Biomechanical Transport
Processes

NATO ASI Series

Advanced Science Institutes Series

A series presenting the results of activities sponsored by the NATO Science Committee, which aims at the dissemination of advanced scientific and technological knowledge, with a view to strengthening links between scientific communities.

The series is published by an international board of publishers in conjunction with the NATO Scientific Affairs Division

A	**Life Sciences**	Plenum Publishing Corporation
B	**Physics**	New York and London
C	**Mathematical and Physical Sciences**	Kluwer Academic Publishers Dordrecht, Boston, and London
D	**Behavioral and Social Sciences**	
E	**Applied Sciences**	
F	**Computer and Systems Sciences**	Springer-Verlag
G	**Ecological Sciences**	Berlin, Heidelberg, New York, London,
H	**Cell Biology**	Paris, and Tokyo

Recent Volumes in this Series

Volume 189—Free Radicals, Lipoproteins, and Membrane Lipids
edited by A. Crastes de Paulet, L. Douste-Blazy, and R. Paoletti

Volume 190—Control of Metabolic Processes
edited by Athel Cornish-Bowden and María Luz Cárdenas

Volume 191—Serine Proteases and Their Serpin Inhibitors in the Nervous System:
Regulation in Development and in Degenerative and Malignant Disease
edited by Barry W. Festoff

Volume 192—Systems Approaches to Developmental Neurobiology
edited by Pamela A. Raymond, Stephen S. Easter, Jr., and Giorgio M. Innocenti

Volume 193—Biomechanical Transport Processes
edited by Florentina Mosora, Colin G. Caro, Egon Krause, Holger Schmid-Schönbein, Charles Baquey, and Robert Pelissier

Volume 194—Sensory Transduction
edited by Antonio Borsellino, Luigi Cervetto, and Vincent Torre

Volume 195—Experimental Embryology in Aquatic Plants and Animals
edited by Hans-Jürg Marthy

Series A: Life Sciences

Biomechanical Transport Processes

Edited by

Florentina Mosora

Université de Liège
Sart-Tilman, Belgium

Colin G. Caro

Imperial College
Physiological Flow Studies Unit
London, United Kingdom

Egon Krause

Aerodynamisches Institut
Aachen, Germany

Holger Schmid-Schönbein

RWTH Aachen
Lehrstuhl für Physiologie
Aachen, Germany

Charles Baquey

INSERM
Université de Bordeaux II
Bordeaux, France

and

Robert Pelissier

Université d' Aix Marseille II
Marseille, France

Plenum Press
New York and London
Published in cooperation with NATO Scientific Affairs Division

Proceedings of a NATO Advanced Research Workshop
on Biomechanical Transport Processes,
held October 9-13, 1989,
in Cargèse, France

Library of Congress Cataloging-in-Publication Data

NATO Advanced Research Workshop on Biomechanical Transport Processes
   (1989 : Cargèse, France)
   Biomechanical transport processes / edited by Florentina Mosora
 ... [et al.].
       p.   cm. -- (NATO ASI series. Series A, Life sciences ; v.
 193)
   "Proceedings of a NATO Advanced Research Workshop on Biomechanical
 Transport Processes, held October 9-13, 1989, in Cargese, France"-
 -T.p. verso.
   Includes bibliographical references and indexes.
   ISBN 0-306-43676-0
   1. Blood--Rheology--Congresses. 2. Blood flow--Congresses.
 3. Biomechanics--Congresses. 4. Biological transport--Congresse.
 I. Mosora, Florentina. II. Title. III. Series.
 QP105.15.N38 1989
 612.1'181--dc20   90-45246
                                                       CIP

© 1990 Plenum Press, New York
A Division of Plenum Publishing Corporation
233 Spring Street, New York, N.Y. 10013

Printed in the United States of America

PREFACE

This book contains contributions related to most of the talks given at the European Mechanics Colloquium and NATO workshop : "BIO-MECHANICAL TRANSPORT PROCESSES" held at: "Institut d'Etudes Scientifiques de Cargèse", Corsica France from October 9 - 13, 1989. The following topics were discussed at this meeting:

Transport phenomena, blood vessels and heart thermo-biophysical properties
Arterial, valvular, ventricular and venous flow
Hemorheology and microcirculation
Blood-wall interactions
Instrumentation and hemodynamic investigation

Accordingly, the papers presented in this volume are related to these five topics and reflect the interdisciplinary character of this colloquium.

The meeting brought together a group of active and eminent scientists from theoretical, experimental and clinical disciplines of various countries and provided a forum where major recent discoveries were discussed. The contributions which make up this book are based on those presented at the colloquium and which have been completed after as a result of discussion at the sessions.

The aim of the workshop was to analyse some problems of bio-mechanical transport processes of the cardio-vascular system, to facilitate an understanding of their applications, and ultimately to lead to a cross-fertilization of ideas and the development of new approaches. To this extent, in our opinion, this colloquium was highly successful and a rewarding experience for all the participants. It is our belief that readers of this book will obtain benefits similar to those obtained by all of us during the five days of our meeting.

Florentina Mosora
Professor of Biophysics and Biomechanics
University of Liège, Belgium

Editor and Colloquium Chairperson

CONTENTS

HEMORHEOLOGY AND MICROCIRCULATION

BLOOD-WALL INTERACTIONS

OBJECTIVES AND PERSPECTIVES OF THE COLLOQUIUM:

"BIO-MECHANICAL TRANSPORT PROCESSES"

Florentina Mosora

Institut de Physique, Université de Liège
4000 Sart-Tilman par Liège 1, Belgique

Ladies and Gentlemen, Dear Friends and Colleagues,

I am very glad that you all arrived here at the Institut of Research Studies of Cargèse to participate in this European Mechanics Colloquium which is at the same time a NATO workshop.

I should like to thank all members of our Scientific and Organizing Committee: Professor C.Caro from Imperial College, Professor Egon Krause and Professor Holger Schmid-Schönbein from RWTH Aachen, Dr. Charles Baquey from Université de Bordeaux II and Professor Robert Pelissier from Université de Marseille for their help in the overall planning and in executing the details of this colloquium and especially Professor Robert Pelissier who accepted to join me for the local organization.

I am very pleased that the participants and session chairmen accepted my invitation and I want to thank the European Mechanics Committee for the patronage of this meeting and the "Université de Liège", the "Région Corse", the "Association Universitaire de Mécanique" and in particular NATO for their economic and financial support. Without them this meeting could not have been held.

On behalf of all of us I would like to express our gratitude to the General Secretary of this Institute: Mrs. Marie-France Hanseler for her devotion and work in the local organization of this workshop.

The aim of this meeting is to discuss some fundamental processes and their modelling related to the behaviour of blood, its flow properties and the structures with which blood or its components come into direct contact, that is the vessel wall and the surrounding tissues and spaces. Interactions of blood or its components with materials of biological significance originating from the human organism or with materials of non-biological origin employed in the biological system or setting, will be also discussed.

These surface phenomena and the different aspects related to the heat and mass transfer are becoming more and more important in bio-mechanical transport processes of the cardiovascular system. For this reason, I feel certain that the discussions and the contributions of our meeting will generate progress in our knowledge of this field.

There are many scientists with a keen interest in this subject who cannot be with us here this week. Professor Nicolaides from St. Mary's Hospital, London and Professor Geert Schmidt-Schönbein from University of California, San Diego, wrote me at the last moment that they are unable to be present at our colloquium. They expressed their regret and send their best wishes to our meeting. Many others scientists are not present because, as you see, here the Institute has a limited capacity and it was impossible to invite the hundred persons interested in the field. Our Scientific Committee was forced to select the participants. Considering that the progress in this field was done thanks to interdisciplinary researches we decided to choose scientists from different specialties.

It is fortunate that the participation at this meeting is an interdisciplinary one. Engineers, surgeons, physicians, physicists and mathematicians are present. Accordingly, the contributions and discussion will reflect this interdisciplinary character and, I hope for a deep understanding between quite different views of the same subject.

It has long seemed to me that a mathematical and an experimental approach to biologically interesting questions are quite insufficient. The real *sine qua non* in the bio-fluids game is insight into the operative physical processes. We must second-guess Nature in an area where physical complexity presents her endless opportunities to be clever.

The simplest of our systems makes the most complex type of flow. Very little has been said about physiological flows. Not that we know little about them: in fact we know so much that the present state of the art almost defies summarization.

Blood flow is clearly the most important bio-mechanical transport system and is enormously complex. I only dwell on the non-steady flow of a non-isotropic, non-Newtonian fluid through non-rigid pipes without speaking of the boundary conditions which, in almost all biological cases, are not exactly known. Moreover, the different hormonal and neurological factors influence and change all the characteristics of the blood flow following an unknown law. Nevertheless, something of the order of one hundred papers appear on the subject of blood flow each year. This proves that this field of interest is far from being entirely known.

Why ?

Are our mathematical, physical models and experimental approaches inadequate , or partially inadequate ?

Or would a complete non-linear treatment of blood flow be necessary to explain some crucial points ?

If this is the case, the future in this field probably belongs to applications of the chaos theory to the blood flow, and the formulation of new field equations dealing with the unique properties of the cardio-vascular system.

Certainly, I hope that our discussions and presentations will contribute to confirm or infirm my opinion and impel valuable recommendations for future research. I hope also that our colloquium will permit a mutually beneficial exchange of information and lay the bases of new scientific collaboration. If at the end of this workshop we can decide about the usefulness of organizing another meeting in a few years time and can establish now the major topics of this , that is, the directions for further research, we can say that one of the aims of the present colloquium has been accomplished.

I want to repeat what Professor Alfred L. Copley said at the occasion of the Second Conference of the International Society of Hemorheology in 1969 : "Being a research worker, of course, I could never have stopped being a student". So, in my two capacities of student and chairperson of the colloquium: "Bio-mechanical transport processes" I extend to you my very best wishes for the success of our workshop.

The members of the Scientific and Organizing Committee of this meeting hope that your days in Cargèse will be scientifically fruitful and that you and the members of your families will have enjoyable and sunny days here.

TRANSPORT PHENOMENA, BLOOD VESSELS AND HEART

THERMO-BIOPHYSICAL PROPERTIES

TRANSPORT OF MATERIALS THROUGH THE WALLS OF DIFFERENT BLOOD VESSELS

M. John Lever and Mark T. Jay

Physiological Flow Studies Unit
Imperial College of Science, Technology and Medicine
London SW7 2AZ, U.K.

INTRODUCTION

The blood vessels exhibit wide variations in structure from one site in the body to another. The marked differences in the composition, the ultrastructural organization of tissue components and the thickness of the vessel walls presumably reflect differences in the mechanical forces and chemical environment which prevailed during their growth. As the vessels age, differences also appear in their susceptibility to atherosclerosis. Veins are normally spared from the disease, except when they take on the features of arteries following the develoment of an arterio-venous fistula, or when they are used as arterial grafts. The pulmonary arteries also tend to be affected by atherosclerosis only if there is pulmonary hypertension. Even amongst the systemic arteries there is marked variability in susceptibility; vessels smaller than 1mm are rarely affected and the ascending portion of the aorta is normally affected to a lesser extent than the abdominal region.

Atherosclerosis is thought to result partly from abnormalities in lipid biochemistry, but many other processes also appear to play a role in the development of the disease, including haemodynamic factors, the interactions of blood leucocytes and platelets with the vessel wall, and changes involving vascular smooth muscle cells and the interstitial components of the wall tissue. None of these latter contributory factors is likely to be uniform throughout the cardiovascular system, and their effects at a particular site are likely to depend on the local tissue architecture.

A significant feature of atherosclerotic lesions is the accumulation of materials within the intimal tissue of the wall. Some of these materials including lipids, fibrinogen and fibrinogen degradation products are usually assumed to have originated from the blood plasma and so another factor which is likely to be of importance in the development of the disease is the transport of these materials between the plasma and the vessel wall and within the wall tissue itself.

Many studies have shown that virtually all plasma proteins are able to enter vessel walls, and there is increasing evidence that most normally undergo continuous flux into, and out of the tissues. Their accumulation within blood vessel walls, may therefore result from a failure of the normal mass balance homeostasis of the wall tissue. In this chapter, consideration will be given to differences which have been observed between the transport properties of various blood vessels in the circulation and whether some of these differences might help to explain the variability in susceptibility of the vessels to atherosclerosis.

VARIATION IN PROTEIN UPTAKE BY BLOOD VESSEL WALLS IN VIVO

Many studies performed over the past thirty years, have demonstrated that proteins enter vessel walls at rates which vary from site to site. Evans blue dye, when injected intravenously is found to be taken up preferentially in certain regions rather than others, and since the dye binds strongly to plasma albumin, it has been assumed that its pattern of uptake reflects varying degrees of permeability to the protein (Packham et al 1967, Somer and Schwartz 1971). Uptake of the dye is most pronounced in rather localized regions around arterial branches and also in larger patches often at the most proximal part of the aorta. Experiments with intravenously injected labelled plasma protein tracers have confirmed that these "blue" areas are indeed sites at which the wall has an elevated permeability to such molecules. Techniques with higher spatial resolution have also shown that there may be additional small areas of high permeability in regions which do not take up observable quantities of blue dye (Stemerman et al 1986). Most studies of protein transport in vessel walls have been concerned with different parts of the aorta and have usually shown greater rates of uptake in the proximal part of this vessel than in the distal parts (Duncan et al 1963). A few studies have been performed on other vessels; Christensen et al (1982), for example, demonstrated that the uptake of various materials is greater in the pulmonary artery than in systemic arteries.

The extent to which tracers are taken up by the vessel wall in the experiments described above, is dependent on several factors including:

a) the permeability of the intimal surface,
b) the affinity and extent of binding of the transported material to components of the wall tissue,
c) the volume of distribution of the material within the wall tissue,
d) the rate of loss of the material from the tissue, by metabolism or by efflux either back across the intimal surface or through the whole thickness of the wall to the adventitia.

The relative importance of each of these factors will depend very strongly on the type of tracer studied, the duration of the uptake experiments and the particular transport properties of the blood vessel under consideration. If short duration experiments are performed with an inert, non-binding tracer, it will enter the wall tissue both across the luminal surface and also across the adventitial surface from the vasa vasorum. Figure 1 shows the probable distribution of such a tracer across the thickness of the wall at the conclusion of an experiment of this type.

We have attempted such studies by injecting radio-iodinated albumin into the circulation of rabbits, and perfusion-fixing the vessels in situ after a period of 15 minutes. Albumin was chosen for these studies since it is relatively inert, and unlike other plasma proteins such as lipoproteins which are of greater importance in atherosclerosis, it does not bind to a significant extent to components of the wall. The transmural distribution of the tracer was measured by en face sectioning of flattened segments of the walls of the vessels, at 10 μm intervals from the luminal to the adventitial surfaces (Tedgui and Lever 1987). In some of the thicker-walled vessels including the ascending aorta and the thoracic portion of the dorsal aorta, the concentration profiles observed were similar to that shown in Fig 1, with the concentration of tracer falling to zero in the middle of the wall. In the other vessels studied, a finite concentration was measurable at all sites across the wall. In thin-walled vessels such as the inferior vena cava and the carotid artery this could be attributed to the short distances over which the tracer had to travel within the duration of the experiment. In other thicker walled vessels, such as the pulmonary artery, it may be due to higher transport rates into, and within the tissue. An estimate of the permeability of the luminal surface was obtained by measuring the quantity of tracer in those sections cut between the luminal surface and the mid-point of the media, and expressing this as a ratio of the product of the plasma concentration of tracer, the intimal surface area and the time during which the tracer was circulating. These estimates of "luminal permeability" are shown in Table 1.

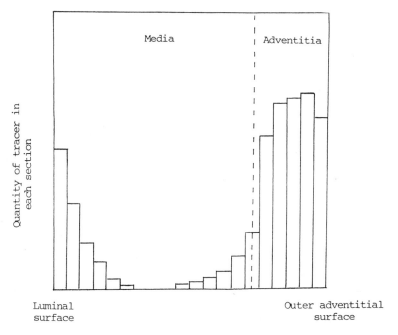

En face sections cut through the wall

Figure 1. The distribution of labelled tracer in en face sections cut through the thickness of the vessel wall, after a relatively short uptake experiment.

Because of the finite quantities of tracer measured in the middle of the media of some of the vessels, there must be some uncertainty concerning the calculated permeability values for these. Although there could have been an over-estimate of the luminal permeabilty if much of the tracer within the medial tissue had originated from the adventitial surface, it is more probable that the values were under-estimates, since data to be presented later shows very rapid transit times for albumin across the wall from the luminal to the adventitial side, particularly in the more porous vessels. In such vessels, some of the tracer which had entered the wall across the luminal surface may have been lost from the tissue before the end of the experiment.

Table 1. Various properties of rabbit blood vessels. See text for definitions of the terms used.

	Ascending Aorta	Dorsal Aorta	Carotid Artery	Pulmonary Artery	Inferior Vena Cava
Luminal permeability $(cm. \ s^{-1} \ x \ 10^{-8})$	27.1 ± 6.3	7.2 ± 3.7	3.5 ± 1.4	21.3 ± 6.8	6.5 ± 3.9
Medial uptake in 3 hours in vivo	0.029 ± 0.006	0.009 ± 0.004	0.005 ± 0.005	0.047 ±.0.022	0.051 ± 0.019
Medial distribution volume	0.095 ± 0.023	0.107 ± 0.021	0.014 ± 0.017	0.234 ± 0.038	0.52 ± 0.08
Endothelial cell density cells/mm^2	3006 ± 284	2690 ± 536	3133 ± 301	3098 ± 496	1598 ± 1594

POSSIBLE FACTORS DETERMINING INTIMAL PERMEABILITY

The high resistance of the intimal surface to the transport of macromoleules appears to be a feature of the endothelium, since removal of this layer greatly increases the rate of movement of such materials into the wall. Various factors have been proposed to account for the spatial variability observed in the permeability of the endothelium including cell morphology, cell turnover rates and local variation of mechanical factors.

Cell morphology varies widely around arterial bifurcations where local variation in blue dye accumulation is also commonly observed (Nerem et al 1981). To investigate the gross endothelial morphology of the vessels considered in this study, we made Hautchen (thin layer) preparations of their intimal surfaces, after they had been fixed at physiological pressure in situ, and stained with haematoxylin. It was apparent that endothelial cells from the various vessels displayed considerable structural differences. The nuclei in cells from the ascending aorta and the pulmonary artery are slightly ovoid but irregularly orientated, whilst those in the dorsal aorta and the carotid artery were much more elongated, and were aligned with the long axis of the vessel. The nuclei in the vein endothelial cells were almost discoid in shape. As shown in Table 1, the surface concentration of endothelial cells is also much lower in the inferior vena cava than in the arteries. How much these differences in morphology account for the apparently wide range of permeabilities is not clear.

It has been suggested that within the intimal surface, highly localized regions of high permeability may occur around endothelial cells undergoing mitosis (Lin et al 1988). Data obtained in the rat, however, suggest that the rate of thymidine incorporation into the endothelial layers of the thoracic aorta, the pulmonary artery and the proximal portion of the inferior vena cava are not significantly different (Schwartz and Benditt 1973). It can therefore be assumed that in each of the vessels there will have been a similar number of mitotic cells during the uptake experiments and that the markedly different permeabilities cannot readily be accounted for by differences in cell turnover rates.

The high permeabilities of the ascending aorta and the pulmonary artery could also be asssociated with the mechanical properties of these vessels. They are the most compliant vessels within the circulation and the endothelial cells may experience considerable strain during each cardiac cycle. They are also thought to experience large, and probably non-axial shear stresses as blood is ejected from the heart. Further work is needed before we have a proper understanding of how these mechanical factors might affect endothelial permeability.

TRANSPORT PROPERTIES OF THE DEEPER LAYERS OF THE VESSEL WALLS

While the endothelial layer may control the rate of entry of a material into the wall across the luminal surface, the transport properties of the deeper layers of the wall will also be important in determining the quantity that accumulates within the tissue. The distribution volume of the material within the tissue indicates the capacity of the tissue for it, and in general, the ease with which the material passes through the tissue, and may drain away, across the wall to the adventitial surface, either by diffusion or by convective transport. Tissue with a high distribution volume will be able to accommodate a large quantity of the material and will also have a high porosity, allowing high rates of outward transport. The dense medial layers of blood vessels have relatively low distribution volumes for larger molecules, much lower than those of the adventitia, and it has been proposed that the low medial porosity may contribute to the intimal accumulation of material during the development of atherosclerosis (Tedgui and Lever 1987). We have estimated the distribution volumes for albumin in different vessels by incubating excised segments in solutions containing the labelled protein. Incubation proceeded for a sufficient length of time for the protein to diffuse into the tissue and to reach saturation levels. The concentration in en face sections was estimated as a fraction of the concentration in the incubating bath, and average values were obtained for the whole of the medial tissue. These values are also given in Table 1. From these data, it would seem that the vein and the pulmonary artery are both considerably more porous than the systemic arteries, which all appear to have similar, but lower porosities. Since protein distribution volumes are decreased by stretching the tissues, we have also estimated distribution volumes in vessels pressurized to normal physiological levels. Although the values obtained were smaller in these experiments than those obtained with unpressurized, relaxed wall tissues, similar differences in porosity were observed between the different vessels.

To investigate the reasons for these differences, we undertook chemical analyses of the tissues. There were marked differences in the concentrations of elastin and collagen in the different vessels and to a lesser extent, differences in the concentration of hyaluronic acid, a component thought to be important in determining tissue distribution volumes. However there was no apparent correlation between any aspect of composition and distribution volumes. It is therefore presumed that the structural organization of the components may be more important than their concentrations. Histological sections certainly demonstrate very large variation in the structure of the different vessels, particularly with regards to the degree of organization of lamellar structures within the wall.

The suggestion has been made that the internal elastic lamella, which separates the intima from the media in certain vessels, may be a transport barrier to macromolecules. Doubt may be cast on this assumption however, because firstly, this layer is normally fenestrated and secondly, intimal accumulation associated with atherosclerosis often occurs in vessels in which the layer is highly fragmented. However in the series of vessels investigated in these studies, there does appear to be an inverse correlation between the prominance of the internal elastic lamella and the "luminal permeability". The layer is not an histologically obvious structure in the ascending aorta, the pulmonary artery or the inferior vena cava, which exhibit higher permeabilities and is most highly developed in the carotid artery which exhibits the lowest permeability amongst this group of vessels.

TRANSMURAL FLUX OF MATERIALS

In addition to vessel wall uptake studies, we have begun to undertake studies on the rate of transport of tracers across the whole thickness of the vessel wall. Vessels are cannulated in anaesthetized rabbits using a procedure which maintains a fully functional endothelial cell layer. They are then transferred to a bath and the lumen is continuously perfused with labelled solutions. Transmural flux is assessed by measuring the rate of appearance of the tracer in aliquots of solution taken from the outer bath. The studies have so far demonstrated that the transit time for the proteins across the wall are relatively short. In the case of the carotid artery which has a medial thickness of aproximately 45 μm, labelled albumin has crossed the wall within 5 min of its introduction into the lumen and a steady state transport rate is achieved within 15 min. With the inferior vena cava, which has a medial thickness of approximately 15 μm, transmural flux achieves a steady state value within 5 min. From these observations, it would seem that the quantity of tracer measurable within the luminal half of the media following the 15 min vessel wall uptake studies performed in vivo, may be an underestimate of the total quantity that has traversed the luminal surface during the period of the experiment. Despite this, the steady state transmural flux of albumin across the carotid artery (5.6 ± 2.4 cm. s^{-1} x 10^{-8} at a transmural pressure of 52 mm Hg) was only slightly greater than that estimated from uptake studies, across the luminal surface in vivo (3.5 ± 1.4 cm. s^{-1} x 10^{-8}). In contrast the steady state transmural flux of albumin aross the inferior vena cava (2.8 ± 1 cm. s^{-1} x 10^{-7} at a transmural pressure of 7 mm Hg) was over 4x greater than that estimated from the in vivo study. This is almost certainly because within the 15 min time-course of the in vivo study, a very large proportion of the albumin which had entered the wall across the luminal surface will have been lost from the tissue to the outside.

Even at the same transmural pressures, albumin flux across the vein was roughly ten times greater than across the artery. This may be due in part to the three fold difference in medial thickness between the two vessels, but is more probably due to structural differences, particularly those of the endothelium.

Transmural pressure was an important determinant of the protein fluxes across both the vessels. Rather surprisingly the greatest increases in flux occurred at pressures toward the upper ends of the relevant physiological ranges (above 10 mm Hg for the vein, and above 100 mm Hg for the artery). At both these pressures, the respective vessels had become relatively stiff and so the increments in flux at higher levels occurred with little tissue strain. Further work is required to test whether this pressure dependency is due to an alteration of the diffusive permeability coefficient or enhanced convective flux of the protein. The latter could result from increased fluid flow across the tissue, and a reduction in the reflection coefficient for the protein at the intimal surface

CONCLUSION

These studies demonstrate that there is a large degree of variability between the transport properties of several different large blood vessels in the rabbit. It would appear that account must be taken of intimal permeability, wall thickness and medial porosity in determining the rate of exchange of materials between the circulating blood and the wall tissue. Other factors which might be important but have not been considered in the present studies include the density and organization of the vasa vasoral capillaries and the adventitial lymphatic vessels.

It is interesting to speculate on the relevance of these findings to atherosclerosis. The pulmonary artery and the inferior vena cava appear to have both a more permeable intima and a more porous media than any of the systemic arteries studied. It can therefore be presumed that in vivo there is a brisk flux of many plasma solutes across the whole thickness of the walls of these vessels. Thus, despite the ease of entry, the apparent ease of egress will tend to hinder the accumulation of any of the plasma components within the tissue. The systemic arteries have uniformly lower medial porosities, and it may be assumed that the ease of transport of materials through this layer at different locations will depend on its thickness. The question therefore arises as to why atherosclerotic material is more likely to accumulate in the carotid artery which is a preferred site for the disease compared with either of the aortic regions studied, which are both considerably thicker. One possible explanation arises from the apparently higher intimal permeability of the aorta, partiularly that of the thickest, most proximal portion. A low resistance to the influx of material into the wall may indicate that there is also a low resistance to efflux back across the luminal surface, and therefore a reduced likelihood of excessive intimal accumulation.

ACKNOWLEDGEMENT

This work was supported by the British Heart Foundation

REFERENCES

Christensen S., Stender S., Nyvad O. and Bagger H., 1982, In vivo fluxes of plasma cholesterol, phosphatidylcholine and protein into mini-pig aortic and pulmonary segments. Atherosclerosis, 41:309-319.

Duncan L.E., Buck K. and Lynch A., 1963, Lipoprotein movement through canine aortic wall. Science, 142:972-973.

Lin S.J., Jan K.M., Schuessler G., Weinbaum S. and Chien S., 1988, Enhanced macromolecular permeability of aortic endothelial cells in association with mitosis. Atherosclerosis, 73:223-232.

Nerem R.M., Levesque M. and Cornhill J.F., 1981, Vascular endothelial morphology as an indicator of blood flow. ASME J. Biomech. Eng., 103:172.

Packham M.A., Rowsell H.C., Jorgensen L. and Mustard J.F., 1967, Localized protein accumulation in the wall of the aorta. Exp. Mol. Pathol., 7:214-232.

Schwartz S.A. and Benditt E.P., 1973, Cell replication in the aortic endothelium: a new method for study of the problem. Lab. Invest., 28:699-707.

Somer J.B. and Schwartz C.J., 1971, Focal 3H-cholesterol uptake in the pig aorta. Atherosclerosis, 13:293-304.

Stemerman M.B., Morrel E.M., Burke K.R., Colton C.K., Smith K.A. and Lees R.S., 1986, Local variation in arterial wall permeability to low density lipoprotein in normal rabbit aorta. Arteriosclerosis, 6:64-69.

Tedgui A. and Lever M.J., 1987, Effect of pressure and intimal damage on the 131-I albumin and 14C-sucrose spaces in aorta. Am. J. Physiol., 253:H1530-1539.

SHUNTING OF HEAT IN CANINE MYOCARDIUM IS CONSIDERABLE

N. Westerhof, J.H.G.M. Van Beek, P. Duijst, G.H.M. Ten Velden and
G. Elzinga

Laboratory for Physiology
Free University of Amsterdam
The Netherlands

ABSTRACT

Shunting of heat in the heart was studied by sinusoidal infusion (0.005-0.1 Hz)
of cold saline into the left canine coronary bed. The resulting temperature change in
the coronary sinus was measured. A model of heat transport in the heart containing
a mixing chamber, a countercurrent exchanger and heat conduction across endocardium
to the ventricular lumen and across epicardium to the thoracic cavity was developed.
The model showed that shunting in the countercurrent exchanger alone does not affect
the mean transit time of heat, but heat loss by conduction shortens this characteristic
time. In the presence of heat loss across endocardium and epicardium, shunting results
in an enhanced reduction of mean transit time. The measured mean transit time for
heat (14.3 $\pm$ 4.5 s, mean $\pm$SD) was only 22% of the theoretical mean transit time (65
$\pm$ 1 s) predicted under the assumption that no heat conduction to the environment
takes place, and which is the inverse of flow rate (87.4 $\pm$ 1.5 ml min^{-1} per 100 gram).
The amount of heat leaving via the coronary sinus (heat recovery) was 76 $\pm$ 11% of
that injected in the coronary artery. Using the model it is estimated from these data
that 62% of the injected heat is shunted. Injection of cold saline in an epicardial vein
gave no temperature change distally (peripherally) in the adjacent artery, so that
shunting between large epicardial vessels is small. The maximum temperature change
in the myocardium following an atrial injection of cold indicator was only 25% of that
in the coronary sinus, indicating that heat shunts before reaching the capillaries. We
conclude that shunting of heat is of considerable quantitative importance, and occurs
at the arteriolar level.

INTRODUCTION

Extravascular shunting, i.e. diffusion from artery (or arteriole) to vein (or venule)
without proper transport into the tissue may exist in heart muscle. Extravascular
shunting has been invoked to explain the finding that tissue pO_2 is observed to be
lower than coronary sinus pO_2 (Schubert et al., 1978; Piiper et al., 1984). We have
chosen to study extravascular shunting of heat since this indicator has an extremely
high diffusivity, and heat shunting might therefore be large and quantifiable. Shunting
of heat (Bassingthwaighte et al., 1984) and hydrogen (Roth and Feigl, 1981) has been
studied by comparing arrival times of an intravascular (non-diffusible) indicator and
a diffusible indicator (heat or hydrogen) following simultaneous bolus injection of both
indicators. However, arrival times of indicators are difficult to measure accurately and
do not make it possible to quantitate shunting. We therefore studied shunting by

sinusoidal infusion of cold saline into the left coronary circulation of the anesthetized dog and measured the resulting sinusoidal temperature variations in the coronary sinus. To analyze the data we developed a model of myocardial heat transport that contains arterio-venous shunting, a mixing chamber, and heat loss to thoracic cavity and ventricular lumen.

METHODS

Experiments were conducted in seven anesthetized mongrel dogs using pentothal sodium 0.5 ml/kg, i.v. after sedation with dehydrobenzperidol (0.6 mg/kg) and methadone (0.4 mg/kg). Anesthesia was maintained by supplemental venous infusion of dehydrobenzperidol (0.5 mg/kg/hr), methadone (0.3 mg/kg/hr), and lidocaine (0.6 mg/kg/hr)). The dogs weighed between 18.5 and 27 kg and were ventilated with a mixture of 30% oxygen and 70% nitrous oxide. Details of the methods have been given elsewhere (Duijst et al., 1987; Duijst, 1988). Anticoagulation was accomplished with sodium heparin (initial dose 300 U/kg, maintained by infusion of 200 U/kg/hr). A thoracotomy was performed in the fifth intercostal space. A thermistor catheter (with linear temperature response) was positioned in the coronary sinus via the left jugular vein, under fluoroscopic guidance. A balloon tipped Gregg cannula was wedged into the left main coronary artery. Coronary artery infusion was accomplished by means of a pump using blood from the femoral artery. The system contained a heat exchanger to obtain a stable baseline temperature, an electromagnetic flow probe (with Skalar, Delft, Holland flow meter) and a bubble trap. The seal between the Gregg cannula and coronary artery wall was checked regularly by stopping perfusion flow for about 10 seconds. If pressure at the cannula tip decreased to 20 mmHg or less the seal was considered satisfactory (Roth and Feigl, 1981). Infusion of cold indicator into the perfusion line was accomplished via a special infusion pump, made in our own workshop, that drives a syringe in any desired pattern. Injection flow was in the present experiments controlled by a sinus generator and was small compared to mean coronary blood flow: about 4 ml/min. A thermistor at the tip of the Gregg cannula measured temperature of the infused blood.

In three separate experiments (same anesthesia) left atrial bolus injections of 20-30 ml saline at room temperature were administered. The temperature distribution over the free wall of the left ventricle was, in these experiments, measured with a thermistor mounted on a needle (Ten Velden et al., 1982 and 1984).

In three further experiments (same anesthesia) cold saline (0.1-0.15 ml, 0 °C) was injected into a small side branch of an epicardial vein and temperature was measured just distally (downstream) in the adjacent artery (artery and vein ran parallel over about 5 cm), using a thermistor mounted in a small needle. Temperature was also measured in the coronary sinus in these experiments.

MODEL OF MYOCARDIAL HEAT TRANSPORT

The model for extravascular shunting is shown in figure 1. It consists of afferent and efferent vessels that provide coronary perfusion. These vessels give rise to a transport delay (t_d) in the response. The more distal vessels run in close proximity giving rise to countercurrent exchange of heat (Piiper et al., 1984), with an effective geometric shunt parameter A_s that accounts for exchange area and diffusion distance. The blood vessels carry blood to and from the tissue which can be represented by a single mixing chamber since longitudinal diffusion in the capillary exchange region is fast (Perl and Chinard, 1968). The mixing chamber volume (V) comprises the entire myocardium, blood flow is F and the chamber's time constant for washout is thus τ_m = V/F, assuming the same specific heat capacity for myocardium and blood. Since we found earlier that there is heat loss to the environment (Ten Velden et al., 1982), we assume that heat can diffuse directly from the tissue (mixing chamber) to the environment, with effective geometric exchange parameter A_1 proportional to exchange area, inversely proportional to diffusion distance. The derivation of the model has been given by Duijst et al. (1988). We define a dimensionless parameter describing the

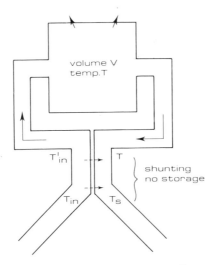

Fig. 1. Model of heat balance of myocardium consisting of a countercurrent exchanger, representing shunt diffusion (without storage of blood) and a mixing chamber, representing the tissue (volume V, temperature T). Arterial and venous blood temperatures are T_{in} and T_s. Heat is lost to the environment.

relative importance of shunt diffusion to transport by perfusion α_s = k.A_s/F, with k denoting heat diffusivity ($1.3.10^{-3}$ cm^2 s^{-1}, according to Grayson, 1973) and A_s the geometric parameter for shunt diffusion. The dimensionless heat loss parameter (α_l) is defined as α_l = k.A_l/F, with A_l the geometric parameter quantifying heat loss. Using a heat balance equation and assuming that heat production by metabolism in the myocardium is not affected by the infusion of the cold indicator we derived the impulse response ($h_m(t)$) of the temperature in the coronary sinus following an arterial bolus injection (impulse) of cold saline (Duijst et al., 1987; Duijst, 1988).

The impulse response is given by:

$$h_m(t) = F_s \, \delta(t-t_d) + 1/[\tau_m.(1 + \alpha_s)^2] \cdot \exp[-(t-t_d)/\tau_m.f)] \qquad (1)$$

where $F_s = \alpha_s/(1 + \alpha_s)$ gives the fraction of infused heat that shunts directly in the countercurrent exchanger. The factor $f = (1 + \alpha_s)/(1 + \alpha_l + \alpha_l \, \alpha_s)$ and we designate $\tau'_m = \tau_m.f$. The apparent mean transit time of this model (t_m) can be calculated to be:

$$t_m = \tau_m/((1 + \alpha_s\alpha_l)\cdot(1 + \alpha_l + \alpha_l\alpha_s)) + t_d \qquad (2)$$

It can be seen from this relation that in case the indicator is neither shunted nor lost by conduction (i.e. $\alpha_s = \alpha_l = 0$) the model's mean transit time is equal to the sum of the delay time (t_d) and the time constant of the mixing chamber ($\tau_m = V/F$, and with tissue density ρ, $\tau_m = 1/(\rho \cdot \text{Flow per gram tissue})$). Diffusional shunting in the absence of heat loss to the environment ($\alpha_s > 0$, $\alpha_l = 0$) does not affect the mean transit time. Loss of heat alone results in a reduction of mean transit time. The combination of shunting and loss to the environment results in an augmented reduction of the mean transit time.

For further calculations it is convenient to introduce the normalized apparent mean transit time: $t'_m = (t_m-t_d)/\tau_m$, so that:

$$t'_m = 1/[(1 + \alpha_l\alpha_s)\cdot(1 + \alpha_l + \alpha_l\alpha_s)] \qquad (3)$$

17

The dimensionless ratio of recovered heat in the venous effluent to the arterially infused heat, R, is given by Duijst (1988):

$$R = (1 + \alpha_s \alpha_l)/(1 + \alpha_l + \alpha_l \alpha_s) \tag{4}$$

From the experimentally determined values of t'_m and R, using equations (3) and (4), it can be derived what the α_s, α_l and the shunt fraction, $F_s = \alpha_s/(1 + \alpha_s)$, are. The shunt fraction is:

$$F_s = (\sqrt{R} - \sqrt{t'_m})/(1/\sqrt{R} - \sqrt{t'_m}) \tag{5}$$

In summary, from the determination of the mean transit time, the delay time t_d, and flow per gram tissue the t'_m is found, subsequently from t'_m and the recovery ratio, R, the shunt fraction, F_s, can be calculated.

DATA ANALYSIS

The mean transit time, t_m, and recovery ratio, R, were obtained from sinusoidal temperature changes in the arterial blood infused into the left main stem coronary artery and measurement of the sinusoidal response in the coronary sinus. As a first step drift correction was carried out on the coronary sinus temperature oscillations. Subsequently cross-correlation between input and output temperatures was performed to obtain the phase differences. Amplitude ratios of output and input temperature were derived from the waveform amplitudes. The sinusoidal input-output relations as a function of frequency were fitted to the model in the frequency domain using a complex Marquardt (1963) nonlinear fit (real and imaginary parts are fitted simultaneously). The model in the frequency domain follows directly from the Fourier transform of $h_m(t)$, and takes the form:

$$H(j\omega) = \exp(-j\omega t_d) [A + B \, \tau'_m/(j\omega\tau'_m + 1)] \tag{6}$$

with:

$$A = F_s \text{ and } B = 1/[\tau_m(1 + \alpha_s)^2] \text{ and } j = \sqrt{-1}.$$

The fit of this four parameter model allows the calculation of the mean transit time and the recovery ratio.

The mean transit time is derived via $t_m = -\lim_{\omega \downarrow 0} (d \lim H(j\omega)/d(j\omega))$ so that:

$$t_m = t_d + B \cdot \tau'^2_m/(A + B \, \tau'_m) \tag{7}$$

The recovery ratio is $R = \lim_{\omega \downarrow 0} H(j\omega)$, which results in $R = A + B \, \tau'_m$.

Together with knowledge on coronary flow per gram tissue, the shunt fraction can then be calculated.

RESULTS

Sinusoidal changes in temperature were studied in 7 dogs. An example of the sinusoidal temperature changes in coronary artery and coronary sinus are given in figure 2. In figure 3 an example of the modulus and phase data of the input-output relation are given. The line through the data points is the fit to the model (figure 1, eq. 6). The fit produces the model parameters. The average results of all seven dogs are (mean ± SD): delay time t_d = 2.80 ± 1.19 s, decay time τ'_m = 13.3 ± 2.0 s, mean transit time t_m = 14.3 ± 4.5 s. Coronary flow is 87.4 ± 1.5 ml min^{-1} per 100 gram. Using this flow value τ_m is calculated to be: 64.7 s. The average recovery ratio R was found to be 0.76 ± 0.11 i.e. 76% of arterially infused heat is recovered in the

freq.=0.04Hz

Fig. 2. Example of sinusoidal excitation. Cold saline is sinusoidally infused in the left main coronary artery (T_{in}) and the resulting coronary sinus temperature is measured (T_{out}).

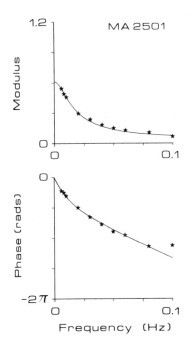

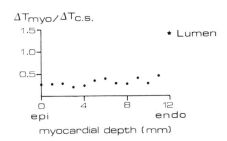

Fig. 3. Example of input-output relation in the frequency domain of coronary sinus and arterial temperature in terms of modulus and phase. The drawn line is the model (figure 1, eq. 6) fit to the data.

Fig. 4. Example of maximum temperature change in the myocardium relative to the maximum temperature change in the coronary sinus resulting from injection of a bolus of saline into the left atrium. The most right hand point (asterisk) is the maximum temperature change in the left ventricle.

coronary venous effluent. The shunt fraction can now be calculated (eq. 5) and is found to be 62%.

An example of the maximum temperature change at various depths in the myocardium relative to the temperature change in the coronary sinus following an atrial bolus injection of cold saline is given in figure 4. Similar results are found in the two other experiments.

A cold bolus of saline given in an epicardial vein did not result in a detectable response in the adjacent artery indicating little countercurrent exchange between epicardial arteries and veins (figure 5). When the bolus arrives in the coronary sinus, temperature does change there, indicating a major change in epicardial venous blood temperature (figure 5).

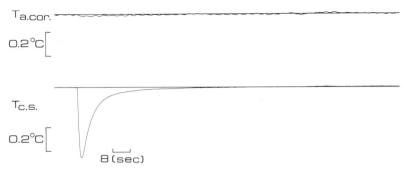

Fig. 5. Example of the temperature changes in the coronary sinus (Tc.s.) and in an epicardial coronary artery (Ta.cor.) after a bolus injection of cold physiologic saline in an epicardial vein adjacent to the artery. The change in temperature in the artery is negligible, indicating minimal countercurrent exchange in epicardial vessels.

DISCUSSION

The objective of the study was to quantitate extravascular shunting. We have chosen heat (cold) as indicator since diffusivity of heat is large, which makes it possible to quantitate shunting. The use of sine wave excitation has the advantage that drift correction can easily be performed. Measurement inaccuracy at high frequencies is not of major concern since the calculation of mean transit time and recovery ratio both make use of the low frequency part of the spectrum (see above). With the use of impulse responses it is the tail of the response that contributes greatly to the mean transit time and drift may strongly affect its value. Drift correction thus not only plays an important role, but the correction is also more difficult than in the cases of sine waves. The introduction of the cold indicator into the coronary artery results in only very small changes in coronary sinus temperature (about 0.1 °C, figure 2). The myocardial temperature (see figure 4) changes are only 25% of those in the coronary sinus. Effects of changes in temperature on cardiac metabolism can therefore be neglected.

The observation that a small cold bolus injected into a coronary vein does not result in a measurable temperature change in the adjacent epicardial artery suggests little extravascular shunting between the large epicardial vessels. The temperature change following an atrial bolus injection of cold saline results in a maximal myocardial temperature change of only 25% of that in the coronary sinus, suggesting that heat shunts before arrival in the capillaries. We conclude that extravascular shunting of heat mainly occurs at the arteriolar-venular level.

From anatomical data of Bassingthwaighte et al. (1974) we calculate an intracapillary mean transit time of 2.4 s for the perfusion rate used in our experiments. For an intravascular indicator a total coronary intravascular mean transit time of 6.8 $\pm$ 1.7 s was found in the same group of dogs (Duijst, 1988). Thus the pre- and postcapillary vessels have a mean transit time of about 4.4 s. The delay time of our model fit (t_d = 2.8 s) indicates the mean transit time of the conduit vessels where no countercurrent exchange takes place. Thus the mean transit time of the vessels where extravascular shunting is supposed to take place is 1.6 s. Thus on basis of mean transit times it is suggested that most shunting takes place at the arteriolar level. The three mean transit times (large vessels (2.8 s), small shunt vessels (1.6 s) and capillaries (2.4 s) are not much different.

The use of a single mixing chamber in the model is an oversimplification. However, the diffusion of heat is fast. If heat were deposited locally at a certain point at t = 0, the root mean square distance of diffusion is $\sqrt{6kt}$ (Carslaw and Jaeger, 1978). With diffusivity of heat in cardiac tissue k = $1.3 \cdot 10^{-3}$ cm^2.s^{-1} (Grayson, 1973) a distance of 0.9 mm per sec is found. This distance is about equal to distances between arterioles and venules (Bassingthwaighte et al., 1974), so that, to a first approximation no temperature differences are expected in the capillary exchange unit. In figure 4 it is also seen that no large temperature gradients are found in the myocardium. Therefore, the assumption of a single mixing chamber seems not unreasonable. The rather large root mean square distance of diffusion in heart tissue and blood also implies that no distinction can be made between axial and shunt diffusion of heat from the comparison of the arrival times of the diffusible and non-diffusible indicators.

The assumption of uniform myocardial perfusion may also be challenged since it was shown that coronary flow is quite inhomogeneously distributed (King et al., 1985). We are presently investigating the effect of flow heterogeneity using a model with a fractal flow distribution.

Since the diffusion constant of oxygen is about 100 times smaller than that of heat a shunt fraction of less than 1% is expected for oxygen. The results imply that the shunting of oxygen is small, which is in agreement with the results published by Van Beek and Elzinga (1987).

REFERENCES

Bassingthwaighte, J.B., Yipintsoi, T., and Harvey, R.B., 1974, Microvasculature of the dog left ventricular myocardium. Microvasc. Res., 7: 229.
Bassingthwaighte, J.B., Yipintsoi, T., and Knopp, T.J., 1984, Diffusional arteriovenous shunting in the heart. Microvasc. Res., 28: 233.
Carslaw, H.S., and Jaeger, J.C., 1978, "Conduction of heat in solids," Oxford Univ. Press, Oxford, reprinted from the second edition (pp. 256).
Duijst, P., 1988, "Cardiac metabolism and coronary flow," Ph.D. Dissertation, Free Univ. of Amsterdam (pp. 42-72).
Duijst, P., Van Beek, J.H.G.M., Ten Velden, G.H.M., Elzinga, G., and Westerhof, N., 1987, Shunting of heat in the canine myocardium. Pflügers Arch. (Europ. J. Physiol.), 410: S29.
Duijst, P., Elzinga, G., and Westerhof, N., 1987, Temperature distribution cannot predict local cardiac metabolism. Am. J. Physiol., 252: H529.
Grayson, J., 1973, Thermal conductivity of normal and infarcted heart muscle. Nature, 215: 767.
King, R.B., Bassingthwaighte, J.B., Hales, J.R.S., and Rowell, L.B., 1985, Stability of heterogeneity of myocardial blood flow in normal awake baboons. Circ. Res., 57: 285.
Marquardt D.W., 1963, An algorithm for least-squares estimation of nonlinear parameters. J. Soc. Ind. Appl. Math., 11: 431.
Perl, W., and Chinard, F.P., 1968, A convection-diffusion model of indicator transport through an organ. Circ. Res., 22: 273.
Piiper, J., Meyer, M., and Scheid, P., 1984, Dual role of diffusion in tissue gas exchange: blood-tissue equilibration and diffusion shunt. Respir. Physiol., 56: 131.
Roth, A.C., and Feigl, E.O., 1981, Diffusional shunting in the canine myocardium. Circ. Res., 48: 470.
Schubert, R.W., Whalen, W.J., and Nair, P.P., 1978, Myocardial pO_2 distribution: Relationship to coronary autoregulation. Am. J. Physiol., 234: H361.
Ten Velden, G.H.M., Elzinga, G., and Westerhof, N., 1982, Left ventricular energetics. Heat loss and temperature distribution of canine myocardium. Circ. Res., 50: 63.
Ten Velden, G.H.M., Westerhof, N., and Elzinga, G., 1984, Heat transport in the canine left ventricular wall. Am. J. Physiol., 247: H295.
Van Beek, J.H.G.M., and Elzinga, G., 1987, Diffusional shunting of oxygen in isolated rabbit heart is negligible. Pflügers Arch. (Europ. J. of Physiol.), 410: 263.

STIFFENING OF THE CARDIAC WALL BY CORONARY BLOOD VOLUME INCREASE:

A FINITE ELEMENT SIMULATION

Jacques M. Huyghe[1], Theo Arts[2], Dick H. van Campen[3], and
Roberts S. Reneman[4]

Departments of Movement Sciences[1], Biophysics[2] and Physiology[4]
University of Limburg, Maastricht, the Netherlands and Department of
Mechanical Engineering[3], Eindhoven University of Technology
Eindhoven, the Netherlands

ABSTRACT

A porous medium finite element model of the beating left ventricle is
used to simulate the influence of the intracoronary blood volume on left
ventricular mechanics. The spongy material is composed of incompressible
solid (myocardial tissue) and incompressible fluid (coronary blood). The
model is axisymmetric and allows for finite deformation, including torsion
around the axis of symmetry. The total stress in the tissue is the sum of
the intramyocardial pressure, effective passive stress due to myocardial
deformation and the contractile fiber stress. The model is able to simulate
a full cardiac cycle. Three-dimensional end-systolic deformation computed
relative to the end-diastolic state is shown to be consistent with
experimental data from the literature. The direction of maximal shortening
varied less than 30° from endocardium to epicardium while fiber direction
varied by more than 100°. It is shown that the ventricular model exhibits
diastolic stiffening following an increase of intracoronary blood volume.
End-diastolic left ventricular pressure increases from 1.5 kPa to 2.0 kPa
when raising intracoronary blood volume from 9 to 14 ml per 100 g
myocardial tissue. The model simulation suggests that the mechanism
underlying the increase in end-diastolic pressure at higher coronary blood
volumes, is an increase in passive stiffness of the myocardial fibers. This
increased stiffness is the combined result of an overall increase in strain
in myocardial tissue and the non-linear stress-strain relationship of
myocardial tissue.

Keywords
Left ventricle / porous medium / mixture / erectile properties / diastole /
coronary perfusion.

INTRODUCTION

The cardiac wall is a complex biological structure composed of
different components: muscle cells, coronary vessels, collagen fibers,

intracellular and interstitial fluid, lymph and blood. For the sake of simplicity, many authors of cardiac models assume the cardiac wall to be a homogenous continuum of solid matter. The aim of this study is to investigate to which extent a more detailed model description of the myocardial tissue, including a solid and a fluid component, is able to describe the stiffening of the cardiac wall by coronary blood volume increase.

METHODS

Material model. We assumed that myocardial tissue was a spongy structure filled with intracoronary blood. The material of which the spongy structure was made and the intracoronary blood were both assumed to be incompressible. Therefore, changes in volume of the solid fluid mixture were equal to the amount of blood being squeezed out or sucked in. The stress in the mixture was the sum of the intramyocardial pressure p (present in both fluid and solid), effective passive stress due to the deformation of the porous structure and contractile fiber stress. As the collagen weave is three-dimensional, we assumed that the passive effective stress was three-dimensional. The contractile fiber stress, however, acted only in the fiber direction, which changed across the wall. The contractile stress was time, strain and strain rate dependent. The redistribution of intramyocardial blood in the coronary bed was modelled by Darcy's law: the flow of intracoronary blood was proportional to the intramyocardial pressure gradient. The proportionality constant is the permeability of the medium. In order to account for the changes in vascular resistance when intracoronary blood volume changes, the permeability of the medium was adapted proportionally to the square of the intramyocardial coronary blood volume in the course of the computation.

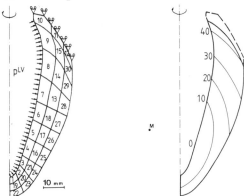

Fig. 1 Left: cross-section of the rotationally symmetric finite element mesh of a canine left ventricle. All elements contain contractile fibers except elements 10, 11 and 12 which represent the annulus fibrosus. Right: in a cross-section of the wall of the left ventricle, the sequence of onset of contraction is simulated to radiate from a point M (time in ms).

Numerical approach. The myocardial wall was subdivided into thirty ring-shaped elements (fig. 1). Each element had eight nodes. The change in position of each node relative to the reference situation was described by a radial, axial and circumferential displacement at any time during the cardiac cycle. Quadratic interpolation yielded displacements at intermediate points. The intramyocardial pressure field was obtained by linear interpolation from pressure values at the corner nodes. The total number of degrees of freedom was 395 (351 displacements and 44 pressures). finite deformation was accounted for by means of a total Lagrangian approach. An implicit-explicit time integration scheme was used. Within each time step a modified Newton-Raphson iterative procedure was used to account for the non-linearities included in the model.

The axial and circumferential displacement of the 7 top nodes of the mesh were suppressed. No blood was allowed to cross the endocardial surface. At the endocardial side of elements 1 to 9 a uniform intraventricular p^{LV} was applied as an external load. The loads exerted by the papillary muscles and by the pericardium were neglected. Along the epicardial surface we allowed free exchange of blood between the intramyocardial coronary vessels and the epicardial coronary vessels. Initiation of contraction was not simultaneous for all sarcomeres (fig. 1). The depolarisation wave moved from endocardium to epicardium and from the apical region toward the basal region. The wave needed about 40 ms to reach the whole left ventricular wall. The initial permeability of the porous medium was derived from data on time constants of the coronary circulation and equals 2 mm^2 $kPa^{-1}s^{-1}$. The initial porosity of the medium (= the percentage of intramyocardial space occupied by coronary blood at 0 kPa perfusion pressure) is 6%.

The transmural variation of fiber angle was derived from experimental data of Streeter and Hanna (1973). The sarcomere model used in the simulations is described elsewhere (Huyghe, 1986). The passive constitutive behaviour of the myocardial tissue was a quasi-linear viscoelastic law with an exponential elastic response, which was fitted to experimental data of van Heuningen et al (1982), and Yin et al (1987).

Three computations were performed. Each computation started with an increase of intracoronary blood volume. The increase equalled 3% of the myocardial volume in the first computation, 5,5% in the second computation and 8% in the third computation. After the increase of intracoronary blood volume, the intraventricular pressure was increased up to 2.5 kPa in all three computations. Finally the third computation was repeated, loading the ventricle only up to 1 kPa, and then consecutive beats were initiated. The duration of the cardiac cycles was 0.55 s.

A commercial post-processing package I-DEAS (Structural Dynamics Research Corporation) produced color-coded plots of different local output variables. To facilitate the comparison of computed strains with experimental strain data, local three-dimensional Green strain tensors were computed with reference to the end-diastolic state and were interpreted in terms of their eigenvalues (the principal strains) and eigenvectors (the principal axes of strain). The three principal strains were ranked from smallest (most negative) to largest (most positive).

RESULTS

Diastolic stiffness. The increase in intracoronary blood volume induced an increase of wall thickness. The pressure-volume curves resulting from the three computations (fig. 2) show that diastolic stiffness increased with increasing coronary vascular volume. At a given left ventricular volume left ventricular pressure increased from 1.5 kPa to 2.0 kPa when raising intracoronary blood volume from 9 to 14 ml per 100 g left ventricle.

The cardiac cycle. Ejection fractions for the three cardiac cycles were 59%, 55% and 54%. The time course of the radial and axial displacement component of all nodes of the mesh is shown in fig. 3.

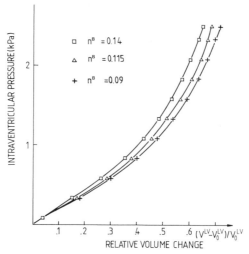

Fig. 2 Simulated pressure-volume relations of the passive left ventricle at different levels of intramyocardial blood volume. The ventricular wall stiffens at increased intramyocardial blood volume. V_0^{LV} and V^{LV} are the initial and current intracavitary volume of the left ventricular model respectively. n^B is the ratio of coronary vascular volume over total myocardial volume.

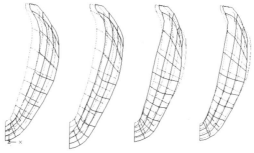

Fig. 3 Simulated successive states of deformation of the left ventricle. A computer generated picture is shown of the deformation of a meridional section of the model (dotted line = reference state, continuous line = deformed state). From left to right: end diastole, beginning and end of ejection and beginning of diastole.

The circumferential displacement component (not shown) showed rotation of the apex relative to the base in counterclockwise direction during the ascending limbs of the ventricular pressure, while the opposite happened during the descending limb. The model computed increasing end-systolic values of the three principal strains with increasing depth (fig. 4).

End-systolic principal strains equalled 0.45; -0.01 and -0.24 at 2/3 of the wall thickness from the epicardium and 0.26, 0.00 and -0.19 at 1/3 of the wall thickness from the epicardium. To analyse the transmural variation in the orientation of the principal strain axes at the end of ejection, we computed the angle between the first principal strain axis (i.e. the axis of maximal shortening) and the circumferential coordinate

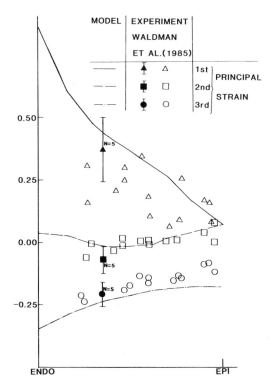

Fig. 4 Transmural distribution of end-systolic principal strains as predicted by the model and measured by Waldman et al (1985). In both model and experiment strains are given with respect to the end-diastolic state.

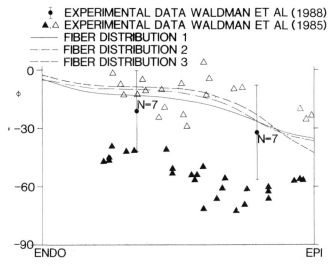

Fig. 5 The angle ϕ is the projection onto the epicardial plane of the angle between axis of maximal shortening and the circumferential direction. The variation of angle ϕ with depth is plotted for three fiber angle distributions, defined in table 1. The two sets of experimental data (▲ and △) are from two animals representing the range observed in five dogs by Waldman et al. (1985).

direction. The projection of this angle on the epicardial tangent plane is refered to as the angle ϕ and is plotted as a function of depth at the interface of the elements 6, 18, 27 and the elements 7, 13, 28 (fig. 5). The angle ϕ shifted from $-37°$ epicardially to $-6°$ endocardially.

DISCUSSION

This study shows that (1) a biphasic model of left ventricular mechanics is able to compute an increase of diastolic stiffness by coronary blood volume increase, provided that the non-linear nature of the constitutive behaviour of myocardial tissue is taken into account and (2) the axisymmetric finite element model computes systolic triaxial strains consistent with experimental data from the literature.

Diastolic stiffness. When using quasi-linear viscoelasticity, stiffness increases with increasing coronary vascular volume (fig. 3). This result is consistent with experimental data of Olsen et al. (1981) and Vogel et al. (1982) (fig. 6).

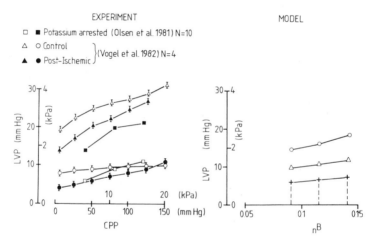

Fig. 6   The left panel shows the dependency of end-diastolic left ventricular pressure (LVP) on the coronary perfusion pressure (CPP) at a given cavity volume. The right panel shows model results of left ventricular pressure as a function of intracoronary blood volume at constant cavity volume. Stiffening of the cardiac wall by increase of coronary vascular volume is more pronounced at higher cavity volumes, in both experiment (left) and model (right).

On the basis of the experimental quantification of the relationship of coronary perfusion pressure and intracoronary blood volume by Morgenstern et al. (1973), we infer that the three values of intracoronary blood volume chosen in fig. 3 correspond with coronary perfusion pressures of 6, 11, and 14 kPa. This implies that the model predicts that at constant cavity volume, left ventricular pressure increases from 1.5 kPa to 2.0 kPa when raising perfusion pressure from 6 to 14 kPa. The same shift of coronary perfusion pressure causes an increase in left ventricular pressure from 1.5 to 2.4 kPa in potassium arrested hearts according to the experimental data of Olsen et al. (1981), while Vogel et al. (1982) find an increase of left ventricular pressure from 1 to 1.3 kPa and from 3 kPa to 3.8 kPa in intact hearts indicating that model results are within the range of values measured experimentally. The increase in wall thickness induced by the increased intracoronary blood volume is consistent with the experimental data of Morgenstern et al. (1973). One might be tempted to attribute the shift of the pressure-volume curve to the left in fig. 3 to the increase in wall thickness, on the basis of Laplace's law. However, it is not evident that Laplace's law applies to biphasic materials. In a single phase solid material, increase in thickness of the shell induces an increased stiffness of the shell, because more material is available to take up shell forces. In the case of a solid-fluid mixture, adding more fluid to the mixture, does not result in additional material to take up shell forces. Tensile forces can only be borne by the solid. The increase in intracoronary blood volume resulted in increased wall thickness, but not in increased stiffness when we replaced the quasi-linear viscoelastic law by isotropic linear elasticity while maintaining Darcy's law as a description for the redistribution of intracoronary blood. These results show that stiffening of the diastolic left ventricle by coronary blood volume increase should not be interpreted in terms of Laplace's law, but rather as the combined result of an overall increase in strain in myocardial tissue and the non-linear stress-strain relationship of the myocardial tissue.

The cardiac cycle. The model computes increasing values of strain with increasing depth. Many experimental data from the literature point in the same direction. In fig. 4 we have used the experimental data of Waldman et al. (1985) to assess the transmural distribution of principal strain as predicted by the model. Although the direction of maximal stress almost coincides with the fiber direction, the direction of maximal shortening (i.e., the third principal strain axis) does not. Across the wall the computed direction of maximal shortening does not vary nearly as much as the muscle fiber direction (fig. 5). This finding is consistent with experimental data of Prinzen et al. (1984) and Waldman et al. (1985). The latter investigators found that the above defined angle ϕ of maximal shortening equaled $-22\pm21°$ in the inner half of the wall ($65\pm9\%$ of the wall thickness from the epicardium). A common feature of the model prediction and Waldman's data is the progressive rotation of the principal axis of shortening towards the circumferential direction with increasing depth.

CONCLUSION

An axisymmetric two-phase finite element model is used to simulate myocardial deformation during the cardiac cycle. Computed transmural strain distribution is in agreement with experimental data from the literature. The model indicates that the increase in diastolic stiffness by increase of coronary vascular volume should be interpreted as a combined effect of an overall increased strain in the myocardial fibers and the non-linear stress-strain relationships of the myocardial tissue.

Question from the audience: Isn't axisymmetry a rather rough approximation of the real geometry of the left ventricle?

Huyghe: We chose for an axisymmetric model in order to reduce computation time. This approximation is indeed a limitation of this model. In december, our group will present a three-dimensional model of the left ventricle at the Winter Annual Meeting of the ASME.

Question from the audience: How do you analyse torsional deformation with an axisymmetric model?

Huyghe: The finite element code that we use has been written specially for finite deformation including torsion. Except a radial and axial displacement, there is also a circumferential displacement, which is also axisymmetric.

Oddou: There is an essential difference between the porous medium approach of your group and our group, because fluid in your model is the coronary blood, while in our analysis it is the interstitial fluid.

Huyghe: This is very right. The two phenomena have also very different time constants and this is why we feel it is acceptable to neglect interstitial fluid flow in our analysis.

REFERENCES

Huyghe, J.M., 1986, "Non-linear finite element models of the beating left ventricle and the intramyocardial coronary circulation". Ph.D.-thesis. Eindhoven University of Technology, the Netherlands.

Huyghe, J.M., Oomens, C.W., Van Campen, D.H. and Heethaar, R.M., 1989, Low Reynolds steady state flow through a branching network of rigid vessels: I. A mixture theory, Biorheology, 26:55.

Huyghe, J.M., Oomens, C.W. and Van Campen, D.H., 1989, Low Reynolds number steady state flow through a branching network of rigid vessels: II A finite element mixture model, Biorheology, 26:73.

Morgenstern, C., Holtes, V., Arnold, G., Lochner, W., 1973, The influence of coronary pressure and coronary flow on intracoronary blood volume and geometry of the left ventricle, Pflueg. Arch., 340:101.

Olsen, C.O., Attarian, D.E., Jones, R.N., Hill, R.C., Sink, J.D., Lee, K.L., Wechsler, A.S., 1981, The coronary pressure-flow determinants of left ventricular compliance in dogs, Circ. Res., 49:856.

Prinzen, F.W., Arts, T., Van der Vusse, G.J., and Reneman, R.S., 1984, Fiber shortening in the inner layers of the left ventricular wall as assessed from epicardial deformation during normoxia and ischemia, J. Biomech., 17:801.

Streeter, D.D. Jr. and Hanna, W.T., 1973, Engineering mechanics of successive states in canine left ventricular myocardium: II. Fiber angle and sarcomere length, Circ. Res., 33:657.

Van Heuningen, R., Rijnsburger, W.H. and Ter Keurs, H.E.D.J.. 1982, Sarcomere length control in striated muscle, Am. J. Physiol., 242:H411.

Vogel, W.M., Apstein, C.S., Briggs, L.L., Gaasch, L. and Ahn, J., 1982, Acute alterations in left ventricular diastolic chamber stiffness: role of the erectile effect of coronary arterial pressure and flow in normal and damaged hearts, Circ. Res., 51:465.

Waldman, L.K., Fung, Y.C. and Covell, J.W., 1985, Transmural myocardial deformation in the canine left ventricle; normal in vivo three-dimensional finite strains, Circ. Res., 57:152.

Waldman, L.K., Nosan, D., Villarreal, F. and Covell, J.W., 1988, Relation between transmural deformation and local myofiber direction in canine left ventricle, Circ. Res., 63:550.

Yin, C.P., Strumpf, R.K., Chew, P.H. and Zeger, S.L., 1987, Quantification of the mechanical properties of non-contracting canine myocardium under simultaneous biaxial loading, J. Biomech., 20:577.

ARTERIAL, VALVULAR, VENTRICULAR AND VENOUS FLOW

STEADY FLOWS AND INSTABILITIES IN COLLAPSIBLE TUBES

O. E. Jensen and T. J. Pedley†

Department of Applied Mathematics and Theoretical Physics
University of Cambridge, Silver Street
Cambridge CB3 9EW, UK

† Department of Applied Mathematical Studies
The University
Leeds LS2 9JT, UK

ABSTRACT

The flow through a finite length of externally pressurised collapsible tube is described using a one-dimensional model incorporating the effects of longitudinal wall tension and energy loss through flow separation. The existence and stability of steady flows are analysed to determine under what external conditions self-excited oscillations may occur.

KEYWORDS: Physiology / Fluid dynamics / Blood flow / Korotkoff sounds / Starling resistor / Collapsible tubes / Self-excited oscillations / Flow separation / Nonlinear dynamical system

INTRODUCTION

When the transmural pressure in a blood vessel is negative (either because of low internal pressure, such as that due to hydrostatic pressure drop in the veins above the heart, or because of high external pressure, such as that imposed by a blood pressure cuff on the brachial artery) the vessel may "collapse", with its cross section undergoing both a substantial reduction in area and a large change in shape. Simple laboratory experiments using thin-walled rubber tubes, mounted between two rigid tubes and enclosed in a pressurised chamber (Conrad 1969; Brower & Scholten 1975; Bonis & Ribreau 1978; Bertram 1982, 1986; Bertram *et al.* 1990; see figure 1), have all demonstrated that in this state the interaction between the internal flow and the distorted tube walls can be a dynamic one: even under steady external conditions, a remarkably wide variety of self-excited oscillations can arise which involve variations of the tube area and the flow. When the inertia of an elastic-walled vessel is significant compared with the inertia of the internal fluid, such as the airways in the lung, such oscillations take the form of small amplitude wall-flutter, leading to wheezing. When the fluid inertia dominates, however, they generally have larger amplitude and lower frequency, and may be responsible for the Korotkoff sounds produced during sphygmomanometry, for

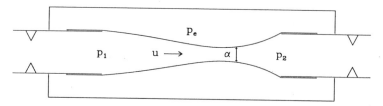

Fig. 1 The conventional laboratory apparatus: a rigid tube is inserted between two rigid tubes and is enclosed in a chamber held at pressure p_e; the flow is driven from a constant head reservoir and is controlled by the variable resistances and inertances of the upstream and downstream rigid tubes; downstream the system is open to the atmosphere.

example. To develop an understanding of the mechanism of these oscillations, and to obtain a description of the very rich bifurcation structure of this nonlinear dynamical system, a one-dimensional mathematical model of the flow in a finite length of elastic-walled tube is presented briefly here; for a more complete description of the model see Cancelli & Pedley (1985) and Jensen & Pedley (1989). A broader review of collapsible tube flow is given by Kamm & Pedley (1989).

THE MODEL

The elastic properties of the tube. Since most experiments use uniform, thin-walled rubber tubes we follow Shapiro (1977) by describing the elastic properties of the tube through a local relationship between the tube area and the transmural pressure,

$$p - p_e = \mathcal{P}(\alpha) = \begin{cases} k(\alpha - 1) & \text{if } \alpha > 1 \\ 1 - \alpha^{-3/2} & \text{if } \alpha \leq 1. \end{cases} \tag{1}$$

$\alpha(x, t)$ is the nondimensionalised tube area at a distance x along the axis of the tube at time t, defined so that the tube is unstressed when $\alpha = 1$; $p(x, t)$ and p_e are the pressures inside and outside the tube respectively, nondimensionalised by the tube bending stiffness; k is a large positive constant. This "tube law" (1) agrees well with experimentally determined pressure-area relations for uniform tubes which are not subject to longitudinal tension; it is plotted on figure 2, together with the approximate shape of the tube cross-section for various transmural pressures.

Ignoring all dissipation, the equations describing the axial flow in an elastic walled tube are those of conservation of mass and momentum,

$$\frac{\partial \alpha}{\partial t} + \frac{\partial (u\alpha)}{\partial x} = 0 \tag{2}$$

$$\frac{\partial u}{\partial t} + u \frac{\partial u}{\partial x} = -\frac{\partial p}{\partial x}. \tag{3}$$

(2) and (3) are valid provided that any transverse velocities are small compared to the axial velocity; $u(x, t)$ is the axial velocity averaged across the cross-section of the tube. It is easily

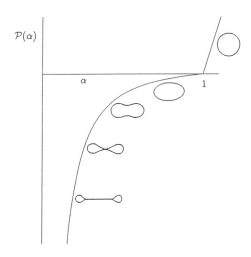

Fig. 2 The tube law: the transmural pressure is shown as a function of the nondimensional cross-sectional area of a uniform tube; the shape of the cross section is indicated for varying pressures.

shown (Shapiro 1977; Cancelli & Pedley 1985) that if a frictional term is included in (3), then under steady conditions these equations break down at a point where the flow speed u equals c, the dimensionless speed of propagation of small amplitude pressure waves in the tube wall, given by $c^2 = \alpha \mathcal{P}'(\alpha)$. At such a point the (negative) gradient of tube area with x becomes infinite, the model therefore breaks down and the tube is said to "choke".

Longitudinal Wall Tension. This difficulty may be overcome by introducing constant longitudinal tension in the tube wall. Considering the tube as a pair of parallel membranes, McClurken *et al.* (1981) showed that longitudinal tension T may be modelled by modifying the tube law (1) to be

$$p - p_e = \mathcal{P}(\alpha) - \frac{T}{R}. \tag{4}$$

$1/R$ is the longitudinal curvature, which can be approximated by $\frac{1}{2}\partial^2\alpha/\partial x^2$ provided that α varies slowly with x; we can rescale the tube length so that T is unity.

Boundary conditions. If we now calculate the pressure gradient in (3) using (4), we obtain a system of equations which require four boundary conditions. To model the experimental conditions (figure 1) we have to ensure that at the upstream and downstream ends of the collapsible segment the tube area matches that of the supporting rigid tubes, and that the internal pressures p_1 and p_2 at these junctions match those determined by the external parts of the apparatus.

Fully Attached Flow. As a first step in predicting time-dependent behaviour it is necessary to calculate the steady solutions of (2), (3), (4) and the boundary conditions described above. Such a calculation, based on a phase plane analysis, was performed by Reyn (1987) (but using a more complicated tube law than described here) and by Jensen & Pedley (1989). We summarise the conclusions of the latter below.

There are two natural parameters describing steady flow in a collapsed tube, the flow-rate Q and the negative downstream transmural pressure $P = p_e - p_2$. Since we are interested in situations in which the tube is collapsed, we suppose that P takes some positive value. For small non-zero Q two steady solutions exist with the tube collapsed ($\alpha < 1$) along its length, one with smaller minimum area and larger maximum velocity than the other. As Q

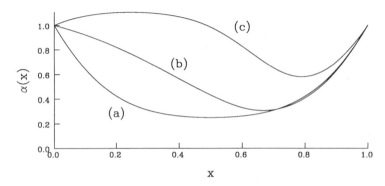

Fig. 3 (a),(b) and (c): Three steady solutions when the flow is separated beyond a constriction in the tube, for $P > 0$ and increasing values of Q.

is increased these two solutions become more alike until at some critical flow rate Q_c they are identical. For larger Q no collapsed solutions exist, although at very large Q (greater than Q_d, say) a fully dilated ($\alpha > 1$) solution exists. All of these solutions are symmetric about the mid-point of the tube; there is no pressure drop across the collapsible segment since we have neglected all dissipation. More recently, Jensen (1989) has shown that of the two collapsed solutions which exist for small Q, that with the larger minimum area is stable to arbitrarily small, time-dependent disturbances, whereas the other is unstable. Thus one would expect only the stable one to be attainable in practice, as small experimental fluctuations would always destabilise the other. Furthermore, recent numerical calculations (consistent with those reported by Cancelli & Pedley 1985) show that starting from some arbitrary initial state with $Q_c < Q < Q_d$, the tube area decreases to zero at some point along the tube in a finite time. Therefore we see that although longitudinal tension allows the tube to assume a stable collapsed state for $Q < Q_c$, choking still occurs for an intermediate range of flow-rates. Since longitudinal tension makes this system dispersive there is no simple relation between flow speed and wave speed determining the critical flow-rate Q_c.

Flow separation beyond a constriction. At high Reynolds numbers the flow downstream of a constriction in a collapsed tube will form a gradually broadening jet, surrounded by regions of separated flow in which there is substantial dissipation of energy. Using a lumped-parameter model, Bertram & Pedley (1982) showed that a description of this energy loss was necessary for the prediction of oscillations, so it is clearly desirable to include it in the present distributed model.

To do so we use an approximation proposed by Cancelli & Pedley (1985), which involves a modification of the inertia term in the momentum equation:

$$\frac{\partial u}{\partial t} + \chi u \frac{\partial u}{\partial x} = -\frac{\partial p}{\partial x}. \tag{5}$$

χ is chosen to equal 1 upstream of the point of flow separation, which is assumed to be the point of minimum tube area, and it is set to a value between 0 and 1 downstream of this point. If $\chi = 1$ in this downstream region we recover (3), and this of course represents fully attached flow in which there is no energy loss; $\chi = 0$ corresponds to a parallel jet beyond the constriction, along which there is no pressure recovery. We assume arbitrarily that $\chi = 0.2$ in

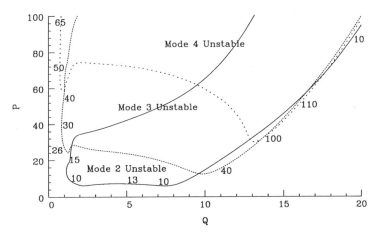

Fig. 4 Neutral curves for modes 2 (solid line), 3 (dotted line) and 4 (dashed line): a steady solution with flow rate Q and negative downstream transmural pressure P will be unstable to a mode 2 oscillation, say, if (Q, P) lies within the mode 2 neutral curve; the figures adjacent to the curves show the nondimensional frequency of the oscillations.

the region of separated flow, allowing for energy loss but partial pressure recovery; the nature of the solutions predicted by this model does not depend qualitatively on the precise value of χ that is choosen.

The introduction of dissipation has a substantial effect on the steady solutions predicted by (1), (2) and (5), as was shown in Jensen & Pedley (1989). For fixed positive P, at small flow-rates there exists a single solution with the tube collapsed along its length (curve (a) in figure 3); as Q is increased the tube begins to bulge at its upstream end (curve (b)), eventually becoming dilated along much of its length but remaining collapsed across a narrow neck at its downstream end (curve (c)). Further increases in flow-rate lead to a reduction in the constriction at this neck until a dilated solution with the flow fully attached is recovered. We see that the introduction of energy loss through flow separation inhibits choking, and leads to highly asymmetric tube profiles. Comparison of the pressure drop across the elastic tube as a function of flow-rate shows good qualitative agreement with the experiments of Bonis & Ribreau (1978), except when the tube is either substantially collapsed along its length or is dilated everywhere, and direct frictional energy losses become significant.

INSTABILITIES OF STEADY SEPARATED FLOW

Linear stability. Having determined the steady flow which exists for a given set of parameters, it is possible to establish whether it is stable to small time-dependent perturbations. The details of this procedure are given in Jensen (1989), but briefly it involves linearising (2), (4), (5) and the boundary conditions about the steady state, allowing the space and time behaviour of the perturbation to be separated; one may then assume that the perturbation is proportional to $\exp(\tau + i\omega)t$, where τ and ω are the growth-rate and frequency of the perturbation; these "eigenvalues" can be calculated numerically, together with the spatial form of the corresponding eigensolution. In general a perturbation consists of a linear combination of an infinite set of discrete eigenmodes, each with a different growth rate τ_j and frequency ω_j;

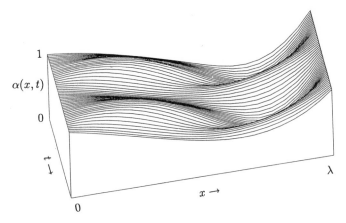

Fig. 5 A mode 2 oscillation: the tube area $\alpha(x,t)$ is plotted at successive instants in time as a function of x.

each mode is labelled $j = 0, 1, 2, \ldots$ according to the magnitude of its frequency. The higher frequency components (with larger j) have spatial structures with shorter lengthscales. Those components which have $\tau_j < 0$ will decay exponentially with time, and are therefore classed as linearly stable; those with $\tau_j > 0$ are linearly unstable, and these will grow rapidly with time until the assumption that they are small breaks down. The locus of points in parameter space along which the intervening condition holds, i.e. $\tau_j = 0$ for $j = 0, 1, 2, \ldots$, are called "neutral curves". A set of such curves has been calculated numerically for fixed values of k, χ, the tube length and the resistances of the upstream and downstream rigid tubes; they are plotted on figure 4. These should be used as a general guide for behaviour at other parameter values. Clearly for small Q (when the tube is collapsed along its length as in figure 3(a)) and small P (when the tube is only mildly constricted), the steady solutions are linearly stable; modes 2, 3 and 4 become progressively unstable as Q and P are increased. The frequencies of the unstable modes are indicated along the corresponding neutral curves, and it is clear that each different mode has frequencies in a distinct range.

Weakly nonlinear stability. Suppose a parameter such as Q or P is varied so that we cross a neutral curve and a linearly stable steady solution becomes unstable to one of these oscillations. In general the oscillation will develop in one of two ways: either "supercritically", with a stable disturbance of gradually increasing amplitude growing as the steady solution becomes linearly unstable; or else "subcritically", with large experimental fluctuations perturbing the linearly stable steady solution to some unsteady nonlinear state. Using weakly nonlinear analysis, Jensen (1989) has determined whether oscillations develop sub- or supercritically for the range of parameters shown in figure 5. In the subcritical case the neutral curve is a poor indication of the stability of the steady state to large perturbations, and the transition to unsteady behaviour is likely to display hysteresis. Jensen (1989) also examined the interactions of two modes which become simultaneously unstable, and demonstrated the existence of stable quasiperiodic motion, for example.

The mechanism of the oscillations. A mode 2 oscillation which has been computed numerically (using the technique described in Cancelli & Pedley 1985) is presented as a function of space and time in figure 5. This is a clear demonstration of the mechanism proposed in that paper: as the minimum area of the tube decreases and the constriction (and therefore the separation point) move downstream, the accelerating attached flow sends a high pressure wave towards the downstream end of the collapsible tube, from where it is reflected and returns to force the constriction open, pushing the separation point upstream; the reverse process begins and so the oscillation continues. For further examples of numerically calculated nonlinear oscillations see Cancelli & Pedley (1985) and Matsuzaki & Matsumoto (1989).

COMPARISON WITH EXPERIMENT

At present the experiments by Bertram *et al.* (1990) provide the most detailed and complete data about the dynamical behaviour of collapsible tubes. Unfortunately, Bertram found it necessary to use a relatively thick-walled tube; the model presented above is based upon the assumption that the tube has a thin wall. In addition, constraints of the numerical method used in the linear stability calculations do not allow comparison using Bertram's exact parameter values. Nevertheless, many significant qualitative features of the experiments are predicted.

The tube is observed to become unstable to a number of different modes of oscillation. Three of these (which Bertram classified by their low, intermediate and high frequencies) correspond approximately to the oscillations predicted by the linear stability calculations, at least in terms of their relative frequencies: conclusive comparison of their distribution in parameter space is not possible. Bertram suggested that the additional class of high frequency "noisy" fluctuations arose from some internal fluid-dynamical instability (such disturbances are inevitable at high Reynolds numbers), and so it is not surprising that similar oscillations are not predicted by this model. However, the different classes of oscillation can modulate one another in a highly complex manner — for a description of simpler mode interactions see Jensen (1989) — giving rise to nonlinear harmonics and disturbances which often have a broad-band freqeuncy structure. The interpretation of this last feature using ideas from dynamical systems theory is of considerable interest.

CONCLUSIONS

This model demonstrates that an essential component in a description of the self-excited oscillations which arise in collapsed, elastic-walled tubes is a good representation of the unsteady flow separation process. Even using the simplest assumptions, much of the complicated dynamical behaviour observed in experiment can be qualitatively predicted.

REFERENCES

Bertram, C.D. 1982 Two modes of instability in a thick-walled collapsible tube conveying a flow. *J. Biomech.* **15**, 223-224

Bertram, C.D. 1986 Unstable equilibrium behaviour in collapsible tubes. *J. Biomech.* **19**, 61-69

Bertram, C.D. & Pedley, T.J. 1982 A mathematical model of unsteady collapsible tube behaviour. *J. Biomech.* **15**, 39-50

Bertram, C.D., Raymond, C.J. & Pedley, T.J. 1990 Mapping of instabilities during flow through collapsed tubes of differing length. *J. Fluids & Structures* (to appear)

Bonis, M. & Ribreau, C. 1978 Etude de quelques propriétés de l'écoulement dans une conduite collabable *La Houille Blanche* **3/4**, 165-173

Brower, R.W. & Scholten, C. 1975 Experimental evidence on the mechanism for the instability of flow in collapsible vessels. *Med. Biol. Engg.* **13**, 839-845

Cancelli, C. & Pedley, T.J. 1985 A separated-flow model for collapsible-tube oscillations. *J. Fluid Mech.* **157**, 375-404

Conrad, W.A. 1969 Pressure-flow relationships in collapsible tubes. *I.E.E.E. Trans. Bio-med. Engg.* **BME-16**, 284-295

Jensen, O.E. & Pedley, T.J. 1989 The existence of steady flow in a collapsed tube. *J. Fluid Mech.* **206** 339-374

Jensen, O.E. 1989 Instabilities of flow in a collapsed tube. *J. Fluid Mech.* (submitted).

Kamm, R.D. & Pedley, T.J. 1989 Flow in collapsible tubes: a brief review. *Trans. A.S.M.E. K: J. Biomech. Engg.* **111**, 177-179

Matsuzaki, Y. & Matsumoto, T. 1989 Flow in a two-dimensional collapsible channel with rigid inlet and outlet. *Trans. A.S.M.E. K: J. Biomech. Engg.* **111**, 180-184

McClurken, M.E., Kececioglu, I., Kamm, R.D. & Shapiro, A.H. 1981 Steady, supercritical flow in collapsible tubes. Part 2. Theoretical studies. *J. Fluid Mech.* **109**, 391-415

Reyn, J.W. 1987 Multiple solutions and flow limitation for steady flow through a collapsible tube held open at the ends. *J. Fluid Mech.* **174**, 467-493

Shapiro, A.H. 1977b Steady flow in collapsible tubes *Trans. A.S.M.E. K: J. Biomech. Engg.* **99**, 126-147

FLOW IN PIPES WITH VARYING CROSS SECTIONS

Arnold F. Bertelsen

Section of Mechanics
Department of Mathematics
University of Oslo
P.O.Box 1053, Blindern
0316 Oslo 3, Norway

KEYWORDS

Arterial blood flow. Mathematical modelling. Varying pipe cross
sections. Steady. Oscillating. Secondary flows.

ABSTRACT

Viscous incompressible flow in pipes of variable cross sections is
considered with a view to physiological conditions. Steady and oscillatory
inlet flow conditions are discussed. Asymptotic solutions of the basic
equations, with respect to inherent parameters, are given. Weakly non-linear
effects are included. The predicted flow fields are generally in accordance
with observations.

INTRODUCTION

Steady and unsteady flow in pipes with varying cross section, both in
area and shape, are quite common in physiological flow problems, e.g. as
aneurisms or stenosis. Several mathematical models depicting such flow
fields have been published, see for example Hall (1974), Doffin, Chagneall,
Borzeix, Bordon & Ripert (1977), Kimmel & Dinnar (1983), Grotberg (1984)
and Wille (1984). In the four first papers, asymptotic analytical methods
are applied, while numerical simulations are used in the last. Hall considered slowly varying cross sections and included one term in the angular
change of the boundary (i.e. $r=a_M \cos M\theta$, see figure 1 in this paper).
Doffin et al. investigated the flow in a stenosis. Grotberg discussed the
flow in a slowly converging cone. The simulations of Wille were limited
to Reynolds numbers smaller than 10. Perktold (1989) and Reuderink,
Schreurs & van Steenhoven (1989) have improved and extended numerical
simulations to cover realistic physiological flow conditions.

Because of limitations in geometrical shape covered by analytical
investigations and the complexity of three dimensinal numerical models, it
may be worth while to extend the analytical calculations to cover more
complex cross sections. So is done in this report by adopting an approach
similar to Hall (1974). Both steady and unsteady inlet conditions are
considered and weakly non-linear effects are included.

FORMULATION OF THE PROBLEM

The flow problems mentioned above will be depicted in a cylindrical coordinate system (r,θ,z) as sketched in figure 1. The pipe wall is supposed to be given by

$$r = \sum_{n=0}^{N} a_n(z) \cos 2n\theta \tag{2.1}$$

which implies symmetry about the planes $\theta=0$ and $\theta=\pi/2$. The inlet section is assumed to be a straight circular pipe where

$$a_0 \Big|_{z\to-\infty} \to a_\infty = \text{const}$$

$$a_n \Big|_{z\to-\infty} \to 0 \qquad (n=1,2,\ldots,N) \tag{2.2}$$

We also suppose

$$a_0 > |a_1| > |a_2| > \ldots > |a_N|$$

$$\frac{da_n}{dz} = 0(\gamma a_n) \ll 0(\frac{a_n}{a_\infty}) \tag{2.3}$$

The continuity equation

$$\frac{\partial u}{\partial r} + \frac{u}{r} - \frac{1}{r}\frac{\partial v}{\partial \theta} + \frac{\partial w}{\partial z} = 0 \tag{2.4}$$

and the vorticity equation

$$\frac{\partial}{\partial t}(\nabla x \vec{v}) + \nabla x(\vec{v}\cdot\nabla\vec{v}) = \nu\Delta(\nabla x\vec{v}) \tag{2.5}$$

is used as the basic equation (ν is kinematic viscosity and standard notations else).

Steady Inlet Conditions

In this case we suppose a parabolic profile has developed in the straight inlet section, i.e.

$$w\Big|_{z\to-\infty} \to W_\infty[1 - (\frac{r}{a_\infty})^2]$$

$$u\Big|_{z\to-\infty} \to 0$$

$$v\Big|_{z\to-\infty} \to 0 \tag{2.6}$$

The boundary conditions are

$$u = v = w = 0 \quad \text{on} \quad r = \sum_{n=0}^{N} a_n \cos 2n\theta \tag{2.7}$$

The basic equations are non-dimensionalized using W_∞ and a_∞ as velocity and length scales, respectively, giving the continuity equation

$$\frac{\partial u}{\partial t} + \frac{u}{r} + \frac{1}{r}\frac{\partial v}{\partial \theta} + \frac{\partial w}{\partial z} = 0 \tag{2.8}$$

and the vorticity equation

$$\nabla \times (v \cdot \nabla \vec{v}) = \frac{1}{Re}\Delta(\nabla \times \vec{v}) \tag{2.9}$$

where the Reynolds number Re is

$$Re = \frac{W_\infty a_\infty}{\nu} \tag{2.10}$$

It is well-known that the non-linearity of the vorticity equation (2.9) causes unsurmountable difficulties with respect to obtaining exact analytical solutions and the complex shape of the pipe wall make it hard to impose the boundary conditions in a convenient way. These difficulties are here resolved by assuming weak non-linearity and introducing a slowly varying scale in the radial direction by defining

$$R = \frac{r}{\sum_{n=0}^{N} a_n \cos 2n\,\theta} \tag{2.11}$$

as the scaled radial coordinate. In accordance with equation (2.3) we also introduce a scaled axial coordinate

$$s = \gamma z \tag{2.12}$$

Introduction of these new variables leads to

$$\frac{\partial}{\partial r} = \frac{1}{\sum_{n=0}^{N} a_n \cos 2n\,\theta} \frac{\partial}{\partial R}$$

$$\left(\frac{\partial}{\partial\theta}\right)_r = \left(\frac{\partial}{\partial\theta}\right)_R + 2\frac{\sum_{n=1}^{N} na_n \sin 2n\theta}{\sum_{n=0}^{N} a_n \cos 2n\theta} R\frac{\partial}{\partial R} \tag{2.13}$$

$$\left(\frac{\partial}{\partial z}\right)_r = \gamma\left[\frac{\partial}{\partial s} - \frac{\sum_{n=1}^{N} a_n' \cos 2n\theta}{\sum_{n=0}^{N} a_n \cos 2n\theta} R\frac{\partial}{\partial R}\right]$$

$$a_n' \equiv \frac{da_n}{ds}$$

which are used to rewrite the basic equations (2.8 & 2.9).

Asymptotic solutions of the rewritten equations subject to the boundary conditions (2.7) are discussed in section 3.1.

Oscillatory Inlet Conditions

In this case the non-dimensional basic equations read

$$\frac{\partial u}{\partial r} + \frac{u}{r} + \frac{1}{r}\frac{\partial v}{\partial\theta} + \frac{\partial w}{\partial z} = 0 \tag{2.14}$$

$$\frac{\partial}{\partial\tau}(\nabla \times v) + \varepsilon\nabla \times (v \cdot \nabla\vec{v}) = \frac{1}{2}\beta^2\nabla^2(\nabla \times \vec{v}) \tag{2.15}$$

43

where

$$\beta = \frac{1}{a_\infty} \sqrt{2\nu/\omega}$$

$$\varepsilon = W_\infty/\omega a_\infty$$

$$\tau = \omega t \qquad\qquad (2.16)$$

ω = frequency of oscillation

W_∞ = typical velocity in straight inlet section

The thin Stokes layer approximation $\beta \ll 1$ will be considered subject to which

$$w|_{z \to -\infty} \longrightarrow \sin\tau - e^{-\eta}\sin(\tau-\eta) \qquad\qquad (2.17)$$

where

$$\eta|_{z \to -\infty} = (1-r)/\beta \qquad\qquad (2.18)$$

Other boundary conditions are

$$u|_{z \to -\beta} \longrightarrow 0$$

$$v|_{z \to -\infty} \longrightarrow 0 \qquad\qquad (2.19)$$

$$u = v = w = 0 \quad \text{on} \quad r = \sum_{n=0}^{N} a_n \cos 2n\theta$$

The slowly varying scaling of the radial coordinate

$$R = \frac{r}{\sum_{n=0}^{N} a_n \cos 2n\theta} \qquad\qquad (2.20)$$

and the scaling

$$s = \gamma z \qquad\qquad (2.21)$$

are also introduced.

Solutions of equations (2.14) & (2.15) subject to boundary conditions (2.17) and (2.19) will be discussed in section 3.2.

APPROXIMATE SOLUTIONS

Steady Inlet Conditions

The conditions specified by equations (2.3) & (2.4) and the restriction

$$\gamma Re < 1 \qquad\qquad (3.1)$$

justify the following perturbation expansion of the solution of the basic equations (2.6) and (2.7)

$$u = \gamma W_0 \left(a_0' u_{010} + \frac{a_0' a_1}{a_0} u_{110} + a_1' u_{210} + \frac{a_0' a_1^2}{a_0^2} u_{310} \right) + \ldots \tag{3.2}$$

$$v = \gamma W_0 \left(a_0' v_{010} + \frac{a_0' a_1}{a_0} v_{110} + a_1' v_{210} + \frac{a_0' a_1^2}{a_0^2} v_{310} \right) + \ldots \tag{3.3}$$

$$w = W_0 \left[w_{000} + \frac{a_1}{a_0} w_{100} + \left(\frac{a_1}{a_0}\right)^2 w_{200} + \frac{a_2}{a_0} w_{300} + \ldots \right]$$

$$+ \gamma Re W_0^2 \left[a_0 a_0' w_{010} + a_0' a_1 w_{101} + a_0 a_1' w_{201} + a_1 a_1' w_{301} + \ldots \right] \tag{3.4}$$

$$X = \frac{W_0}{a_0} \left\{ X_{000} + \frac{a_1}{a_0} X_{100} + \left(\frac{a_1}{a_0}\right)^2 X_{200} + \frac{a_2}{a_0} X_{300} + \ldots \right\} \tag{3.5}$$

$$\Psi = \frac{W_0}{a_0} \left\{ \Psi_{000} + \frac{a_1}{a_0} \Psi_{100} + \left(\frac{a_1}{a_0}\right)^2 \Psi_{200} + \frac{a_2}{a_0} \Psi_{300} + \ldots \right\} \tag{3.6}$$

$$\Omega = \frac{\gamma W_0}{a_0} \left\{ \Omega_{010} + \frac{a_0' a_1}{a_0} \Omega_{110} + a_1' \Omega_{210} + \frac{a_0' a_1^2}{a_0^2} \Omega_{310} + \ldots \right\} \tag{3.7}$$

These expressions inserted into the basic equations (2.9) and (2.10), give subject to $\gamma \to 0$, $\gamma Re \to 0$, (2.3) and the boundary conditions, the asymptotic solutions we are looking for.

It is important to be aware of the prefactor W_0 appearing in the perturbation expansion (3.1)-(3.4). this prefactor is determined by claiming

$$\int_0^1 \int_0^{2\pi} W_0 (1-R^2) \left(\sum_{n=0}^{N} a_n \cos 2n\theta \right)^2 R dR d\theta = \int_0^1 \int_0^{2\pi} (1-R^2) R dR d\theta \tag{3.8}$$

which is an integral continuity condition which has to be applied. The condition gives

$$W_0 = \frac{1}{a_0^2 + \frac{1}{2}\sum a_n^2} \tag{3.9}$$

Oscillatory Inlet Conditions

In this case the conditions

$$\beta \ll 1 \tag{3.10}$$

$$\gamma \varepsilon \ll 1 \tag{3.11}$$

justify the approximation of an outer (leading order) potential flow field associated with the oscillatory motion which has to be matched by an inner Stokes layer solution. In the outer region we solve the equations

$$\nabla \times \vec{v} = 0 \tag{3.12}$$

$$\nabla \cdot \vec{v} = 0 \tag{3.13}$$

subject to the boundary conditions

$$\vec{N} \cdot \vec{v} = 0 \quad \text{on} \quad R = 1 \tag{3.14}$$

$$\vec{v}\big|_{z \to -\infty} \to \vec{i}_z \sin \tau \tag{3.15}$$

The outer asymptotic expansions can be written

$$u = \gamma W_0 [a_0' U_{0100} + a_0' \frac{a_1}{a_0} U_{1100} + a_1' U_{2100}$$

$$+ a_0' \frac{a_1^2}{a_0^2} U_{3100} + a_0' \frac{a_2}{a_0} U_{4100} + a_1' \frac{a_1}{a_0} U_{5100} + \ldots] \tag{3.16}$$

$$v = \gamma W_0 [a_0' V_{0100} + a_0' \frac{a_1}{a_0} V_{1100} + a_1' V_{2100} + a_0' \frac{a_1^2}{a_0^2} V_{3100}$$

$$+ a_0' \frac{a_2}{a_0} V_{4100} + a_1' \frac{a_1}{a_0} V_{5100} + \ldots] + \ldots \tag{3.17}$$

$$w = W_0 W_{0000} + \gamma^2 W_0 [(3a_0'^2 - a_0 a'') W_{0100} + \ldots]$$

$$+ \gamma \varepsilon W_0 [W_{0101} + \ldots] + \ldots \tag{3.18}$$

where weakly nonlinear effects are included. These expansions inserted
into equations (3.12, 3.13) give subject to $\gamma \ll 1$ and (2.3), the differential equation of each term in the expansion.

The integral continuity condition

$$\int_0^1 \int_0^{2\pi} W_0 (\textstyle\sum a_n \cos 2n\theta)^2 R dR d\theta = \pi \tag{3.19}$$

determines

$$W_0 = \frac{1}{a_0^2 + \frac{1}{2}\sum a_n^2} \tag{3.20}$$

The linearized outer solution gives on $R = 1$ and $\tau = \pi/2$ a velocity vector
$\vec{\Sigma}_B$ which is tangent to the pipe wall. We find

$$\Sigma_B = \gamma W_0 [\vec{i}_r (a_0' U_{0100} + a_0' \frac{a_1}{a_0} U_{1100} + \ldots)$$

$$+ \vec{i}_\theta (a_0' V_{0100} + a_0' \frac{a_1}{a_0} V_{1100} + \ldots)]_{\substack{R=1 \\ \tau=\pi/2}}$$

$$+ \vec{i}_z W_0 \{W_{0000} + \gamma^2 [(3a_0'^2 - a_0'') W_{01000} + \ldots) + \ldots\}]_{\substack{R=1 \\ \tau=\pi/2}}$$

$$\equiv W_0 [\gamma(\vec{i}_r U_B + \vec{i}_\theta V_B) + \vec{i}_z (1 + \gamma W_B)] \tag{3.21}$$

In order to simplify the boundary layer calculations on the pipe wall, we introduce a coordinate system (λ,ϕ,σ) where $\vec{i}_\lambda = \vec{N}$, $\vec{i}_\sigma \| \vec{\Sigma}_B$ and $\vec{i}_\phi = \vec{i}_\sigma \times \vec{i}_\lambda$ where

$$\vec{i}_\sigma \approx \vec{i}_z + \gamma(\vec{i}_r U_B + \vec{i}_\theta V_B) - \vec{i}_z \frac{\gamma^2}{2}(U_B^2 + V_B^2) \tag{3.22}$$

The Stokes layer solution is then simply,

$$\Sigma_{00} = \Sigma_B [\sin\tau - e^{-\eta}\sin(\tau-\eta)] \tag{3.23}$$

where

$$\eta = -\beta\lambda \tag{3.24}$$

and

$$\Sigma_B = W_0[1 + \frac{1}{2}\gamma^2(u_B^2 + V_B^2 + 2W_B)] \tag{3.25}$$

The velocity component normal to the pipe wall is found using the continuity equation and it turns out to have a series expansion

$$\Lambda = \beta\gamma\Lambda_{10} + \beta^2\gamma\Lambda_{20} + \cdots \tag{3.26}$$

The leading term is governed by the equation

$$-\frac{\partial\Lambda_{10}}{\partial\eta} + \left[\frac{\partial W_0}{\partial s} + \kappa W_0\right][\sin\tau - e^{-\eta}\sin(\tau-\eta)] = 0 \tag{3.27}$$

with the boundary condition

$$\Lambda_{10}(\eta=0,\tau) = 0 \tag{3.28}$$

The solution is

$$\Lambda_{10} = \frac{1}{2}(\frac{dW_0}{ds} + \kappa W_0)\{\cos\tau - \sin\tau$$
$$+ 2\eta\sin\tau - e^{-\eta}[\sin(\tau-\eta) - \cos(\tau-\eta)]\} \tag{3.29}$$

where κ is essentially the curvature of the pipe wall along the ϕ-coordinate curve. The linearized boundary layer solution $(\Lambda_{10},\Sigma_{00})$, is substituted into the nonlinear terms of the boundary layer equations. The solution of these equations gives among other things a drift velocity of order $O(\gamma\varepsilon)$.

$$\Sigma_{01}^{(s)} = 2\{W_0 \frac{dW_0}{ds}[-\frac{3}{4}e^{-\eta}\sin\eta + \frac{1}{4}e^{-\eta}\cos\eta$$
$$-\frac{\eta}{4}e^{-\eta}(\cos\eta - \sin\eta) + \frac{e^{-2\eta}}{8}]$$
$$-\kappa W_0^2[\frac{1}{4} - \frac{e^{-\eta}}{4}(\cos\eta + 2\sin\eta) + \frac{\eta}{4}e^{-\eta}(\cos\eta - \sin\eta)]\} \tag{3.30}$$

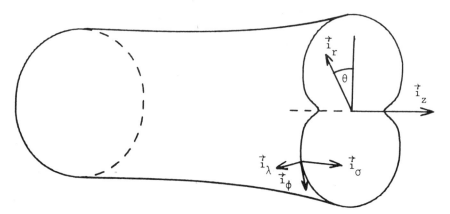

Fig. 1. A sketch of the coordinate systems referred to in the
         text.

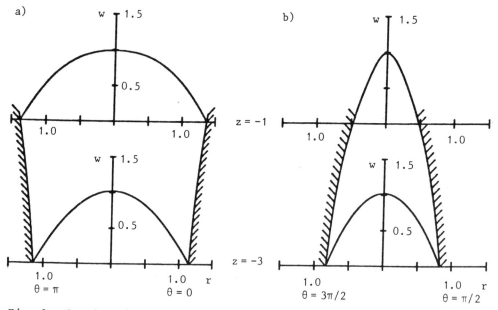

Fig. 2. Steady axial velocity profiles at $z = -1$ and $z = -3$ in the plane
         $\theta = 0$ are shown in figure 2a and in the plane $\theta = \pi/2$ in figure 2b.

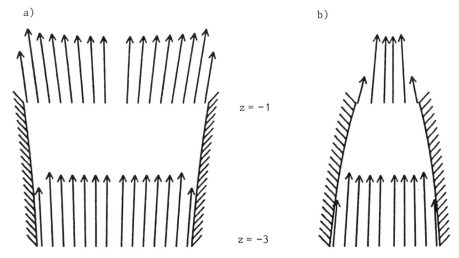

Fig. 3. Oscillatory velocity vector plots (time $\tau = \pi/2$) at $z = -1$
and $z = -3$ in the plane $\theta = 0$ are shown in figure 3a and
in the plane $\theta = \pi/2$ in figure 3b.

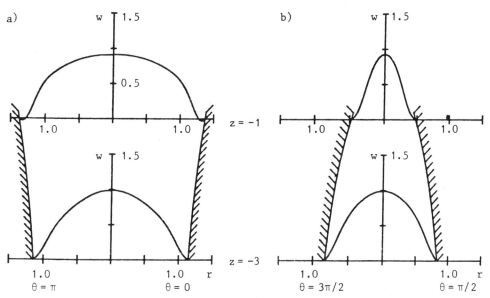

Fig. 4. Time averaged pulsatile axial velocity profiles at $z = -1$ and
$z = -3$ in the plane $\theta=0$ are shown in figure 4a and in the plane
$\theta = \pi/2$ in figure 4b.

This drift velocity will, because of the viscosity, induce streaming effects outside the boundary layer. The outer flow proportional to $W_0 \dfrac{dW_0}{ds}$ has has been calculated.

DISCUSSION

Many (28) terms of the perturbation expansions (3.2), (3.3) and (3.4) have been calculated. Blood vessel data from an aortic bifurcation has been used to estimate the coefficients a_n (n=0,...,7). Using a Reynolds number Re = 100 we get the axial velocity profiles shown in figure 2. Weakly non-linear effects are observable in z=-1 (z=R=0 is the apex point).

Velocity vector diagrams based on the oscillatory solution (19 terms) (3.16), (3.18) and the boundary layer solution (3.23) are shown in figure 3. The velocity field close to the wall are fairly parallel to the wall indicating an accurate outer solution.

Time averaged pulsatile profiles are probably most relevant to flow induced wall lesions. Such profiles have been constructed by adding the steady solution and the steady streaming induced by an oscillatory inlet flow. Using Re=100, β=0.1 and ϵ=2.5 we get the profiles shown in figure 4. We notice that back-flow is predicted close to the wall in the cross section z=-1. The calculations are formaly valid for $\gamma Re \lesssim 1$; $\gamma\epsilon<1$; $\beta<<1$ and $\gamma\epsilon^2/\beta^2<1$.

CONCLUSION

A mathematical model of blood flow including weakly nonlinear effects seems to predict flow features of importance for a discussion of flow induced lesions in the vessel wall. Modified wall shear stress and secondary flows are found.

ACKNOWLEDGEMENT

Mr. M. Brattberg is acknowledged for his most helpful check of the calculations above.

REFERENCES

Doffin, J., Chagneau, F., Borzeix, J., Bordon, P., and Ripert, G., 1977, INSERM-Euromech 92 Cordiovascular and pulmonary dynamics, Sept. 1977, 71, pp. 43-52.
Gaver, Donald P., III and Grotberg, James B., 1986, J. Fluid Mech., 172, pp. 47-61.
Grotberg, J. B., 1984, J. Fluid Mech., 141, pp. 249-264.
Hall, P., 1974, J. Fluid Mech., 64(2), pp. 209-226.
Kimmel, E., and Dinnar, U., 1983, J. Biomech. Eng., 105, pp. 112-119.
Perktold, K., 1989, Proc. 2nd Int. Symp. on Biofluid Mech. and Biorheology (Munich, FRG), pp. 685-694.
Reuderink, P., Willems, P., Schreurs, P., and van Steenhoven, A., 1989, Proc. 2nd Int. Symp. on Biofluid Meck. and Biorheology (Munich, FRG), pp. 455-462.
Wille, O., 1984, J. Biomech. Eng., 6, pp. 49-54.

QUASI-STEADY LAMINAR VISCOUS FLOW IN CIRCULAR CURVED PIPE

M. Thiriet [1,2,3], J.M.R. Graham [2] and R.I. Issa [4]

[1]LBHP URA 343 CNRS, Université Paris VII
2, Place Jussieu - 75251 Paris Cedex 05 France
[2]PFSU and [3]Departement of Aeronautics - Imperial College
[4]Royal School of Mines, London SW 7 - UK

ABSTRACT

A numerical study of a pulsatile flow of an incompressible viscous
fluid in a circular-sectioned tube of given curvature (90° bend, curvature ratio of 1/10) has been carried out. The flow rate, generated by a
non-zero mean sinusoidal pressure variation remains positive throughout
the whole cycle. The flow is characterized by the following set of values
of the governing parameters : Womersley parameter of 4, amplitude ratio
of 1.25, Reynolds number range 40 - 360, Strouhal number range 0.05 -
0.45. In such conditions, the flow is always unidirectional and a single
secondary motion occurs in the half cross-section. The flow pattern
varies during the deceleration phase, the cross flow becoming very weak
when the axial flow reaches its minimal value. Consequently the shear
stress undergoes great changes during the cycle, the circumferential
component being smaller than the axial component. The low-shear region is
located near the inner edge, except in a short entrance segment of the
curved tube.

KEY WORDS : CURVED TUBE - NUMERICAL MODEL - PULSATILE FLOW

INTRODUCTION

The flow behaviour in curved pipes have practical importance in biomechanics, because the network of anatomical conduits (blood vessels,
tracheo-bronchial tree, ...) are characterized by numerous sites of curvature and branching. Besides, Jan et al. (1989) have observed, from
flow-visualization experiments of oscillatory flow in a symmetrical rigid
bifurcation, that many of the flow features are mainly due to the effect
of curvature. The knowledge of the temporal and spatial distribution of
the flow velocities and the shear stresses may provide a greater understanding of the mass transfer from the tube lumen to the tube wall. In
curved pipes, the secondary motions induce an increase in diffusional
deposition of solid particles and the mass radial transport is thus
greater than in straight ducts. However, the present study is focused on
the fluid-mechanic aspect. The classical fluid dynamic equation (mass and
momentum conservation) are solved for a quasi-steady flow in a curved
tube.

This study employs an uniform rigid curved tube of circular cross-section. The curvature ratio, k = a/R (a : tube radius, R : radius of curvature), is 1/10. The angle of curvature is 90°. The fluid is assumed to be incompressible, homogeneous in composition and newtonian. The flow is considered laminar.

The temporal variations of the axial velocity profiles in the straight pipe upstream from the bend are imposed by a pulsatile (sinusoidal with non-zero mean) pressure variation (amplitude of 2.7 kPa, mean of 13.3 kPa, frequency of oscillation 1 Hz). The Womersley parameter α is 4 and the amplitude ratio γ (ratio of the amplitude to the mean of the velocity waveform) is 0.8. From the beginning of the acceleration period to its end, the Reynolds number Re increases from 40 to 360, the Dean number De from 12 to 113, and the Strouhal number St decreases from 0.45 to 0.05 (Fig.1)

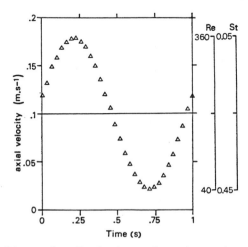

Figure 1. Variations throughout the entire cycle of the cross-sectional average of the axial velocity with the range of values of both Reynolds Re and Strouhal St numbers.

The numerical model uses the finite-volume method. The domain is discretised into hexahedra; the pressure node is located at the center of the hexahedron, while the components of the flow velocity are computed at the center of the faces of the cell to which they are normal.

The curved section is interposed between 2 straight pipes of length 3.6 d (d : tube inner diameter) and 8.1 d upstream and downstream from the bend respectively. A coarse mesh (7x7x48) was used because of the limit in available memory size.

Linear interpolations with spatial weighting factors are used to calculate the physical quantities at the required locations from the values stored at the grid nodes. The convection and diffusion terms in the transport equations are represented by an hybrid upwind/central differencing scheme (Caretto et al., 1972). The set of the finite-difference forms of the transport and continuity equations is solved with the PISO algorithm (Issa, 1982). This predictor-corrector scheme is regulated by 2 "pressure-increment" equations. The velocity field is computed with a line-by-line counterpart of Gauss-Seidel iteration, using the tri-diagonal matrix scheme. The pressure field is calculated with the Stone

implicit method. The values of the residuals are compared to the specified maximum tolerable thresholds for each equation solved over the whole
field. New iterations are applied repeatedly until the prescribed level
of convergence is satisfied.

The classical boundary conditions are applied at the wall
($u = v = w = o$, where u, v, w are respectively the circumferential,
radial and axial velocity components) and at the centerplane
($u_x = v = w_x = o$, where the subscript denotes partial differentiation,

X : circumferential direction) of the pipe model. The velocity field is
given at the tube inlet. At the tube outlet, the flow is assumed unidimen-
sional and insensitive to downstream conditions.

RESULTS

Comparison between experimental results, data from the literature
and the numerical results were done in order to validate the computational model . The numerical tests were carried out (i) with entrance
Poiseuille flow conditions ($140 < De < 440$) (ii) with constant injection
velocity conditions, either in the upstream straight pipe (De = 216) or
directly in the curved tube (De = 183), (iii) with turbulent flow régime
(De = 22091) using the $k - \varepsilon$ model of turbulence of L aunder and S palding
(1972).

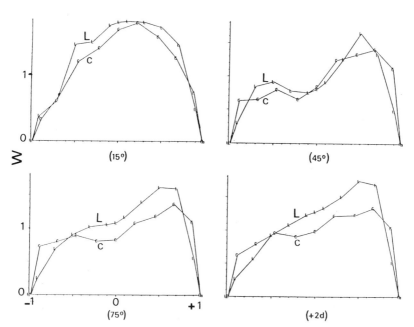

Figure 2. Comparison between the measured
velocity profile (L) and the numerical
extrapolated values (C) in a plane located
at 0.1 a from the centerplane, at different
stations along the pipe model (at 15°, 45°,
75° in the 90° bend and at + 2 d in the
downstream straight tube).

53

Fig. 2 shows the axial velocity profiles with condition (i)
(De = 433 , k = 1/10) in a plane located at a distance 0. 1 a of the
centerplane. A reasonable good agreement is obtained between the laser-
doppler measurements and the computational results ; however the shear
stress is overestimated, especially at the inner tube edge. This overes-
timation in shear stress magnitude is also observed with condition (ii).
The comparison with velocity measurements made by Agrawal et al. (1978)
in a 180° bend and the present numerical results for a 90° bend is displayed in Fig. 3. As for the finite-difference model of Soh and Berger
(1984), taking a constant entrance dynamic pressure as inlet boundary

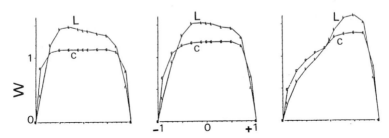

Figure 3.   Comparison between the laser-doppler
measurements (L) of Agrawal et al (1978) and the
computed velocity profiles (C) at 3 differents
stations within the bend ($\sim$ 15°, $\sim$ 30°, $\sim$ 60°)
from left to right.

condition, the agreement between the calculated axial velocity profiles
and the laser-doppler measurements are better in the downstream segment
than in the entrance region of the bend.

Fig. 4 illustrates the velocity profiles in the mid-vertical plane
obtained in the present study and the laser-doppler measurements of
Azzola et al. (1984) in a 180° bend. At least for this position, the difference in shear rate is smaller.

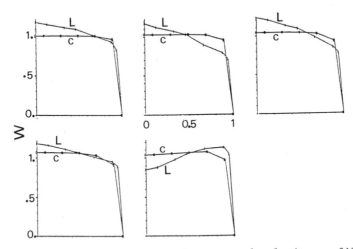

Figure 4.   Comparison between the computed velocity profiles
(0) and the laser-doppler measurements (1) of Azzola et al.
(1984) at different stations in the upstream straight pipe
($\sim$ – 2d, $\sim$–1d,) and in the bend ($\sim$ 3°, $\sim$ 45°, $\sim$ 90°) from top
left to bottom right.

The axial velocity profiles in the centerplane for given cross-sections are plotted at different times of the cycle in Fig.5. In presence of a strong steady component, the velocity profiles undergo a similar distorsion as those observed when the steady laminar flow at the entrance section of the curved tube is fully-developed. However, the outer shift of the peak velocity tends to disappear at the end of flow deceleration and at the beginning of the acceleration phase. The huge change of the profiles of the axial velocity is confirmed by the isovelocity contours in the half cross-section and suggests strong variations in secondary motions. The magnitude of the cross velocity becomes indeed negligible during the phase of minimal flow (in Fig.5, the scaling factor of the secondary-velocity vectors is three times greater at the beginning of flow acceleration than at the beginning of flow deceleration).

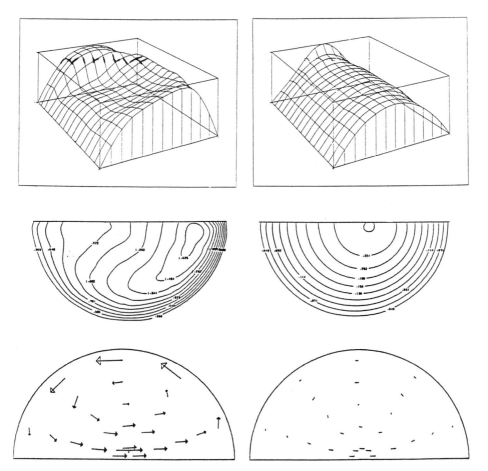

Figure 5. From top to bottom : axial velocity profiles in the centerplane, iso-axial-velocity contours in the half cross-section at 63° from the bend inlet and vector plot of the secondary motion in the cross-section located at 61.2° from the bend inlet. Left : beginning of flow deceleration ; right : beginning of flow acceleration.

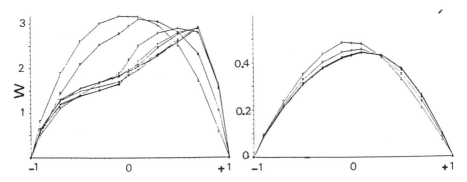

Figure 6 . Development of the axial velocity profiles in the center plane
from the bend inlet to the bend outlet.
Left: beginning of flow deceleration;   Right: beginning of flow acceleration.

The flow does not reach a fully-developed regime, because the velocity profiles of the developing axial flow are never completely superim-
posed in a 90° bend. However only small changes in velocity profiles are
observed in the upstream segment of the curved tube and their variations
become negligible in the downstream segment (Fig.6).

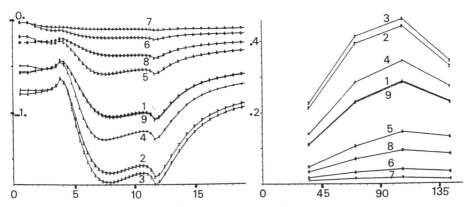

Figure 7 . Left : axial shear stress τ_a (N.m^{-2}) versus the ratio of the
axial distance to the tube diameters at time intervals of 1/8 of the
period from the beginning (1) to the end (9) of the cycle (see Fig.1),
near the outer wall of the pipe model. Right   : circumferential shear
stress τ_c versus the circumferential angle X in the cross-section at
48.6° from the bend inlet.

Due to the change in shear rate, the magnitude of the axial shear
stress at a given point in the bend can increase more than 15 times
during the acceleration phase ; the circumferential component of the
shear stress is always smaller than the axial component.

The maximum of the circumferential shear stress is more often located at the upper (lower) wall toward the outer edge, while the maximum
of the axial component is located near the outer wall, except in the
entrance region of the curved tube (Fig.8)

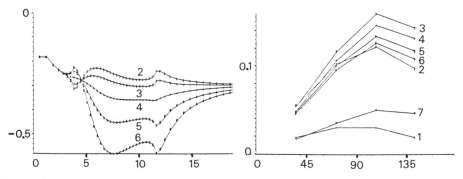

Figure 8. Left : axial shear stress versus the normalized axial distance at different circumferential location (from the inner wall (1) to the outer wall (6). Right, circumferential shear stress versus circumferential angle in different cross-sections within the bend (from 1.8° (1) to 88.2° (7)). Non-dimensional time of 0.52 (16th point of Fig. 1).

DISCUSSION

This preliminary study has been carried out with a given set of governing parameters and fixed entrance flow conditions. The aim of this work was at presenting the main characteristics of simple pulsatile flow in curved pipe. Moreover, a strong steady component was used in order to avoid any flow reversal because this computational model cannot work with bidirectional flow conditions. Another main limitations of the present study is the coarse grid used because of the memory size available on the 855 CDC computer under the NOS operating system. However reasonably good agreements have been found when the flow behaviour is not too complex. The downstream effect of the curved tube can be under-estimated by the outlet boundary conditions, set at 8.1 d of the straight pipe connected to the exit section of the bend. Nevertheless the secondary motions can remain strong in a section located +2d from the bend outlet.

The sinusoidal with non-zero mean flow does not model the highly variable and complex blood flow, even if the usual assumptions (newtonian fluid, rigid wall,...) are made. The Poiseuille component appears to be unrealistic in a network of more or less curved vessels with numerous branch points. The blood pressure wave is known to produce reversed flow. Moreover the flow parameters undergo great changes with the sites in the arteries and with the biochemical environment. Consequently further studies, both numerically and experimentally, are required. The finite-element method is easier to adapt to complicated geometries as in physiological flows than the finite-difference method. The model of bend flow could produce a classification of the major flow regimes defined by the range of values of the main governing parameters. For each flow pattern, the knowledge of the spatial and temporal variations of the shear stress should lead to a better understanding of the deposition of solid particules on the arterial wall.

CONCLUSION

The simplest numerical approach of pulsatile flow, here carried out with a finite-volume technique, reveals the huge variability of the shear stress components, both along the curved pipe and with the time during the cycle. Not only the magnitude of the shear stress changes, but also the location of its peak value. However, the low shear region is in general situated at the inner wall of the curved tube.

ACKNOWLEDGEMENTS

   Financial support of this work was provided by the Royal Society
and the Medical Research Council. One of us acknowledges the receipt of
grants from NATO, Fondation pour la Recherche Médicale and Association
Claude Bernard.

REFERENCES

Agrawal Y., Talbot L. and Gong K., 1978, Laser anemometer study of flow
      development in curved circular pipes. J. Fluid Mech., 85 : 497.
Azzola J., Humphrey J.A.C. and Launder B.E., 1984, Developing turbulent
      flow in a 180° curved pipe and its downstream tangent. 2nd Int.
      Symposium on applications of laser anemometry to fluid mechanics,
      Lisboa, 3.5 : 1.
Caretto L.S., Gosman A.D., Patankar S.V. and Spalding D.B., 1972, Two
      calculation procedures for steady, three-dimensional flows with
      recirculation, Proc. Third Int. Conf. on Numerical Methods in Fluid
      Dynamics, 60.
Issa R.I., 1982, Solution of the implicity discretised fluid flow equa-
      tions by operator splitting, Imperial College Report n° FS/82/15.
Jan D.L., Shapiro A.H. and Kamm R.D., 1989, some features of oscillatory
      flow in a model bifurcation. J. Appl. Physiol., 67 : 147.
Launder B.E. and Spalding D.B., 1972, Mathematical models of turbulence,
      Academic Press, London.
Soh W.Y. and Berger S.A., 1984, Laminar entrance flow in a curved pipe,
      J. Fluid Mech., 148 : 109.

PULSATILE FLOW THROUGH PARTIALLY

OCCLUDED DUCTS AND BIFURCATIONS

J.M.R. Graham[+]   and   M. Thiriet[*]

[+]Dept. of Aeronautics, Imperial College, London S.W.7. 2BY. U.K.
and [*]LBHP. URA. CNRS. 343, Universite Paris VII, 75251 Paris, France

ABSTRACT

Viscous pulsatile flow, of mean plus sinusoidal form, passing through a partially occluded duct and a sharp bifurcation is modelled. The method is based on the Lagrangian (Vortex in cell) technique of solving the two - dimensional Navier - Stokes equations. The computed results are used to examine the effects of the unsteady component of the incident flow on the separations which occur downstream of the restriction and of the bifurcation.

KEYWORDS

Physiological Flows / Blood Flow / Arterial Flow / Atheroma / Secondary Flows.

INTRODUCTION

It has been established (Schroter and Sudlow 1969, Olson 1971, Snyder and Olson 1989) that flow separation may occur downstream of bifurcations in the larger airways of the respiratory system, depending on the local curvature of the outer walls. Similar separated regions occur downstream of stenoses and in the daughter branches of bifurcations in the larger arteries of the body. But in the arteries the flow regime may be significantly pulsatile modifying the separations. Separated regions associated with stenoses and arterial bifurcations are known to correlate, at least partially, with sites at which atheroma develops (Caro et al. 1971). The purpose of the present work has been to examine, using two-dimensional techniques as far as possible, the differences between the behaviour of separation regions at restrictions, representing stenoses, and bifurcations in steady and unsteady (pulsatile) incident laminar flow. Unsteady flow through sharp edged restrictions in pipes has been studied experimentally by Djilali(1979) and comparisons between the numerical predictions and experiments reported by De Bernardinis et al. (1981). In that case the numerical representation of the separated flow forming from the restriction was based on an inviscid axisymmetric discrete vortex model. A similar model has been used by Cassot et al. (1989) to study the haemodynamics of heart valves. Oscillatory flow (of zero mean) generates pairs of vortices during a flow cycle which depending on flow parameters may interact or remain independent. Interaction, together with the effect of the induced velocities of the vortex and its image in the pipe wall may lead to quite rapid motions of the vortex away from its formation region. In pulsatile flow where oscillatory and mean flow are combined a similar effect can occur. Separation on a smoothly curved wall tends to occur followed by the rapid development of a vortex during the decelerating phase of the flow cycle. During the accelerating phase the vortex is swept downstream and the flow reattaches temporarily. Two dimensional computations may well be used to study some aspects of this phenomenon but it must be remembered that the actual flow through a partial occlusion or bifurcation in a pipe is highly three dimensional. In particular the effect of curvature of the streamlines of a flow with significant shear (vorticity) in the direction of the plane of curvature is to generate strong secondary (streamwise) vorticity, as is seen in computations and measurements in curved pipe flow (e.g. Thiriet et al . 1989). The strength of the secondary flow depends on the thickness of the initial shear layers as well as the curvature and viscous diffusion and hence on the Womersley parameter $\alpha = d/2(\omega/\nu)^{1/2}$ and the Dean parameter D $= Re.(d/2R)^{1/2}$

Results of two-dimensional computations may therefore be qualitatively useful only and only for certain flows.

NOMENCLATURE

d   Duct width, pipe diameter
D   Dean number $= (\bar{U}^2 d^3/2R\nu^2)^{1/2}$
h   height of restriction above wall
J   Jacobian of transformation $= |\delta(x,y)/\delta(\xi,\eta)|$
R   Radius of curvature
Re Reynolds number $= \bar{U}d/\nu$
t   time
U,$\bar{U}$,u Velocity, mean velocity, fluctuating velocity component
x, y streamwise, normal coordinates
α Womersley parameter $= (\omega d^2/4\nu^2)^{1/2}$
ζ Vorticity
ν Kinematic viscosity
ξ,η Coordinates in transformed (computation) plane
ψ Streamfunction
ω Frequency of inlet flow (radians/sec.)

NUMERICAL METHOD

The duct is assumed to simulate a large blood vessel where the flow may be assumed to be Newtonian. The flow is described by the Navier-Stokes equations in streamfunction / vorticity form for two dimensional incompressible unsteady flow.

$$\zeta_t + \psi_y.\zeta_x - \psi_x.\zeta_y = \nu \{ \zeta_{xx} + \zeta_{yy} \} \qquad (1)$$

$$\psi_{xx} + \psi_{yy} = -\zeta \qquad (2)$$

Use of equation (1) avoids the pressure term which can be inconvenient to deal with in incompressible flow. Equation (2) is the Poisson relation between vorticity and streamfunction. Equation (1) is solved by conformally transforming the actual flow region (x,y) into a rectangular computation region (ξ,η). The transformed equation

$$J.\zeta_t + \psi_\eta.\zeta_\xi - \psi_\xi.\zeta_\eta = \nu \{ \zeta_{\xi\xi} + \zeta_{\eta\eta} \} \qquad (3)$$

is solved by a split time step approach. The convection part is modelled by the vortex - in - cell Lagrangian technique in which vorticity is convected through the region as circulation on vortex particles. The diffusion part is calculated by an implicit central finite difference scheme applied on a fixed mesh, The method is described in greater detail elsewhere (Graham 1988). The walls of the ducts are transformed into the upper and lower boundaries of the computation region. In the case of the bifurcating duct this transformation necessitates the use of a source to represent the inlet flow. The no-slip condition is applied to the upper and lower ξ,η plane boundaries representing the walls. The transformations used are based on a numerical Schwartz - Christoffel method derived by Davis (1983).

Equation (2) is similarly transformed to the rectangular region, becoming:

$$\psi_{\xi\xi} + \psi_{\eta\eta} = -J.\zeta \qquad (4)$$

This equation is solved by finite difference representation with Fourier transformation in one coordinate direction across the computation region and solution of the resulting tridiagonal set of equations in the other direction. This gives a fast method of solution. In the computations shown the vorticity field was represented by between 3000 - 5000 vortices and the finite difference mesh contained 32 x 50 node points evenly distributed across the computation region.

RESULTS

Figure 1 shows streamlines obtained for a two-dimensional computation of steady flow and Figure 2 for pulsatile flow past a one-sided restriction (on one wall of a duct). In the latter case the pulsatile velocity was:

$$U(t) = 1/2 \, U_0(1 + \sin \omega t) \qquad (5)$$

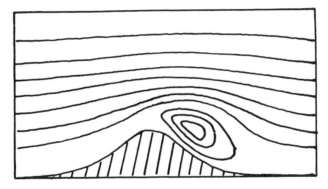

Fig. 1.   Streamlines for steady flow past a two-dimensional restriction.

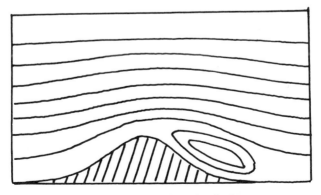

Fig. 2.   As figure 1. for pulsatile flow.

The peak Reynolds number based on duct width of both flows was Re = 1000 and the Womersley parameter of the unsteady flow $\alpha = 7.9$.

Comparison of the two results shows that although the thinner shear region in the unsteady case has a delaying effect on separation, once the flow separates a significant recirculation region is generated which is subsequently swept downstream in a periodic manner by the flow oscillations. The effect of unsteadiness in the incident flow as a tendency to suppress separation has also been observed in flows in curved pipes (Thiriet et al 1989) where secondary flow also tends to inhibit separation. This same behaviour is also seen in flow through a bifurcation, the main object of the present study. Bifurcation flow has several features in common with flow in a curved pipe.

Flow from each half of the parent vessel into either daughter branch follows a similar curved path but the upstream part of the outer wall of the curved pipe representing each half is missing, being in effect replaced by the median plane which divides the bifurcation. Hence the no slip condition is absent from this part of the flow. However, the main source of secondary flow in a curved pipe lies in the shear layers on the top and bottom parts of a pipe curved in a horizontal plane. These are present in the bifurcation and hence secondary motion as observed in the curved pipe (e.g. Thiriet et al 1989) will be expected in the bifurcation. Similarly, considering the results from the curved pipe, pulsatility is expected to weaken this secondary flow. However the computation presented in this paper has been carried out for two-dimensional planar flow and secondary motions will not be present.

The main feature which does occur in the present computed flow is separation on the outer wall of the bifurcation. This has similarities with separation on the inner wall of a curved tube or duct. In the case shown the radius of curvature of the outer wall of the bifurcation is small, equal to half the parent duct width, equivalent to one tube radius. The no-slip condition on the outer wall of a curved pipe which is not present in effect in the upstream part of the bifurcation appears to be relatively passive in its effect. However, when asymmetric flow occurs in a bifurcation, the flow divider (carina) might be expected to cause separation and in extreme cases shed vortices like a restriction. However this is not observed in practice for a sharp carina (Schroter and Sudlow 1969).

The present computations are for a symmetric bifurcation with a sharp flow divider of included angle equal to 45°. The combined area of the two daughter vessels is equal to the area of the parent vessel. Figure 3 shows the streamline pattern and velocity vectors for steady flow through this bifurcation when there is a flow ratio of 2 :1 between the two daughter vessels. This asymmetric flow contains two thin regions of separation on the outer walls of the bifurcation which remain fairly constant throughout the time of the computed flow after a short initial development time.

Figure 4 shows the sequence of streamlines and figure 5 the velocity vectors which result in the case of pulsatile flow through the same geometry. The peak Reynolds number of the flows based on the parent duct width is Re = 1000 in both cases. The ratio of the flows in the daughter vessels is 2 : 1 and the pulsatility is, as for the restriction:

$$U(t) = 1/2 \ U_0(1 + \sin \ \omega t) \qquad (6)$$

The Womersley parameter takes the same value as before.

The effect of pulsatility on the flow is, in this case, to cause the separated regions which form on the outer walls of the bifurcation to be swept downstream regularly with each cycle of the incident flow. This is seen in the velocity vector plot in Figure 5. Because the separated region is now regularly swept downstream, as for pulsatile flow through the restriction, we conclude that the residence time which fluid particles spend in the region of a bifurcation is greatly reduced by pulsatility in comparison with the time for those which enter the separation

62

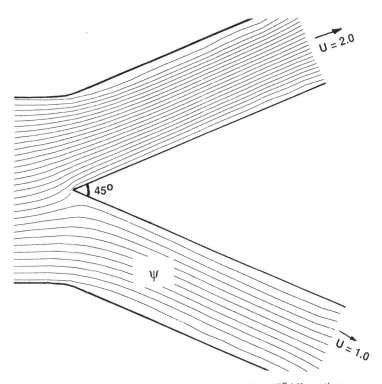

Fig. 3. Streamlines for steady flow through a 45° bifurcation.

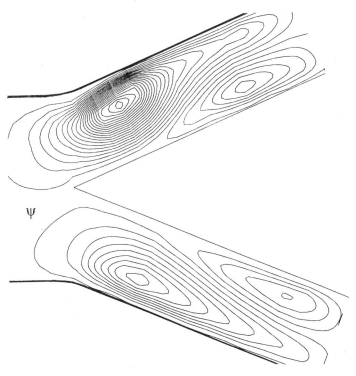

Fig. 4. Streamlines for pulsatile flow through
a 45° bifurcation.   t/T = 2.0

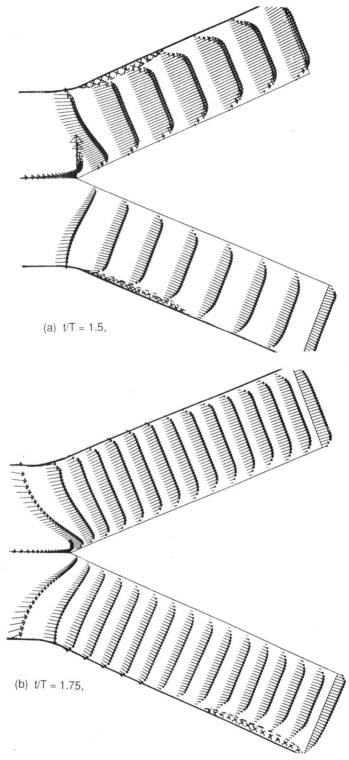

(a) t/T = 1.5,

(b) t/T = 1.75,

Fig. 5.   Velocity vectors as for figure 4.
xxxxx Separation region.

region in steady flow. Long residence times may be associated with higher rates of development of atheroma and strong pulsatility therefore will be an inhibiting factor from this point of view.

The other feature of the bifurcation, which should be considered, is the flow past the flow divider. The situation shown here has been forced to be strongly asymmetric with a ratio of 2 : 1 between the rates of volume flow through the two daughter branches. However, even for this degree of cross-flow past a sharp flow divider, no significant flow separation was observed. Although there are some similarities to flow past an abrupt restriction, the strong component of flow onto the wedge shape of the flow divider should always tend to promote early reattachment downstream of any separation.

It has to be emphasised that the calculations shown here are two-dimensional and therefore cannot model secondary flow effects which are induced in the bifurcation by the skewing of the vorticity as in a curved pipe. However the results do show the effects of pulsatility on the flow separations and hence on the residence times.

CONCLUSIONS

The two dimensional computations carried out for flow past a restriction and through a bifurcation show separation regions forming downstream of the restriction and on the outside walls of the start of each daughter vessel of the bifurcation (for small radii of curvature of these outer walls and no flow expansion or contraction). No separation was observed from the flow divider despite the asymmetry of the flow. The effect of pulsatility is to cause the separation region in both the restriction and the bifurcation to be regularly swept downstream with each cycle of the flow. This reduces the residence times for particles passing through these regions.

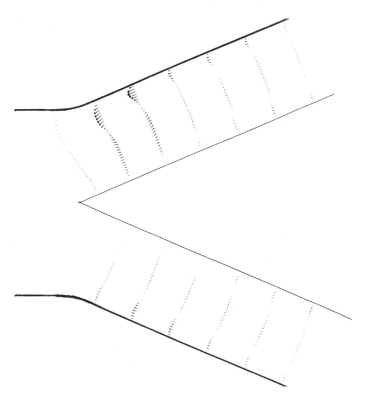

Fig. 5. (continued), (c) t/T = 2.0.

REFERENCES

Caro C.G., Fitzgerald J.M. and Schroter R.C.   1971   Atheroma and arterial wall shear stress. Observation, correlation and proposal of a shear dependent mass transfer mechanism for arterogenesis. Proc. Roy. Soc. (B) 177, p109.

Cassot F., Pelissier R and Tonietto G.   1989   Vorticity formation and transport from mechanical heart valve prostheses. Euromech. Colloqium 259, Biomechanical transport processes. Cargese, Corsica.

Davis R.T.   1983   Numerical methods for coordinate generation based on a mapping technique. "Comp. Methods for Turbulent, Transonic and Viscous Flow".p1. Ed. Essers J. Hemisphere.

De Bernardinis B., Graham J.M.R and Parker K.H.   1981   Oscillatory flow around discs and through orifices. J. Fluid Mech. 102, p282.

Djilali N.   1979   Oscillatory flow through a sharp edged orifice plate. M.Sc Thesis Aeronautics Dept., Imperial College, London University.

Graham J.M.R.   1988   Computation of viscous separated flow using a particle method. "Num. Methods for Fluid Dyn. III", p310. Ed. Morton K.W and Baines M.J., IMA. Conf series 17.

Olson D.E.   1971   Fluid mechanics relevant to respiration - flow with curved or elliptical tubes and bifurcating systems. PhD. Thesis. Univ. London .

Schroter R.C. and Sudlow M.F.   1969   Flow patterns in models of the human bronchial airways. Respir. Physiology 7, p341.

Snyder B and Olson D.E.   1989   Flow development in a model airway bronchus. J. Fluid Mech. 207, p379.

Thiriet M., Graham J.M.R and Issa R.   1989   Quasi - steady laminar viscous flow in a circular curved pipe. Euromech. Colloquium 259, Biomechanical transport processes. Cargese, Corsica.

NUMERICAL AND EXPERIMENTAL ANALYSIS OF CAROTID ARTERY BLOOD FLOW

A.A. van Steenhoven, C.C.M. Rindt, R.S. Reneman[+] and J.D. Janssen

Department of Mechanical Engineering, Eindhoven University of
Technology, Eindhoven, The Netherlands

Department of Physiology, University of Limburg
Maastricht, The Netherlands

ABSTRACT

To obtain more insight into the complex blood flow patterns in the carotid
artery bifurcation, finite element calculations have been carried out
combined with flow visualization studies and laser-Doppler velocity measurements. As curvature effects are expected to be important in a bifurcation,
first the steady flow development in a curved tube was investigated. From a
detailed analysis of steady flow in a three-dimensional model of the carotid
artery bifurcation it is concluded that curvature effects indeed play an
important role in the daughter branches of this bifurcation, but also that
the local geometry of the carotid sinus highly affects the axial and
secondary flow fields. In general a good agreement is found between the
numerical and experimental results.

KEYWORDS

Bend, carotid artery bifurcation, finite element analysis, visualization
studies, laser-Doppler measurements.

INTRODUCTION

In the development of non-invasive detection methods of atherosclerotic
lesions in the carotid artery bifurcation at an early stage of the disease,
insight into the complicated flow field in this bifurcation is indispensable.
The geometry of this bifurcation is shown in figure 1. It consists of a main
branch, the common carotid artery, which asymmetrically divides into two
branches, the internal and external carotid arteries. The internal carotid
artery is characterized by a widening in its most proximal part, the sinus or
bulb. Detailed experimental information about the flow field in this
bifurcation has been obtained for the steady case by Bharadvaj et al.(1982)
and for the unsteady case by Ku et al.(1983).
   The aim of the present study is to develop an accurate numerical model of
fluid flow in the carotid artery bifurcation. With such a model the
influences of interindividual variabilities in geometry and the presence of
small stenoses on the flow field can be quantified. The geometry used in this
study is similar to that used by Bharadvaj et al.(1982); the characteristic
dimensions are shown in figure 1. Both calculations and measurements were
carried out enabling an experimental validation of the numerical model. In
earlier studies, the numerical (Cuvelier et al.,1986 and van de Vosse et

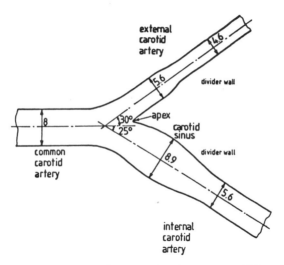

Figure 1. Schematical presentation of the human carotid artery bifurcation
(Adapted from Bharadvaj et al., 1982).

al.,1986) and experimental methods (Corver et al.,1985 and van Steenhoven et
al.,1988) were described. Besides, in an experimental and numerical study by
van de Vosse et al.(1985) steady and pulsatile flow over a two-dimensional
square step was analysed. Rindt et al.(1987) performed steady and unsteady
velocity measurements and calculations of fluid flow in a two-dimensional
model of the carotid artery bifurcation. A study by van de Vosse et al.(1990)
indicated that three-dimensional analysis of the flow field in the carotid
artery bifurcation is necessary to better understand the in vivo flow
situation. To gain more insight into the secondary flow patterns and because
of its geometrical simplicity, first steady entrance flow in a 90-degree bend
was studied. Bovendeerd et al.(1987) performed laser-Doppler velocity
measurements in such a bend, while three-dimensional calculations were
reported by van de Vosse et al.(1989) and Rindt (1989). Next, the steady flow
in a three-dimensional model of the carotid artery bifurcation was analysed,
using experimental (Rindt et al.,1989) and numerical (Rindt et al.,1990)
techniques. In all those studies a reasonable agreement was found between the
numerical and experimental results.

In this paper the numerical and experimental methods used will be briefly
described and some characteristic results for steady flow in a 90-degree bend
and in a 3D-model of the carotid artery bifurcation will be presented. For
the sake of clearness the total flow field will be divided into axial and
secondary flow fields, the latter defined as the flow field in a
cross-section perpendicular to the tube axis. The results will be shown at a
Reynolds number, based on the diameter of the common carotid artery, of about
700. The flow division ratio between the internal and external carotid
arteries is chosen to be about 50/50. Both numbers correspond to the peak
systolic values of blood flow in the carotid artery bifurcation (Ku et
al.,1983)

NUMERICAL METHOD

Flow of an incompressible and isothermal fluid is described by the momentum
and continuity equations. Neglecting gravity effects, for steady flow these
equations read in dimensionless form:

$$\vec{u} \cdot \vec{\nabla}\vec{u} - \vec{\nabla} \cdot \sigma = \vec{0} \quad ; \quad \vec{\nabla} \cdot \vec{u} = 0 \tag{1}$$

with $\vec{u}$ the velocity vector, σ the Cauchy stress tensor and $\vec{\nabla}$ the gradient vector operator. In this study only Newtonian fluids will be considered for which the Cauchy stress tensor is coupled to the velocity field as:

$$\sigma = -\,pI + \frac{1}{Re}\,[\vec{\nabla}\vec{u} + (\vec{\nabla}\vec{u})^C] \quad ; \quad Re = \frac{DU}{\nu} \tag{2}$$

with p the pressure, I the unit tensor and $(\vec{\nabla}\vec{u})^C$ the conjugate of the velocity gradient tensor. Re denotes the Reynolds number and its value is calculated from the diameter (D) and mean velocity (U) at the inlet and the kinematic viscocity ν of the fluid.

The numerical method used is based on a Galerkin finite element approximation of equations 1 and 2, which leads to the following set of non-linear equations (Cuvelier et al.,1986):

$$[N(\underline{U}) + S]\underline{U} + L^T\underline{P} = \underline{B} \quad ; \quad L\underline{U} = \underline{0} \tag{3}$$

Here $N(\underline{U})\underline{U}$ represents the convective acceleration term, $S\underline{U}$ the viscous term, $L^T\underline{P}$ the pressure gradient term and $L\underline{U}$ the velocity divergence term. $\underline{B}$ represents the boundary forces and $\underline{U}$ contains the velocity and $\underline{P}$ the pressure unknowns in the nodal points. To avoid partial pivoting this set of equations is uncoupled with the penalty function method. The non-linear convective term $N(\underline{U})\underline{U}$ is linearised by a Newton-Raphson iteration method. The construction of the system of equations, its solution and the post-processing of the velocity data is carried out with the finite element package SEPRAN (Segal,1984).

The discrete points, in which velocities and pressures are calculated, are determined by partition of the 3D-geometry into elements. The 3D element used is the so-called Crouzeix-Raviart element, shown in figure 2a. It has 27 nodes for the velocity (triquadratic, 81 velocity unknowns) and 1 node for

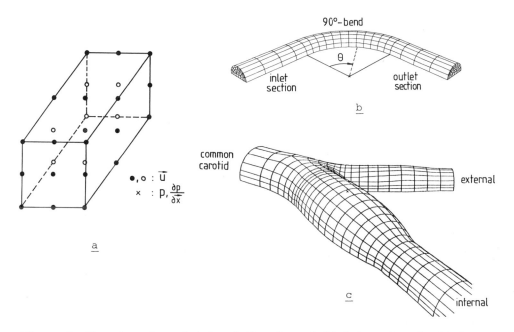

Figure 2. The element used (a) and the element divisions for the 90-degree curved tube (b) and the carotid artery bifurcation (c).

the pressure (linear, 4 pressure unknowns). Due to the triquadratic approximation within the element the velocity has a third order accuracy. Figure 2b shows the element division used for the analysis of the steady entrance flow in a bend. It consists of 20 elements in axial direction and 30 elements per cross-section. Figure 2c shows a plot of the element division of the carotid artery bifurcation, where 1474 elements were used.

For both geometries similar boundary conditions were used. Flow at the inlet was supposed to be fully developed, which means a parabolic axial velocity profile and zero secondary velocities. The velocities at the wall were presumed to be zero, according to the no-slip condition. At the outlet the normal and both tangential stresses were set to zero, while in the plane of symmetry both tangential stresses and the normal velocity component were put to zero.

EXPERIMENTS

In figure 3a the experimental set-up is shown for the visualization experiments and velocity measurements under steady flow conditions. By means of a voltage controlled gear pump P the fluid was pumped from the reservoir R into the measurement section. A long circular pipe upstream of the model was used to ensure a fully developed pipe flow at its entrance. The 3D-models consisted of two halves of perspex, split at the plane of symmetry, in which the 90-degree curved tube and the carotid artery bifurcation were machined out. In figure 3b the models used for the laser-Doppler experiments are presented; for the visualization experiments 5 times enlarged models were used. The ratio of tube radius and curvature radius for the bend was chosen to be 1/6, resembling more or less the geometry of the entrance region of the internal carotid artery.

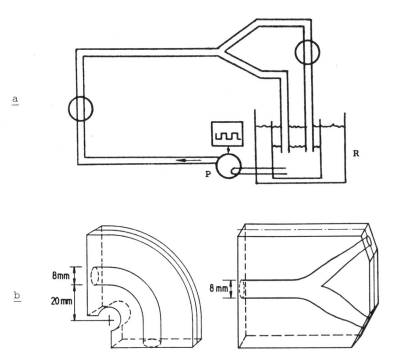

Figure 3. Fluid circuit as used in the experiments (a) together with the perspex models of the 90-degree curved tube and the carotid artery bifurcation (b).

The fluid flow was visualized with the use of the hydrogen bubble technique, as described by Merzkirch (1974). The cathode is a thin (20 μm) platinum wire positioned in the flow field to be studied. The anode is a stainless-steel plate located in such a place that is does not disturb the flow. The fluid used was demineralized water to which a small amount of an electrolyte (0.5% acetic acid) was added. Hydrogen bubbles are produced periodically at the cathode by a square wave generator. The axial flow field in the symmetry plane of the 90-degree curved tube was visualized simultaneously at two locations (0 and 38 degrees) along the bend. To visualize the secondary flow patterns also sheets of hydrogen bubbles were produced applying a dc voltage between the electrodes. In that case the cathode wire was placed out of the plane of symmetry.

The fluid velocity was measured by a one-component laser-Doppler instrument (DISA, measuring volume = 400 μm x 40 μm x 40 μm), based on the forward-scattering reference-beam method. A Bragg cell was used to discriminate velocity directions. Velocity measurements in 3D-geometries require exact matching of the refraction indices of the fluid and of the perspex models. Therefore, as circulating fluid a mixture of oil (Shellflex 214 BG) and kerosine was used. The experiments were performed at 40^{0}C to lower the kinematic viscosity of the oil mixture and to eliminate the influence of ambient temperature variations. For seeding silicagel (Lichrosorb, Si 100, mean particle diameter 5 μm) was used. Three stepper motors were used to traverse the model in three independent directions, through which positioning of the measuring volume at various sites in the model was possible with an accuracy of about 5 μm in each direction. Independently, the axial velocity component and both secondary velocity components were measured in about 100 grid points per cross-section. For each measurement 10 samples were taken from which the mean value and the 95% confidence interval were calculated. A personal computer was used to control the traversing mechanism and the data acquisition.

RESULTS

In figure 4 the results of the flow visualization in the 90-degree bend are shown. In the upper panel the axial flow field is visualized in the plane of symmetry. Due to the centrifugal forces the profiles remain not axial symmetric. The maximum velocities occur near the outside wall, while at the inner wall a so-called velocity plateau is formed. Due to the interaction of centrifugal, pressure, and viscous forces two helical vortices develop; one at each side of the plane of symmetry. This pattern is visualized for the upper half in the lower panel of figure 4.

In figure 5a the axial velocity profiles in the plane of symmetry are presented for both the laser-Doppler measurements and the finite element calculations. A similar shift to the outer bend is found as observed from the visualization experiments. There is a fair agreement between the experimental and numerical data, although in the numerical case the axial velocity plateau at 58.5 degrees is somewhat less developed. In figure 5b the calculated and measured secondary flow fields at 23.4 and 58.5 degrees are presented by velocity profiles of the component parallel to the plane of symmetry (upper half) and the component perpendicular to the plane of symmetry (lower half). As shown in this figure and visualized in figure 4, the secondary velocities near the plane of symmetry are directed from the inner bend towards the outer bend, while near the side wall a circumferentially inward motion of the fluid occurs. Here again a good agreement exists between the experimental and numerical results.

In figure 6 the visualized flow field in a 3D-model of the carotid artery bifurcation is shown. The visualization wires are placed perpendicular to the plane of symmetry. From this figure it is concluded that also at the entrance of the internal carotid artery secondary motions are present.

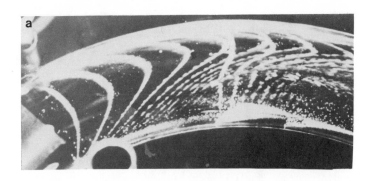

Figure 4. Flow visualization in the plane of symmetry (a) and in the upper half (b) of the 90-degree curved tube (Re = 840).

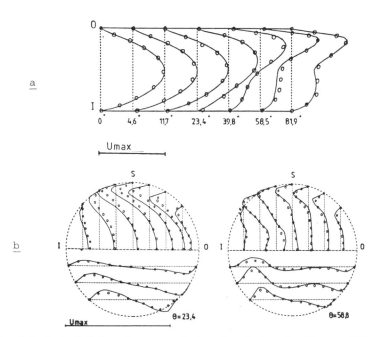

Figure 5. Calculated (-) and measured (ooo) axial velocity profiles in the plane of symmetry (a) and secondary velocity profiles at two cross-sections (b) of the 90-degree curved tube (Re = 640, I: inner wall, O: outer wall, S: side wall).

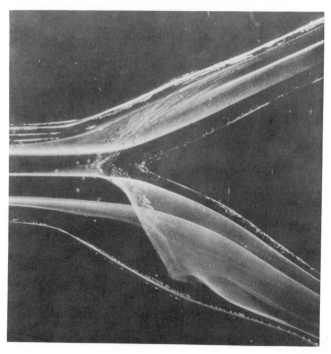

Figure 6. Flow visualization in a 3D-model of the carotid artery bifurcation (Re = 900).

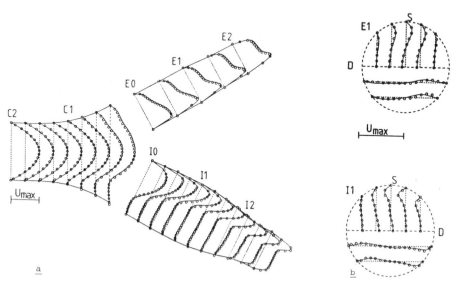

Figure 7. Calculated (-) and measured (ooo) axial velocity profiles in the plane of symmetry (a) and secondary velocity profiles at two cross-sections (b) of the carotid artery bifurcation (Re = 640, D: divider wall, S: side wall). From Rindt et al.(1990), with permission.

In figure 7a the axial velocity profiles in the plane of symmetry of the bifurcation are presented for both the laser-Doppler measurements and the finite element calculations. The characters C, I and E refer to the common, internal and external carotid arteries and the numbers to axial distances to the flow divider expressed in diameters of the main branch. In the model common carotid artery (C1.5) the velocity profile is parabolic. At the entrance of the internal carotid artery high axial velocities are found near the divider wall, which is primarily caused by flow branching. A region with negative axial velocities with a diameter up to 60% of the local diameter of the bulb is found opposite to the flow divider. The agreement between the experimental and numerical data is good. In figure 7b the profiles of the secondary velocity components are shown at positions one diameter downstream in the internal and external carotid arteries. Near the plane of symmetry the secondary velocities are directed towards the divider wall and near the side wall they point circumferentially back towards the non-divider wall. There is again a fair agreement between the numerical and experimental data.

CONCLUDING DISCUSSION

From the present study it is concluded that the finite element method can be used for detailed analyses of fluid flow in complex three-dimensional geometries. The findings in this study also indicate that qualitatively the secondary flow fields in a curved pipe and halfway the carotid sinus show remarkable similarities, although other results (Rindt et al.,1990) suggest that at the entrance and end of the bulb the secondary flow fields are also highly influenced by the specific geometry of the carotid sinus. Both findings support the statement of Olsen (1971) that the flow phenomena in a symmetrical bifurcation with straight daughter branches mainly originate from curvature effects. On the contrary to the flow in a 90-degree curved tube, a large region with reversed axial flow is observed opposite to the flow divider in the internal carotid artery. These negative axial velocities are mainly due to the divergent geometry of the inlet section of the carotid sinus. The width of the region is somewhat larger than the value reported by Bharadvaj et al.(1982), probably due to the higher flow division ratio (70/30) used by them.

In the near future the unsteady flow phenomena in the carotid bifurcation will be investigated. Next, the influence of the wall flexibility on the axial and secondary flow fields has to be evaluated. To that end, the analysis of wave propagation of pressure and flow pulses (van Steenhoven and van Dongen,1986 and Horsten et al.,1989) has to be incorporated in the numerical model. A strategy and some preliminary results have been described by Reuderink et al.(1989).

ACKNOWLEDGEMENTS

We thank W. Vegt and F. Smeets for performing the visualization experiments and J.W.G. Cauwenberg and L.H.G. Wouters for their technical assistance.

REFERENCES

Bharadvaj B.K., Mabon R.F. and Giddens D.P. (1982). Steady flow in a model of the human carotid bifurcation, Part 1- Flow visualization, Part 2- Laser-Doppler anemometer measurements, J. Biomechanics, 15: 349-378.
Bovendeerd P.H.M., van Steenhoven A.A., van de Vosse F.N. and Vossers G.(1987). Steady entry flow in a curved pipe, J. Fluid Mech., 177: 233-246.
Corver J.A.W.M., van de Vosse F.N., van Steenhoven A.A. and Reneman R.S. (1985).The influence of a small stenosis in the carotid bulb on adjacent velocity profiles, in: "Biomechanics Current Interdisciplinary Research", S.M. Perren and E. Schneider eds., 239-244, Martinus Nijhoff Publ., Dordrecht.

Cuvelier C., Segal A. and van Steenhoven A.A. (1986). "Finite Element Methods and Navier-Stokes Equations", D. Reidel Publ., Dordrecht.

Horsten J.B.A.M., van Steenhoven A.A. and van Dongen M.E.H. (1989). Linear propagation of pulsatile waves in viscoelastic tubes, J.Biomechanics, 22: 477-484.

Ku D.N. and Giddens D.P. (1983). Pulsatile flow in a model carotid bifurcation, Arteriosclerosis, 3: 31-39.

Merzkirch W. (1974). "Flow Visualization", Academic Press, New York.

Olson D.E. (1971). Fluid mechanics relevant to respiration: flow within curved or elliptical tubes and bifurcating systems, Ph.D-thesis, University of London.

Reuderink P.J., Willems P.J.B., Schreurs P.J.G. and van Steenhoven A.A.(1989). Fluid flow through distensible models of the carotid artery bifurcation, Proceedings Second International Symposium on Biofluid Mechanics, 455-461, München.

Rindt C.C.M, van de Vosse F.N., van Steenhoven A.A., Janssen J.D. and Reneman R.S.(1987). A numerical and experimental analysis of the flow field in a two-dimensional model of the human carotid artery bifurcation, J.Biomechanics, 20: 499-509.

Rindt C.C.M., van Steenhoven A.A. and Reneman R.S.(1988). An experimental analysis of the flow field in a three-dimensional model of the human carotid artery bifurcation, J.Biomechanics, 21: 985-991.

Rindt C.C.M., van Steenhoven A.A., Janssen J.D., Reneman R.S. and Segal A.(1990). A numerical analysis of steady flow in a 3D-model of the carotid artery bifurcation, J.Biomechanics, in press.

Segal A.(1984). "Sepran User Manual and Programmers Guide", Ingenieursburo Sepra, Leidschendam.

Van de Vosse F.N., Vial F.H., van Steenhoven A.A., Segal A. and Janssen J.D.(1985). A finite element and experimental analysis of steady and pulsating flow over a two-dimensional step, in: "Numerical Methods in Laminar and Turbulent Flow", C. Taylor ed., 515-526, Pineridge Press, Swansea.

Van de Vosse F.N., Segal A., van Steenhoven A.A. and Janssen J.D.(1986). A finite element approximation of the unsteady 2D'Navier-Stokes equations, Int.J.Num.Meth.Fluids, 6: 427-443.

Van de Vosse F.N., van Steenhoven A.A., Segal A. and Janssen J.D.(1989). A finite element analysis of the steady laminar entrance flow in a 90⁰ curved tube, Int.J.Num.Meth.Fluids, 9: 275-287.

Van de Vosse F.N., van Steenhoven A.A., Janssen J.D. and Reneman R.S.(1990). A two-dimensional numerical analysis of unsteady flow in the carotid artery bifurcation, Biorheology, accepted.

Van Steenhoven A.A. and van Dongen M.E.H.(1986). Model studies of the aortic pressure rise just after valve closure, J. Fluid Mech., 66: 93-113.

Van Steenhoven A.A., van de Vosse F.N., Rindt C.C.M., Janssen J.D. and Reneman R.S.(1990). Experimental and numerical analysis of carotid artery blood flow, in: "Blood flow in large arteries: Applications to atherogenesis and clinical medicine", D.W. Liepsch ed., 250-260, Monogr. Atheroscler. vol. 15, Karger, Basel.

INTERNAL FLOWS WITH MOVING BOUNDARIES

AND APPLICATIONS IN VENTRICULAR EJECTION

S. Tsangaris and N. Koufopoulos

National Technical University of Athens
Fluids Section
P.O. Box 64070, 15710 Zografou, Greece

ABSTRACT

Unsteady flows of an incompressible fluid induced by the motion of confining walls are studied numerically for a two-dimensional plane and axisymmetric geometry. The scope of the present study is to predict the distribution of the local flow quantities inside the left ventricle of the heart. The exact inviscid or viscous flow equations are transformed in curvilinear boundary fitted coordinate systems in order to have arbitrarily moving boundaries included in the study. The balance of the energy, which is useful for the estimation of the efficiency of the ventricle is calculated.

Keywords : Blood flow - Ventricular flow - Moving boundaries.

INTRODUCTION

The analysis of the flow inside the heart, that influences the performance of the heart, has been the subject of intensive investigation in recent years. The inviscid mathematical model of flow in a pulsating bulb proposed by Jones (1968) can be considered as the simplest model of ventricular flow (see also Fung (1984)).

The fibers comprising the ventricular wall shorten during systole. An accurate description of the changes in the left ventricular dimensions during contraction has been summarized by Covell (1972) for a dog. During the phase of ventricular ejection according to these results, there is a 10-35% decrease in the minor axis dimension at the endocardium (4.5-4.0 cm) while the distance from the apex to the aortic valve (outflow tract) decreases by an average of 1-5% only.

The duration of the ventricular ejection has a duration of about 0.26 s, that is about 32.5% of the cardiac period, which is assumed in the physiological case to be of a duration of 0.8 s.

During the phase of ventricular ejection the aortic valve is opened and the blood is ejected to the aorta, while the mitral valve is closed. More accurate models for the flow inside the heart during the ejection phase are given by Pedley, Seed (1977) and Wang, Sonnenblick (1979) for the case of a spherical or elliptic ventricle. The above mathematical models are based on the analytical or approximate solutions of the inviscid flow equations and they give variable results for the evaluation of the contractility of the heart. In the last decade more accurate models are introduced for the analysis of flow inside the heart, using numerical methods.

In the present study a solution of the incompressible viscous flow equations for the case of an axisymmetric, arbitrarily shaped ventricle is given. The methodology uses the solution of the Navier-Stokes and continuity equations for arbitrary, curvilinear, body fitted, coordinate system using a modification of the Chorin (1968), Fortin (1971), Temam (1969) method introduced by Hilbert (1987).

METHODOLOGY

The momentum equations used are those for the two dimensional (or axisymmetric), unsteady, incompressible, viscous fluid, which can be written in the following conservative form, according to Viviand (1974):

$$\frac{\partial \tilde{Q}}{\partial \tau} + \frac{\partial \tilde{E}}{\partial \xi} + \frac{\partial \tilde{F}}{\partial \eta} = 0$$

$$\tilde{Q} = \frac{1}{J}\begin{pmatrix} \rho u \\ \rho v \end{pmatrix}$$

$$\tilde{E} = \frac{1}{J}\begin{pmatrix} \rho u U + \xi_x(p - \mu u_x) - \xi_y \mu u_y \\ \rho v U - \xi_x \mu v_x + \xi_y(p - \mu v_y) \end{pmatrix}$$

$$\tilde{F} = \frac{1}{J}\begin{pmatrix} \rho u V + \eta_x(p - \mu u_x) - \eta_y \mu u_y \\ \rho v V - \eta_x \mu v_x + \eta_y(p - \mu v_y) \end{pmatrix}$$

In the above equations (u,v) are the orthogonal cartesian components of the velocity on x and y directions respectively, where x,y are the cartesian independent variables in the physical domain, ρ is the fluid density and μ the viscosity coefficient, U and V are the contravariant velocities along the ξ, η coordinates, ξ_t, ξ_x, ξ_y etc. are the metric coefficients and J ($J = \xi_x \eta_y - \xi_y \eta_x$ the Jacobian of the transformation:

78

$$\tau = t, \xi = \xi(x,y,t), \eta = \eta(x,y,t)$$

In addition the continuity equation for incompressible fluid is taken into account:

$$div\vec{W} = 0$$

$$u_x + v_y = 0$$

The method used for discretization is the projection method. This method was proposed independently by Chorin (1968) and Temam (1969), while an implicit version of such a method was presented by Fortin et al (1971). Another implicit version of the projection method is proposed by Hilbert (1987) and it is adopted here, in the modification of introducing the arbitrary curvilinear coordinate system while Hilbert uses a finite element method:

Predictor step for the velocity field:

$$\frac{\tilde{Q}^* - \tilde{Q}^n}{\delta\tau} + \frac{\partial \tilde{E}^{n,*}}{\partial\xi} + \frac{\partial \tilde{F}^{n,*}}{\partial\eta} = 0$$

where,

$$\tilde{Q}^* = \frac{1}{J}\left(\rho u^* \; \rho v^*\right)$$

$$\tilde{E}^{n,*} = \frac{1}{J}\begin{pmatrix} \rho u^* U^n + \xi_x\left(p^n - \mu u_x^*\right) - \xi_y \mu u_y^* \\ \rho v^* U^n - \xi_x \mu v_x^* + \xi_y\left(p^n - \mu v_y^*\right) \end{pmatrix}$$

$$\tilde{F}^{n,*} = \frac{1}{J}\begin{pmatrix} \rho u^* V^n + \eta_x\left(p^n - \mu u_x^*\right) - \eta_y \mu u_y^* \\ \rho v^* V^n - \eta_x \mu v_x^* + \eta_y\left(p^n - \mu v_y^*\right) \end{pmatrix}$$

$\vec{W}^* = (u^*, v^*)$ is an intermidiate velocity field which is non divergence-free.

Predictor step for the pressure:

$$\Delta q = \frac{\rho}{\Delta\tau} div\vec{W}^*$$

where q is the correction of the pressure.

<u>Corrector step for the velocity</u>:

$$\vec{W}^{n+1} = \vec{W}^* - \frac{\Delta t}{\rho} grad q$$

<u>Corrector step for the pressure</u>:

$$p^{n+1} = p^n + q$$

This method is specifically designed to deal with the incompressibility constraint $div \vec{W}^{n+1} = 0$ and introduces a Poisson equation for the correction pressure q at each time step.

The algorithm, was used here, is based on the M.A.C method. The spatial discretization makes use of the staggered marker-and-cell (MAC) mesh introduced by Harlow and Welch (1965). The mesh is generated by the Sorenson (1981) method. According to this methodology the pressure is defined in the center of the cell while the u and v velocity components in the middle of the east and north faces of the rectangular mesh in (ξ, η) plane. The finite difference approximations are in general centered in space with a parameter a which controls the desired amount of upstream differencing. For instance the inertial $(u^2)_\xi$ can be expressed as:

$$\frac{\partial u^2}{\partial \xi} = \frac{1}{4\delta\xi}[(u_{i,j+1} + u_{i+1,j})^2 + \alpha |u_{i,j} + u_{i+1,j}|(u_{i,j} - u_{i+1,j})$$

$$-(u_{i-1,j} + u_{i,j})^2 - \alpha |u_{i-1,j} + u_{i,j}|(u_{i-1,j} - u_{i,j})]$$

The predictor equation for the velocity as well as the predictor equation for the pressure was solved by using the S.O.R method.

RESULTS AND DISCUSSION

Results are shown for the unsteady (pulsating) flow in semi-infinite duct with pulsating walls, figure 1. The x-axis is the symmetry of the duct. The wall AB is sinusoidal and pulsating in time, having a time dependent radius:

$$R(x,t) = R_0(t) + R_a(t)\cos\left(\frac{\pi x}{L}\right)$$

where

$$R_0(t) = \frac{R_{min} + R_{max}(t)}{2} \qquad and \qquad R_a(t) = \frac{-R_{min} + R_{max}(t)}{2}$$

and

$$R_{max}(t) = R_{max\,2} + \left(R_{max\,1} - R_{max\,2} \right) \cos^2 \left(\frac{\omega t}{2} \right)$$

The rest of the upper wall BC of the duct is straight and non moving. Thus the radial velocity of the upper wall is:

$$v_w = \cos^2 \left(\frac{\pi x}{2L} \right) \frac{R_{max\,1} - R_{max\,2}}{2} \omega \sin(\omega t) x < L$$

$$v_w = 0 \, x > L$$

R_{min} is the straight duct's radius, $R_{max\,1}$ is the radius of the duct for x=0 at t=0 and $R_{max\,2}$ is the radius of the duct for x=0 and $t' = t\omega = \pi$.

Results are shown in figures 2 and 3 for the following values of the parameters:

$$R_{max\,1} = 0.6, R_{max\,2} = 0.4, R_{min\,1} = 0.5, L = 2, L_o = 1$$

The flow field is shown for five different time instants. The results are obtained with an accuracy of 0.0001 using a time step of $\delta t' = 0.01$

A time periodicity of the solution is obtained after some periods of running. For the above examples a periodic convergence of accuracy 0.0001 is reached after four periods. The velocity vector field is shown in figure 2 for five different times ($\phi' = 0.57$ deg, 4.7 deg, 105.3 deg, 184.7 deg, 271.46 deg), after the end of the fourth period, where convergence is reached. In the case of $\phi' = 0.57$ the boundary is nearly non-moving and shows the velocity profile as positive as well negative velocities. In the next time step, very close to the previous one for $\phi' = 4.7$ deg, the negative displacement (contraction) of the moving boundary causes a positive velocity profile at the axis. But a vortex bubble remains in the upstream part of the duct near to the axis. This vortex bubble disappears in the next time steps ($\phi' = 105.3$ deg), where the flow takes a developed axial form at the exit, while it is purely two dimensional in the upstream part of the duct. For a next time step ($\phi' = 184.7$ deg) the velocity profile has changed the direction and it is the phase, where an inflow begins due to the expansion movement of the duct. Remarkable at this time step is that the inflow rate is mainly caused in the outside region of the duct cross section, while in the region near to the axis the flow is rather low. The last shown time instant is $\phi' = 271.46$. At that time instant the velocity profile becomes more or less developed in the entrance region of the duct, that means the maximum velocities are again near to the duct axis.

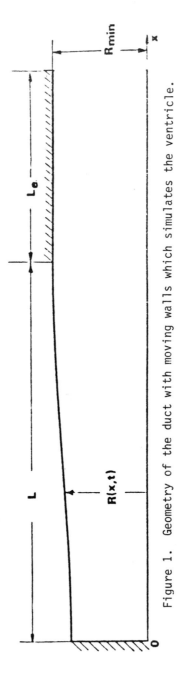

Figure 1. Geometry of the duct with moving walls which simulates the ventricle.

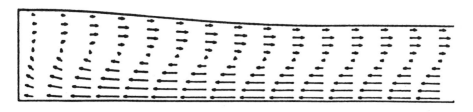

(a) φ'= 0.57 deg

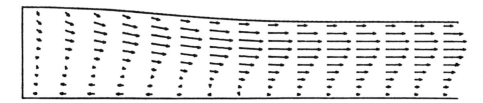

(b) φ'= 4.7 deg

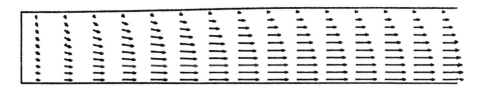

(c) φ'= 105.3 deg

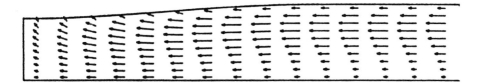

(d) φ'= 184.7 deg

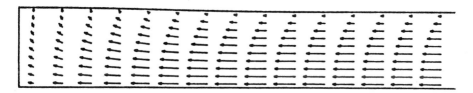

(e) φ'= 271.46 deg

Figure 2.   Velocity field in the duct at different time   instants.
Plotting scale of velocity vectors for (a)(b) and (d) five
times the scale of (c) and (e)

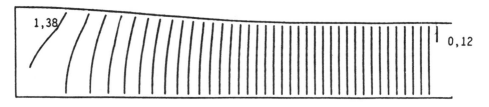

(a) φ′= 0.57 deg

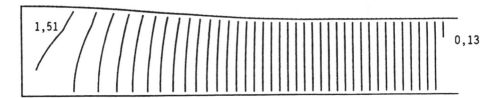

(b) φ′= 4.7 deg

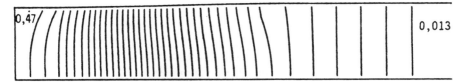

(c) φ′= 105.3 deg

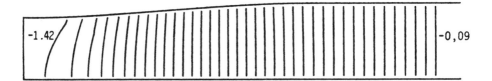

(d) φ′= 184.7 deg

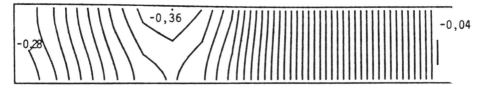

(e) φ′= 271.46 deg

Figure 3. Iso - pressure lines in the duct at different time instants.

Similar remarks can be done for the pressure distributions shown in the next figure 3, with iso-pressure lines.

CONCLUSIONS

In this paper a description of the application of an implicit projection method, in the case of internal flows for two dimensional or axisymmetric ducts with moving walls, was given. The method was applied to the system of momentum and continuity equations for viscous fluid written in an arbitrarily curvilinear, body fitted, coordinate system. For the spatial discretization of the obtaining equations in (ξ, η) plane the M.A.C method is applied. The flow field, in the case of an axisymmetric duct with pulsating walls, was demonstrated. The method should be extended in the future to three dimensional flow problems for arbitrary geometries.

REFERENCES

Chorin, A. J., 1968, Numerical solution of the Navier-Stokes equations, Math. Comput., 22: 745-762.

Covell, J. W., 1972, Mechanics of contraction of intact heart, in: "Biomechanics: Its Foundations and Objectives", (eds.) Fung, Y. C., Perrone, N., Anliker, M., Prentice-Hall, Englewood Cliffs, N. J.

Fortin, M., Peyret, R., Temam, R., 1971, Resolution numerique des equations de Navier-Stokes pour un fluide incompressible, J. Mecan., 10/3: 375-390.

Fung, Y. C., 1984, "Biodynamics - Circulation", Springer Verlag, N.York.

Harlow, F. H., Welch, J. E., 1965, Numerical calculation of time-dependent viscous incompressible flow, Phys. Fluids. 8: 2182.

Hilbert, D., 1987, Zur numerischen Loesung der instationaeren Navier-Stokes Gleichungen, ZAMM, 67/5: T 289-290.

Jones, R. T., 1968, Blood flow, in: "Annual Review of Fluid Mechanics", (eds.) Sears, W. R. and Van Dyke, M., Annual Reviews, Palo Alto, Cal.

Pedley, T. J. and Seed, W. A., 1977, The Fluid mechanics of left venticular ejection, in: "Cardiovascular and pulmonary dynamics", INSERM - Euromech 92. vol. 71: 311-320.

Sorenson, R. S., 1981, A computer program to generate two dimensional grids about airfoils and other shapes by the use of Poisson's equation, NASA TN-81198.

Temam, R., 1969, Sur l' approximation de la solution des equations de Navier-Stokes par la methode des pas fractionnairs II, Archiv. Ration. Mech. Anal., 32: 377-385.

Viviand, H., 1974, Formes conservatives des equations de la dynamique des gaz, La Recherche aerospatiale, 1: 65-68.

Wang, C. Y. and Sonnenblick, E. H., 1979, Dynamic pressure distribution inside a spherical ventricle, J. Biomech., 12: 9-12.

UPSCALING AS A TOOL IN BIOFLUIDMECHANICS - DEMONSTRATED

AT THE ARTIFICIAL HEART VALVE FLOW

K. Affeld, P. Walker, and K. Schichl

Medizintechnik Forschungslabor
der Abteilung für Innere Medizin im
Universitätsklinikum Rudolf Virchow der
Freien Universität Berlin

ABSTRACT

The small size of most geometries investigated in Biofluidmechanics permits the up-scaling of the flow problem. Many benefits can be achieved - the flow is greatly reduced in speed and many methods of flow visualization can be applied. The latter provide an integral picture of the flow and also render quantitative results if the geometry is sufficiently enlarged. The flow through an artificial heart valve is treated as an example.

KEYWORDS

Enlarged model, similarity, artificial heart valve, flow visualization, image processing

INTRODUCTION

The flow through bloodvessels and artificial heart valves has a great degree of complexity - the flow is usually threedimensional, instationary and periodic, has flow separations und often a transition between laminar and turbulent flow. In engineering sciences this is not uncommon, such cases are the rule rather than the exception, for instance in ship hydrodynamics and aircraft engineering. The answer to this complexity of the flow since long has been the modelling of the flow. The actual structure is geometrically scaled down to a manageable size and subjected to the flow. But the engineers often are unable to keep the similarity numbers in the correct relation. The shipbuilders cannot keep the Reynolds number, because they primarily have to abide the Froude number (Relation of boat speed to surface wave speed) and the aircraft engineers run into a similar problem, because they have to pay attention to the Mach number (Relation of aircraft speed to the speed of sound). Both have to sacrifice and have to extrapolate their data to the real size, often with great loss in precision.

Unlike in these fields, in Biofluidmechanics one can <u>scale up</u> as opposed to scaling down, because our real flow is of small geometrical dimensions and in addition slow compared to most technical flows. No surface waves play a role, nor does compressibility of the fluid influence the flow. So if one keeps the Reynolds number constant one can scale up the flow and does not have to sacrifice anything - instead one gains a number of advantages:

the time scale is greatly extended, i.e. the flow can be observed in slow motion,

the lighting is much easier.

Biomechanical Transport Processes, Edited by F. Mosora *et al.*
Plenum Press, New York, 1990

For some reasons these favorable features have been overlooked so far. As an example for the scaling up the flow through an artificial heart valve is described.

One of the most successful artificial heart valves is the Björk-Shiley valve, which has been implanted in several hundertthousand patients. A drawing of this valve is shown in Fig. 1. The valve consists of a ring with a circular tilting disk - the occluder - which by its movement opens the cross-section and closes it when the flow reverses. Fluidmechanics have only partially influenced this design, the main consideration was the hinge mechanism, which is designed not to have hidden pockets to avoid the formation of thrombi. The shape of the disk is circular in order to reduce wear between the ring and the disk. The fluidmechanics of this valve have been extensively investigated, when the Laser Doppler Anemometry (LDA) was introduced into the artificial heart valve field (Affeld et al., Pszolla et al, 1979). Measurements performed with this method do not interfere with the flow, because only laserbeams and no physical probes are present within the flow. The velocity at one point within the flowfield can be measured with a good timewise resolution. The spatial resolution is determined by the accuracy of the traversing unit. With several laserbeams it is possible to measure all three components of the velocity (Yoganathan et al., 1982). From this, turbulent information such as the Reynolds shear stresses can be calculated. A great number of artifical heart valves have been systematically investigated in this way (Reul et al., 1989).

As successful as the LDA method is, it has the great disadvantage that the velocity can only be measured at one particular location at a time. The picture of the flow must be composed from many velocity vectors, which in addition are averaged vectors. Information is lost in this process.

An overview of the flow can be obtained by the use of flow visualization, which has been used for heart valve flow studies since they were first began. The results up until now however, have lacked great detail, mainly due to the small size of the valves and the high velocity of the fluid. By enlarging the valve it is possible to reduce the velocity and therefore to overcome these problems.

MATERIAL AND METHODS

Most methods of flow visualization require a low velocity of around 5 cm/s in water. At higher velocities dyes disperse too quickly and particles are difficult to light properly. Velocities of 150 cm/s which are typical for flow in an artificial heart valve are much too high for good flow visualization. A sufficiently low velocity can be achieved however with upscaling. The model flow will remain similar to the real flow, provided the Reynolds num-

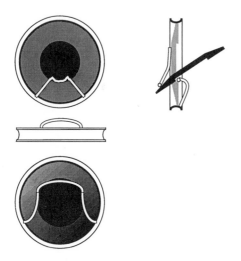

Fig. 1. The Björk-Shiley valve seen from both sides and in a cross section.

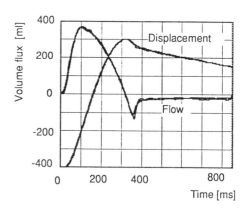

Fig. 2. The flow in the model and the piston displacement as a function of time, calculated for the real size valve.

ber is the same. A scale of 10 to 1 allows the reduction of the velocity in the model flow to 1/35th of the real one, producing a maximum velocity of about 5cm/s.

The flow through the enlarged model must be similar to that through the real valves, for this purpose a water tunnel has been built which can be thought of as a mock circulation. The model valve has a diameter of 220 mm and is mounted in a transparent model of the aortic root. Water is used as the fluid. The tunnel weighs around 600 kg and holds some 400 l of water. The total length is 3.7 m.

Unlike a conventional watertunnel this one produces a pulsatile flow as it is required for artificial heart valves. Another new feature is the way to establish an even and laminar flow to the model: quiet water from a settling chamber is separated from the disturbed fluid by a free floating wall. This wall in the form of a piston moves with the flow from one end of the settling chamber to the other. Once the piston has reached the far end, the flow is stopped and the piston reversed before the next run is started. The volume contained in the settling chamber limits the number of cycles to 3, but this is sufficient for all methods of flow visualization. The floating piston also serves another purpose: since hardly any forces are acting upon it, it floats very precisely with the fluid and can be used to monitor the flow. The displacement of the piston is the same as the displacement of the water and so through differentiation the water flow can be calculated. The piston is connected to a displacement transducer by a thin string. The fluid is moved by a propeller housed in the return section of the tunnel and is driven by a computer controlled electric DC motor. By integrating the aortic flow curve taken from a textbook of physiology a displacement-time curve has been obtained. The electric motor is controlled so that the displacement of the piston follows the physiological curve. In Fig. 2. the real curve and the ideal curve for the flow and the displacement are plotted on top of each other. The proper speed of this displacement is calculated with the help of the similarity laws. Most important is the Reynolds number (Re) similarity. With u being the velocity, d the diameter of the valve and v the kinematic viscosity we obtain:

$$Re = \frac{u\,d}{v} \qquad\qquad 1$$

With a model scale of 10:1 for the geometry and 1:3.5 for the kinematic viscosity - (water: blood) - we obtain a reduction of the velocity of 1:35 and a time expansion of 350:1. This means that a systole of 300 ms lasts in the model more than 100 seconds thereby giving the flow a slow motion appearance.

The second number to be observed is the Strouhal number (S):

$$S = \frac{d}{u\,t} \qquad\qquad 2$$

this number assures that the stroke volume is in the proper relation to the geometry.

The final number to be considered is the Archimedes number (Ar). With ρ_{fluid} being the specific density of the fluid, $\rho_{occluder}$ that of the valve occluder and g the gravity, the Archimedes number is defined as:

$$Ar = \frac{\rho_{occluder} - \rho_{fluid}}{\rho_{fluid}\,u^2}\ d\,g \qquad\qquad 3$$

The number takes into account the buoyancy of the occluder. In the real valve the specific weight of the occluder differs considerably from that of the blood. This has little influence on its movements because the velocities are high and the dynamic pressure easily overcomes the influence of its buoyancy. In the enlarged model this cannot be said because the dynamic forces are so small that the buoyancy plays an important role. With the model scale inserted in the formula one obtains the specific weight of the model occluder as

ρ_{occluder} = 1.00006. This is practically equal to that of the water. To achieve this density the model occluder is made of polyethelene, which is lighter than water and weighted with lead to obtain a neutral buoyancy. The lead is positioned at the occluder´s center of gravity to maintain it´s balance. The movement of the occluder is very sensitive to its specific weight, even some small air bubbles which attach easily when the tunnel is freshly filled will influence its movement. Once the scaled flow and the proper movement of the occluder has be established, the flow can be made visible. A great variety of flow visualization techniques have been developed (Gad-el-Hak, 1988). From these, the most appropriate have been selected and applied. These are:

Particle Method: This method provides an <u>overall picture</u> of the flow, usually of a selected light plane. Small plastic (polystyrene) spheres with a diameter of 0.2 to 0.4 mm are added to the water. The flow is then lit with a light plane so that a cross-section is seen.The spheres move with the flow and show the path lines. On photographs taken with an extended exposure time, each particle produces a streak the length of which is proportional to the velocity. If in addition an electric motor with a little weight attached to its axle is mounted at the camera, a vibration of a known frequency is generated, which results in sawtoothed or cycloidic streaks in the photograph. This can be used for the direct reading of the velocitiy. Fig. 4 shows as an example the trailing edge of the disk. If the flow is recorded upon video then by analysing a series of consecutive pictures the motion of the particles can be obtained and therefore the flowfield (Affeld et al., 1979).

Electrolytic Precipitation Method: This method permits the visualization of a <u>boundary layer</u> and of its separation. A part of the model is covered with a very thin layer of tin solder, this forms with another piece of metal in the water an electrolytic pair. When an electrical potential of about 10 V is applied, the solder undergoes an electrolytic reaction which produces a white tin salt upon the covered surface (Taneda et al., 1979). The salt is carried away by the fluid near the wall and therefore visualizes the boundary layer.

Hydrogen Bubble Method: This method is used to show a <u>velocity profile</u>, which in turn is used to determine the deformation and the shearstress. A thin platinum wire with a diameter of 15 µm and a length of 100 mm is placed in the flow and given a pulsating electrical

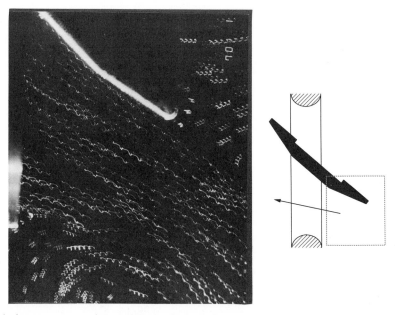

Fig. 3. A time exposure of the particles show their path. By deliberately shaking the camera with a motor a vibration with a known frequency is generated, which deforms the streaks and permits the direct quantification of the velocity. The flow is from left to right, the arrow indicates the position of the frame in relation to the valve.

potential of 10 V (Schraub et al., 1965). This produces electrolysis of the water which leads to the formation of hydrogen bubbles upon the wire, these bubbles are pulled from the wire by the surrounding fluid to form a bubble sheet within the flow which through the correct illumination and pulsation shows the instantaneous velocity profile.

Dye Injection Method: This method shows the path fluidparticles take and is suited for the visualization of <u>vortices</u> and <u>flow separations</u>. By injecting dye into the flow the streak lines can be visualized, with the advantage that the flow in all three dimensions can be seen.

Fluorescent Dye Method: This method is best suited to show the mixing of different flows, in a <u>jet</u> or after a <u>flow separation</u>. Before the valve opens at the start of systole the fluid on the aortic downstream side is mixed with a fluorescent dye, the fluid on the up-stream side is uncoloured. As the valve opens the clear fluid displaces the dyed fluid except in those regions where separation occurs. A clear view showing the separation regions and the mixing of the fluid within them and the main flow is obtained. As in most of these methods a slit lighting is used. The fluorescent dye has the property that it cannot be seen unless strongly lit so that dye between the observer and the light plane does not obscure the view.

Each of these methods has its own merits and will be used according to on what the focus is. Especially useful is the particle method. Still photographs already permit a good impression of the overall flow, the upscaling permits already a quantification, as Fig. 3 shows. The use of a videorecorder makes this method even more impressive - the slow motion effect permits the observation of very fine details of the flow and can be repeated as needed. The timescale of 350 to 1 results in a camera speed which is 25 times 350 equal to 8750 frames per second.

One can also quantify this material if one uses the methods of image processing. The velocity field can be calculated from the displacement of the particles between different video frames, the time between frames is known from the camera shutter speed. The particle displacement has been obtained using the cross correlation method. Once the velocity is obtained the stream function and stream lines can be calculated in order to make the flow patterns more understandable. The flow was filmed using a Panasonic F10 video camera and recorder. The video was digitized by a frame grabber board and software (Quick Capture from Data Translation) which was mounted in a Macintosh II computer from Apple. The resolution of the digitizer is 768 lines by 512 pixels. Standard picture analysis was performed by the Quick Capture software and the specialized cross correlation analysis by in-house programs written in Z-Basic. This method uses standard pattern recognition procedures to match the patterns formed by the particles between pictures. Two consecutive video frames are used, each is divided up into a 19*12 grid, each grid square is 16 by 16 pixels large. The gray scale value of the pixels range from 0 to 255 , let these be represented by f (x,y) for the first video frame and g(x,y) for the second. x is the horizontal pixel number and y the vertical. Within each grid square the white particles upon the black background form a pattern, each grid square has a different pattern. However, corresponding grid squares in the two video frames will have the same pattern except that the position of the pattern within the grid will be different as the particles will have moved because of the velocity of the water. The cross correlation method involves moving the grid square from the first video frame over the corresponding grid square from the second video frame until the patterns of the particles match one another or at least until the best match is made. The difference in position of the pattern between the first and second frames then gives the particle displacement and hence velocity vector. Intrinsic in this process is the assumption that the pattern undergoes a pure solid body translation with no rotation or shearing being involved. This is of course in all but the simplest of motions untrue but can be used if the size of the grid squares are kept small and the time between pictures short relatively speaking. The process of finding the best match between grid squares is mathematically equivalent to the cross correlation as defined in equation 1.

$$f(x,y)\, g(x,y) = \sum_{m=0}^{M=1} \sum_{n=0}^{N=1} f(m,n)\, g(x + m, y + n) \qquad\qquad 4$$

This process is slow if performed using equation 1, but can be vastly accelerated if use is made of equation 2 (Gonzalez et al., 1987). That is the cross correlation of two signals can be calculated by taking the Fourier transform of the two signals, one transform is then complex conjugated and the two transforms are multiplied together to give one complex signal, an inverse Fourier transform is then performed upon this signal to produce a real signal corresponding to the cross correlation.

$$f(x,y)°g(x,y) <=> F^*(u,v).G(u,v) \qquad\qquad 5$$

Where f(x,y) and g(x,y) represent the gray scale values of the two pictures and F(u,v) and G(u,v) their Fourier transforms, the * is the complex conjugate and <=> the forward and inverse Fourier transform. The Fourier transforms are speeded up by using a fast Fourier transform (FFT) routine and long integer (32 bit) numbers. The first step in the formation of the cross correlation is to read in the gray values of the grid squares under consideration, that is to form f(x,y) and g(x,y). These arrays are then expanded up to 64 by 64 pixels to increase the accuracy of the result. A Fourier transform is now performed upon the two arrays to produce the fourier image of the two grid squares. The complex conjugate of one image is taken and the two images multiplied together, the product is then transformed back to the physical plane using an inverse Fourier transform. The resulting array is equal to the cross correlation of the two grid squares. Array element i,j corresponds to the correlation between the two pictures if the first picture is displaced i pixels to the right and j pixels downwards, the largest element in the final cross correlation array therefore corresponds to the position of best correlation between the two grid squares. The displacement of the particles in the grid square is now known in terms of pixels which can be converted to actual displacement in meters and by dividing by the time difference between the two video frames the velocity can be calculated. By performing the above procedure upon all 228 (19*12) grid squares the velocity field over the entire frame at regular intervals is obtained.

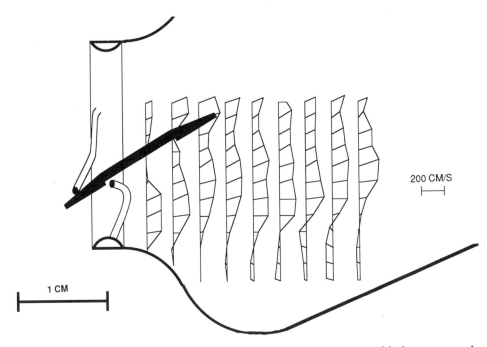

Fig. 4. The flowfield behind the valve as calculated from a video tape with the crosscorrelation method.

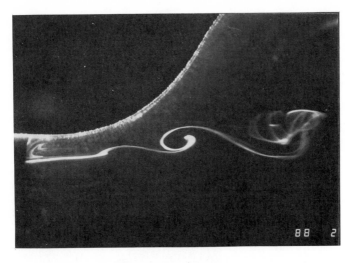

Fig. 5. A flow separation made visible with the electrolytic method. The separation takes place at the ring which is not visible. Note the accumulation of the white salt in the corner between ring and aortic wall.

RESULTS

Artificial heart valves still have a considerable complications rate - valves made out of biological material deteriorate and valves made out of artificial material induce the formation of thrombi. Patients having theses implants die at a rate of about 10% per year of valve related thromboembolytic complications. If the survival rates of patient populations is plotted for the different types of heart valves, one finds a significant difference in performance which can only be due to the design variations and therefore the fluid dynamic characteristics. The influence of the materials can be excluded, since some valves are made of the same materials and still differ in their performance. Despite the many efforts made so far, it has not been possible to specify the flow responsible for thrombus formation in a specific valve. On the other hand one knows from detailed in vitro experiments the flow conditions, which are relevant for the generation of a thrombus. These are:

firstly an area of high shear stress, either in a jet or at a wall, which exceeds

$\tau = 100$ N/m^2 and lasts for more than 30 ms (Wurzinger et al., 1985). No information is yet available on the influence of a repetitive exposure to the shear stress and

secondly an area of recirculation close to a wall of foreign material. The recirculation allows a concentration of platelets. If they are activated, accumulate densely enough and in addition they are in the vicinity of foreign material which they can attach to, then the formation of a thrombus is likely (Hashimoto et al. 1985).

Using the methods applied up until now only an estimate of the position of these areas has been obtained. The enlarged model however, permits the precise location of them - they are found in areas of flow separation. Fig. 4 shows a flow separation region behind the valve ring. The ring is on the right of the picture, but itself not visible. The boundary layer upon the ring has been made visible with the electrolytic method, the solder being directly attached to the ring. The white tin salt shows the boundary layer separating from the ring itself to form a shear layer within the flow. The fluid below the white salt is from left to right and is the major jet of the valve flow. The separation region is found above the white salt where the aorta wall is also visible, the flow here is from right to left. The shear layer becomes unstable further downstream and rolls up to form discrete vortices. The process of vortex rollup depends on the velocity - the frequency of the formation increases with the velocity and ranges between 30 and 200 Hz, if transformed to the real size valve. At the beginning of the cycle within the lower Re-number range we find a periodic laminar flow separation, while within the higher Re-number range the wavelength of the separated boundary layer changes and stochastic turbulent influences prevail. This shows that in this part of the flow a transition from laminar to turbulent flow takes place.

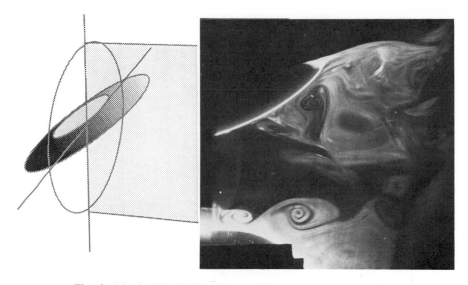

Fig. 6. A jet issuing behind the valve into the aortic root. The mixing of the jet with the surrounding fluid is made visible with fluorescent dye. This picture also confirms the flow separation at the valve ring and behind the disk.

Since the rolling up of the shear layer entrains fluid from within the separation region a flow from the bulbs towards the ring is initiated. There is therefore a mechanism by which fluid which was originally within the free shear layer and is therefore coloured white, can be recirculated back towards the ring where, since the velocity there is very small, it can accumulate. An accumulation of white salt is visible in Fig. 4 on the left side downstream of the ring. If the water were blood, i.e. as in the real valve, it would remain for a relatively long time in close contact with the valve ring which being a foreign material is likely to cause thrombus formation as described in the second condition above.

In order to check if the first condition - the exposure to high shear stress - is met, an evaluation of the shear stress in the shear layer was made. If one evaluates the velocity profiles generated with the hydrogen bubbles method the shear rate can be obtained. The derivative of these curves is calculated and through the use of Newton´s formula the resulting shear stress is found. The shear stress easily exceeds 100 N/m^2 and sometimes comes close to even 200 N/m^2. These values agree with data from other authors (Woo et al., 1985). Moreover it can be seen that very close to an area of flow recirculation an area of high shear stress exists. The recirculation persists in some areas of the ring throughout the whole cycle, so a permanent flow separation exists. From these findings one could predict that some areas of the ring are likely sites for thrombus generation.

Pathologists indeed have found thrombi at these loci: through the examination of artificial heart valves which have been implanted in patients or subjected to bloodflow in artificial bloodpumps, they have noticed that the ring within the minor orifice is a common site of thrombus formation. This gives an indirect proof that areas of thrombus generation and areas of flow separation coincide.

DISCUSSION

The flow through an artificial heart valve is very complex. It is instationary-periodic, three-dimensional, with moving boundaries, with a laminar turbulent transition and in addition the fluid is non-Newtonian. Thus making heart valve flow much more complex than typical engineering flows. No wonder it has so far resisted a complete clarification, despite the sophisticated methods which have been applied. Modelling applied to the artificial heart valve flow permits the application of many flow visualization methods - particles, dye, precipitation, hydrogen bubbles - to be applied. The pictures show a high space and time resolution previously unobtainable.

One of the most interesting aspects of this method is to apply it to the investigation of flow separation: There is flow separation from the disk itself, which is of a transient nature and probably irrelevant for the thrombus formation. More serious is the separation from the ring. There is one region on the downstream side - at the ring within the minor orifice - which is not washed away and persists throughout the whole cycle. Platelets can remain in this region for several cycles. This is an encouraging result, because there are many ways to modify the flow. With the enlarged model we have a way to objectively check the modifications.

CONCLUSION

In engineering one can solve difficult problems by model simulation. Usually one has to scale down, i.e. work with smaller models and has to cope with errors. Unlike in mechanical engineering in biofluidmechanics one can scale up - sacrificing nothing, but gaining greatly in precision and handling.

REFERENCES

K. Affeld, H. Pszolla, B. Lehmann and R. Mohnhaupt, (1979), Measurement of the flow field behind artificial heart valves with the help of the Laser-Doppler-effect, in:"Proc. ISAO II", 439-441.

K. Affeld, P. Walker, K. Schichl, (1989), The Use of Image Processing for the Investiga tion of Artificial Heart Valve Flow, in: "Proceedings ASAIO".

M. Gad-el-Hak, (1988), Visualization Techniques for Unsteady Flows: An Overview, Journal of Fluids Engineering, Vol. 110: 231-243.

R.C. Gonzalez, P. Wintz, (1987), "Digital Image Processing", Addison-Weseley Publishing Company, Reading.

S. Hashimoto, H. Maeda and T. Sasada, (1985), Effect of Shear Rate on Clot Growth at Foreign Surfaces, Artificial Organs 9 (4): 345-350.

H. Pszolla, K. Affeld, B. Lehmann and A. Mohnhaupt, (1979), Messung des Geschwindig-keitsfeldes hinter einer künstlichen Herzklappe (Björk-Shiley-Ventil) mit dem Laser-Doppler Anemometer, in "Biomed. Technik "24, Erg.-Band.

H. Reul, (1989), Qualitätssicherung von Herzklappenprothesen, "Abschlußbericht BMFT Projekt 01 ZQ 0141".

F.A. Schraub, S.J. Kline, J. Henry, P.W. Runstadler and A. Littell, (1965), Use of Hydro-gen Bubbles for Quantitative Determination of Time-Dependent Velocity Fields in Low-Speed Water Flows. Journal of Basic Engineering, Transactions of the ASME: 429-444.

S. Taneda, H. Honji and M. Tatsuno, (1979), The Electrolytic Precipitation Method of Flow Visualization, in: " Proc. Int. Symp. Flow Visualization", Tokyo 1977, Hemi-sphere Publishing Corp.

Y. R. Woo and A. Yoganathan, (1985), In Vitro Pulsatile Flow Velocity and Turbulent Shear Stress Measurement in the Vicinity of Mechanical Aortic Heart Valve Pros-tethes, Life Support Systems 3: 283 - 312.

L. Wurzinger, R. Opitz, M. Wolf and H. Schmid-Schönbein, (1985), Shear induced Plate-let Activation, Biorheology.

A. Yoganathan, A. Chaux, R. Gray, M. De Robertis and J. Matloff, (1982), Flow Charac-teristics of the St. Jude Prosthetic Valve: An In Vitro and In Vivo Study, Artificial Organs, Vol. 6, No 3: 288.

THE STUDY OF INTEGRATED VENTRICULAR AND ARTERIAL FUNCTION

USING THE METHOD OF CHARACTERISTICS

C.J.H. Jones and K.H. Parker

Imperial College of Science
Technology and Medicine
London SW7 2AZ, U.K.

INTRODUCTION

The wave-like nature of flow in the elastic arteries suggests that analysis by means of the method of characteristics may be appropriate and, being non-linear, may be advantageous (Anliker et al., 1971; Parker and Jones, 1990). Assuming one-dimensionality, the equations describing arterial flow, written in terms of U, the mean velocity, and P, the pressure, are hyperbolic with characteristic directions

$$\frac{dz}{dt} = U \pm c$$

where $c = \sqrt{A/\rho \, (dA/dP)}$ is the wave speed which depends upon the density, ρ, the instantaneous local area of the artery, $A(P)$, and its distensibility, dA/dP. As $c > U$ for normal arterial flows the + direction is forward, away from the heart and the − direction is backward, toward the heart. Along each characteristic direction, the equations reduce to a single ordinary differential equation in terms of the Riemann functions (Landau and Lifschitz, 1959),

$$R_{\pm} = U \pm \int_{P_0}^{P} \frac{dP}{\rho c}$$

If the artery is locally uniform and dissipation and dispersion can be neglected (Anliker et al., 1968), the Riemann functions are constant along the characteristic directions and are generally refered to as the Riemann invariants.

Key Words: Left ventricle, waves, aorta, vasodilatation, vasoconstriction.

For the purpose of analysing finite waves, it is convenient to define d as the difference of any variable between two characteristics so that

$$dR_{\pm} = dU \pm \frac{dP}{\rho c}$$

Solving for dP and dU, we can calculate the energy flux per unit area in the wave motion

$$dPdU = \rho c (dR_+^2 - dR_-^2)/4$$

which has the useful property that changes arising from forward running waves, dR_+, contribute positively while those arising from backward running waves, dR_-, contribute negatively. Thus the value and sign of $dPdU$ provide a ready indication of the net magnitude and direction of wave propagation at the time and point of measurement.

RESULTS

Figure 1 shows the (a) ensemble averaged and (b) beat by beat P and U measured using a multisensor catheter (Millar) in the ascending aorta of a patient with normal left ventricular function, treated with vasodilators (diltiazem, a calcium channel blocker, and intravenous nitroglycerine) immediately after coronary angioplasty. Also shown is the $dPdU$ calculated from the ensemble average data which shows two positive peaks of $dPdU$, indicating periods when forward running waves dominate the flow.

The first peak in early systole corresponds to increasing pressure and so represents a forward running compression (distending) wave generated by the contraction of the ventricle. The second peak occurs in late systole when the pressure is decreasing and so represents a forward running expansion (constricting) wave, also generated by the left ventricle, which causes the blood to decelerate. Negative values of $dPdU$ in mid-systole indicate a backward running wave, probably the reflection of the initial systolic forward running wave from an arterial discontinuity close to the left ventricle. Predominant backward running waves were no longer seen when the catheter was withdrawn into the upper descending aorta beyond the origin of the head and neck vessels, the likely site of wave reflection in this patient. The P and U data obtained in the upper descending aorta, with the calculated $dPdU$, are shown in figure 2(a) and (b).

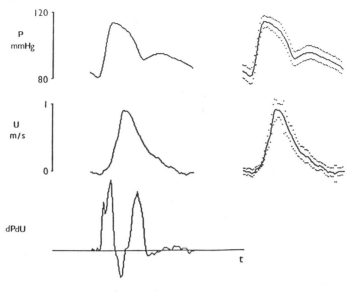

Figure 1 (a)

The ensemble averaged P and U measured in the ascending aorta of a patient maximally vasodilated after angioplasty, with the dPdU calculated from the ensemble averaged data. On the right are shown the mean ± 1 standard deviation. The standard deviation of P is primarily due to low frequency, respiratory variation.

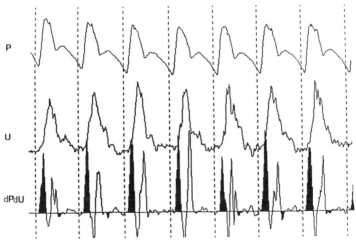

Figure 1 (b)

A portion of the measured P and U from which the ensemble average was calculated. The instantaneous *dPdU* is also shown. Positive values of *dPdU* indicate predominantly forward running waves and negative values indicate backward running waves. Compression waves are shaded in black while expansion waves are unshaded. The initial systolic compression wave is apparently reflected as an expansion wave.

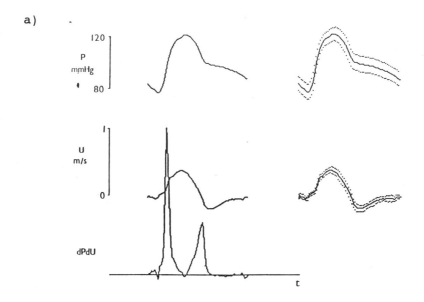

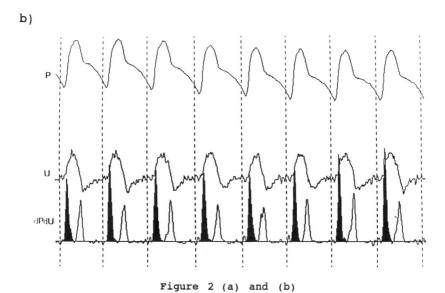

Figure 2 (a) and (b)

Measurements obtained in the same patient in the upper descending aorta just beyond the origin of the subclavian artery. Note the absence of backward running waves, and the change in shape of the forward running waves.

The application of a non-linear analysis enables explanation of the changing shape of the propagating pressure and flow waves. $dPdU$ calculated from data acquired in the upper descending aorta shows peaking and steepening of the forward running compression wave, as has been predicted for non-linear travelling waves on the basis of decreasing distensibility with increasing local pressure. As forward running waves dominate all the flow at the measurement site in the upper descending aorta, peripheral pressure wave amplification may be explicable in terms of non-linear wave propagation, rather than summating distal reflections as are predicted in a linear wave analysis (Milnor, 1982).

As the $dPdU$ calculation indicates periods of uni-directional wave travel, the local wave speed during these periods may be estimated from the general relationship between the dP and dU of travelling waves

$$dP_\pm = \pm cdU_\pm$$

If the forward and backward running waves intersecting at the measurement site are superimposable, knowledge of the wave speed, calculable for a given pressure, enables the forward and backward running waves to be separately calculated

$$dP_\pm = (dP \pm \rho cdU)/2$$

$$dU_\pm = (dU \pm \frac{dP}{\rho c})/2$$

Despite the assumption of linear superimposition, a degree of non-linearity is retained in this calculation as the wave speed remains a function of pressure. Figure 3(a) shows the incremental forward and backward running pressure waves calculated from ascending aortic P and U in a different patient with normal ventricular function (Parker et al., 1988). Although forward running waves are generally predominant in the flow, backward running waves arrive at various times in middle and late systole, indicating diffuse reflecting sites. Figure 3(b) indicates that the backward running waves are increased after vasoconstriction caused by intravenous phenylephrine infusion and are then associated with an accentuated forward running pressure wave which may result from mid-systolic re-reflection of waves within the heart.

a)

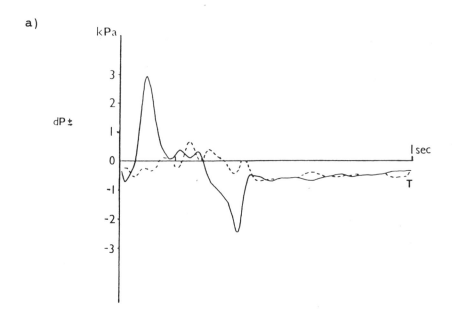

b)

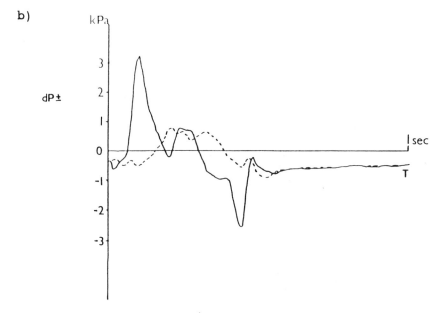

Figure 3

Incremental forward (solid lines) and backward running (broken lines) pressure waves calculated from ascending aortic *P* and *U* in a patient with normal cardiac function (a) in control state (b) after intravenous phenylephrine infusion. The backward running waves are increased and arrive earlier in systole than during the control state.

CONCLUSIONS

In summary, non-linear wave travel in the arterial system may be studied from measurements of P and U at a single point in the artery using the method of characteristics. The value and sign of dPdU indicates the net magnitude and direction of wave travel. Our measurements show that forward waves generated by the left ventricle normally dominate ascending aortic flow. Comparison of the wave pattern with that in the descending aorta suggests that the changing shape of propagating pressure and flow pulses may be a non-linear phenomenon, and that, under conditions of maximal vasodilatation, waves can be reflected at the origin of the head and neck vessels. In addition, the wave speed may be calculated when forward waves predominate, and may be used in the separate calculations of the forward and backward waves from single measurements. This approach, which is performed entirely within the time domain, may enable the separate dynamic assessment of myocardial shortening and arterial reflection in man.

ACKNOWLEDGEMENTS

CJHJ is a British Heart Foundation Junior Research Fellow.
KHP is grateful for the support of the Clothworker's Foundation.

REFERENCES

Anliker, M., Histand, M.B. and Ogden, E., 1968, Dispersion and attenuation of small artificial pressure waves in the canine aorta, <u>Circ. Res.</u>, 23: 539-551.

Anliker, M., Rockwell, R.L. and Ogden, E., 1971, Nonlinear analysis of flow pulses and shock waves in arteries. 1: derivation and properties of mathematical model. <u>Z. ang. Math. Phys.</u> 22: 217-246.

Landau, L.D. and Lifshitz, E.M., 1959 "Fluid mechanics," Vol. 6. "Course of Theoretical Physics," Pergamon Press, London.

Milnor, W.R., 1982, "Hemodynamics," Williams and Wilkins, Baltimore.

Parker, K.H. and Jones, C.J.H., 1990, Forward and backward running waves in the arteries - analysis using the method of characteristics. <u>J. Biomech. Eng.</u> (in press).

Parker, K.H., Jones, C.J.H., Dawson, J.R. and Gibson, D.G., 1988, What stops the flow of blood from the heart?, <u>Heart Vessels</u>, 4: 241-245.

VORTICITY FORMATION AND TRANSPORT FROM MECHANICAL HEART VALVE PROSTHESES
EXPERIMENTAL AND NUMERICAL SIMULATIONS

F. Cassot*, R. Pelissier**, and G. Tonietto**

* Laboratoire de Mécanique et d'Acoustique du C.N.R.S.
31, ch. J. Aiguier, 13402 Marseille Cedex 9
** Institut de Mécanique des Fluides, Marseille, France

ABSTRACT

Unsteady post valvular velocity fields are studied on a cardiovascular simulator which enable to duplicate in vitro the exact conditions of transvalvular flow. Emphasis is put on flow visualizations and on instantaneous velocity maps reconstructed from ultrasonic Doppler measurements. Analysis of vortex formation and dynamics provides documented characterizations and comparisons between artificial valves.

In addition, bidimensional numerical simulation using a vortex method together with an integral approach shows results in good qualitative agreement with experimental data.

KEYWORDS

mechanical heart valves / flow visualizations / flow mapping / ultrasound / vorticity / numerical simulation

INTRODUCTION

A thorough hemodynamic characterization of artificial heart valves involves a detailed study of the tridimensional and unsteady post valvular velocity field. There are many experimental works, on this subject, but it is still necessary to get a detailed picture of the overall instantaneous velocity field as recognized by Hasemkam .

The work presented here deals with _in vitro_ studies of the flow patterns behind different mechanical heart valve prostheses (disk, bileaflet, hemispheric) in aortic and mitral positions. The techniques used are : (i) flow visualizations, and (ii) flow mapping using ultrasonic Doppler velocimetry.

In addition, the results of a bidimensional _numerical simulation_ of the flow in an anatomically shaped left ventricle (LV), behind a disk valve are compared to the in vitro data.

MATERIALS AND METHODS

Cardiovascular simulator

For in vitro studies, we have developed a mock circulatory loop (Cassot et al 1985) which is schematized in figure 1. It includes an

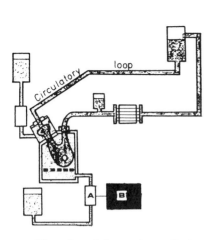

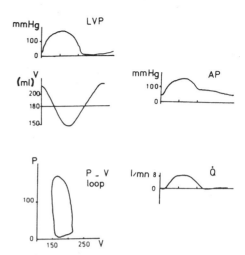

Fig. 1. Schematics of the cardiovascular simulator.

Fig . 2. Pressure, volume, and flow waveforms duplicated in the LV (left) and aortic (right) in vitro models.

elastic, transparent and anatomically shaped LV model pumping into an elastic model of the ascending aorta. The LV volume and pressure waveforms are monitored using an hydrodynamic generator which consists of a pump (A) driven electronically by a synthetiser (B). Such a cardiovascular simulator enables to duplicate in vitro the physiological conditions of transvalvular flow. An example of physiological pressure and flow waveforms generated in the ventricle and in the aorta is given in figure 2. The circulating fluid is a glycerol-water mixture, which has the same viscosity as blood.

Flow visualizations

The trajectories of amberlite particles suspended in the fluid are visualised, and photographed at given instants during the cardiac cycle, using a light plane created by the diffraction of a laser beam on a small glass rod.

Flow mapping using ultrasound Doppler velocimetry

Complete bidimensional velocity maps are reconstructed from an ultrasonic Doppler system (Morvan et al 1984, Farahifar et al 1985). In order to take into account the tridimensional intraventricular flow behaviour we draw up the projections of the velocity field onto three orthogonal sections.

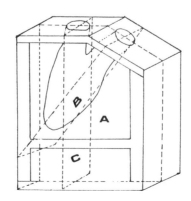

Fig. 3. The projections planes for the LV flow mapping (see text)

Figure 3 shows a sketch of these three planes that we have called :
(i) the longitudinal section A, (ii) the mitral section B, and (iii) the
aortic section C.

The same flow mapping technique is also used downstream aortic valves.

Numerical modelling of left ventricular flow

In order to describe the gross qualitative features of the real flow,
and as a first step towards a more realistic simulation, we have developed
a numerical model of the LV flow behind a disk valve, within the theoretical framework of bidimensional inviscid and incompressible flow with vorticity. We summarize below the main steps of this model which has been presented in more details elsewhere (Cassot et al 1988). The total velocity
field is split into irrotational and rotational parts. The potential flow
problem is solved using a boundary element method. The vorticity formulation of the momentum equation is used to solve for the rotational part of
the flow. The vorticity is approximated as a sum of functions with small
supports (vortex blobs of finite size) which are advected using a Lagrangian scheme, based on a second order predictor-corrector.

At each time step, more vorticity is introduced into the fluid from
the valve extremities and from a point, on the LV wall, where flow separation occurs due to the abrupt enlargement of the LV cavity. A Kutta-
Joukovski condition is applied to determine the locations and the strengths
of the new vortices. This gives the vortex shedding mechanism. This
time-stepping procedure is repeated at the following instant with the new
time-dependent input flow and LV wall normal velocity data.

RESULTS

Flow visualizations for five mechanical valves in aortic position

Using the flow visualization technique described above we have compared the flow patterns behind five mechanical valves, three among the most
commonly implanted, the Bjork-Shiley, the St Jude and the Duromedics and
two prototypes which are currently developped in France (cf figure 4).

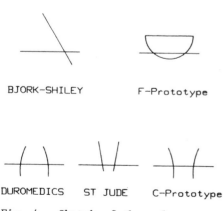

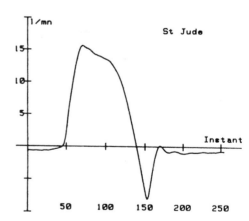

Fig. 4.   Sketch of the valves
    tested in aortic position

Fig. 5.   Instantaneous flow rate for
    the St Jude valve in the aortic model

Note the different shapes of the three bileaflet valves. Note also
that in these experiments the elastic aorta model has been replaced by a
glass one with a more anatomical shape. Nevertheless, the flow conditions
fit accurately physiological ones, as can be shown on figure 5.

Bjork-Shiley. From peak ejection (figure 6a), the wake formed on the unwetted side of the disc, growes and tends to block the jet from the minor orifice of the valve. At mid-systole (figure 6b), immediately downstream the valve, we note, from left to right, a jet focusing at the level of the Valsalva sinuses, a marked tangential velocity discontinuity in the prolongation of the disc, and so a growing vortex sheet, and the flow separated from the unwetted side. We can observe the onset of the vortices formation in the Valsalva sinuses. The mechanism of the vortex shedding (Gerrard 1966) is clearly apparent from the end of the systolic plateau : the two shear layers, issued from the leading and the trailing edges of the disc, interacts in the near wake (figure 6c); the approach of oppositely signed vorticity cuts off further supply of circulation to the growing vortex, which is then shed and moves downstream (figure 6d). At deceleration and during closure the flow pattern evokes a burst turbulence.

At the beginning of diastole, we can see a big vortex zone immediately behind the valve with smaller structures downstream (figure 6e). During diastole, the number of these structures tends to decrease either by coalescence or under the effect of the straining field due to the leakage flow of the valve (figure 6f).

Bileaflet valves. The flow patterns of these valves remain nearly uniform during the major part of the systolic phase: indeed, until the end of the systolic plateau, flow disturbances and, so vorticity formation, are weak and located in the immediate neighbourhood of the leaflets. During deceleration and closure, we observe the partition of the flow into three jets divided by helicoïdal vortex filaments convected from the leaflets (figure 7a). Differences between the Duromedics and St Jude systolic flow patterns are very tiny; the vortex filaments appear later and seem to be less marked for the"C" prototype than for the other valves.

The qualitative evolution of their diastolic flow patterns exhibits a common feature: there occurs a coalescence of vorticity so that stronger, fewer, more distant vortices replace the rather complex but yet organized telesystolic flow structure (figure 7b-c). This progressive amalgamation process is clearly apparent for the Duromedics: it reduces the number of vortices from eight recognizable small ones to two larger ones at telediastole. However, this process is mixed with the leakage flow effect, which is much more important for the St Jude and "C" prototype valves than for the Duromedics.

Caged hemispheric valve. The last flow visualizations here presented concern the "F" prototype (figure 8): vortices form at the sharp edges as early as ejection begins; from peak ejection the flow becomes asymetrical with a divergent jet from one edge and a poorly washed but very disturbed flow pattern on the remaining part of the near wake.

Flow mapping for mechanical valves in aortic position

The flow mapping technique, based on ultrasound velocimetry, described above is, at present, applied to reconstruct the complete velocity field near the same mechanical valves. Figure 9 displays an example of such a reconstruction, for a Bjork-Shiley valve, limited to the axial velocity component on three sections and at four instants during systole. This figure is only to illustrate the complex three-dimensional feature of these flows and, accordingly, the necessity to achieve elaborate studies particularly of the near wake.

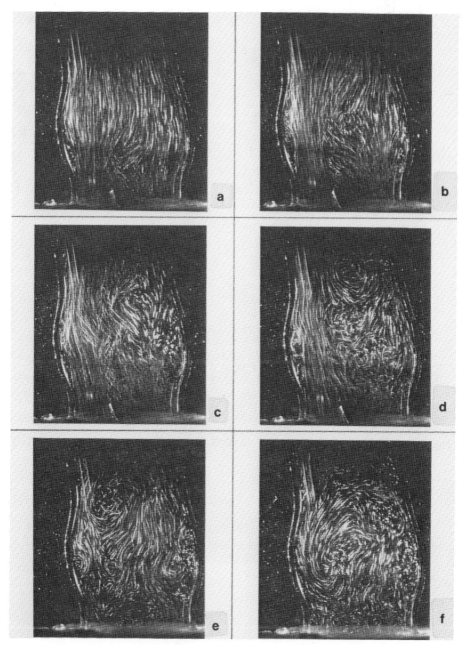

Fig. 6. Flow pattern visualizations behind the Björk-Shiley valve in aortic position (see text).

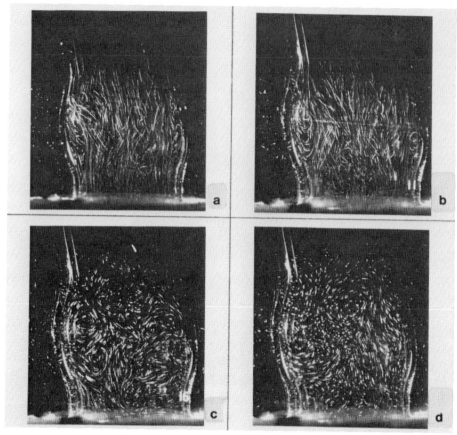

Fig. 7. Aortic flow pattern visualizations behind the bileaflet valves :
(a) "C" prototype (b-d) Duromedics (see text).

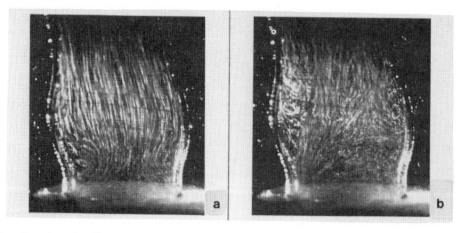

Fig. 8. Aortic flow patterns behind the caged hemispheric prosthetic valve

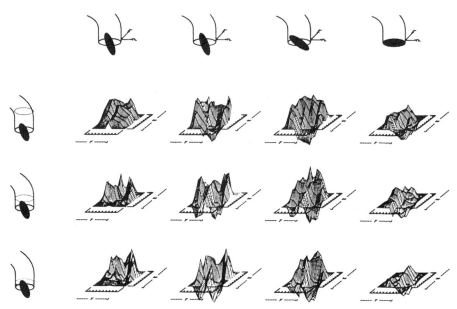

Fig. 9. Aortic flow cross-sections at three levels downstream the Björk-shiley valve (axial component)

We have already shown in previous studies (Farahifar et al 1985, 1987) of the flow development in the aorta that the aortic wall elasticity contributes to the damping of the post-valvular disturbances, and that reverse flow and vortex formation are strongly dependent on the disc orientation.

Flow mapping for a disk valve in mitral position

We summarize here the more important features of the LV flow behind a Bjork-Shiley valve. Figure 10 shows the projection of the velocity field on

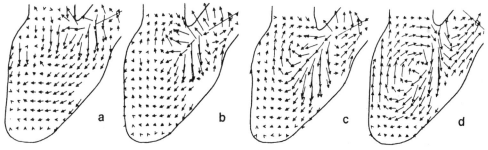

Fig. 10. Velocity field projections on the longitudinal (A) section from early filling (a) to contraction (b)

the longitudinal section A, during the filling phase: it is characterized by the formation of a large vortex flow pattern. The initial development of the flow displays a divergent jet issuing from the major aperture of the valve. Then, from both sides of this jet, much before than its front reaches the apex, the velocity vectors roll up to form fully closed curve flow patterns at mid-diastole. At end filling and during contraction, the main vortex structure under the aortic chamber tends to occupy the major part of the cavity, reinforced by the recirculation of the jet along the wall.

The mitral B-plane corresponds approximately to the central jet direction. We can distinguish two phases (cf. figure 11): during the first half of the filling, , we observe a fairly simple structure with some perturbations only in the near wake of the tilting disc. From mid-diastole, the projections of the velocity vectors near the apex reduces due to the rolling-up process that was already shown on the previous figure. But, we observe also a marked disymetry which may be due to a swirling effect which arises in the atrial model, and to a disymetry of the LV wall motion. At end--filling, the vortex wake engendered at the disk edges seems to remain close to the mitral valve and has a nearly circular shape.

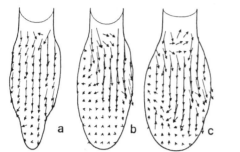

Fig. 11. Velocity field projections on the mitral (B) section during the filling phase.

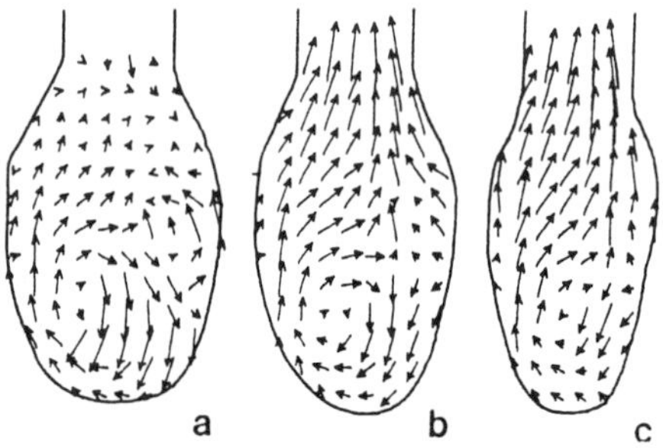

Fig. 12. Velocity field projections on the aortic (C) section during the ejection phase.

The projections of the velocity field in the aortic section (figure 12) show how the vortices are eliminated during the ejection phase. The vortex size decreases very quickly and it seems to be simply unrolled rather than convected towards the aortic orifice. By the end of systole, the velocity vectors have a nearly uniform orientation towards the aortic output.

Bidimensional numerical model of the flow behind a disc valve in mitral position

Figure 13 gives the evolution of the vortex sheets during the filling phase. The vortices are represented as points but they aren't point vortices. The structure of an individual vortex blob consists of three zones: a viscous subcore with a null velocity at the center, an inner region following an analytical model for leading edge vortices, and the outer inviscid finite region.

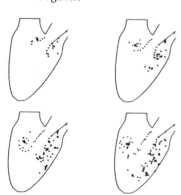

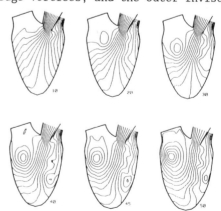

Fig. 13. Distribution of the discrete vortex particles during the filling phase.

Fig . 14. Streamlines pattern evolution during the filling phase.

Figure 14 presents the evolution of instantaneous streamlines during diastole and shows clearly the rolling up process, the interaction between vortex sheets shed from both extremities of the valve and the calculated recirculation zones.

Figure 15 shows a comparison between the computed and experimental flow fields during the filling phase. As expected, only the gross features of the real flow pattern are numerically simulated. There is a good qualitative agreement between the kinematics of the experimental an numerical big vortex zones. On the other hand, the large reverse flow, found experimentally behind the valve, is not present on the numerical simulations.

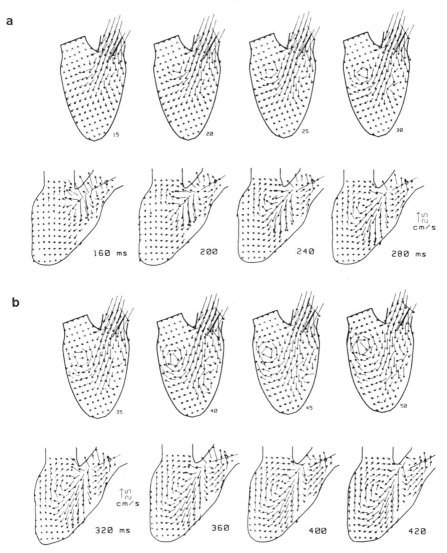

<u>Fig. 15.</u> (a) Comparison between the computed (top) and measured (bottom) velocity vectors during the first half of the filling phase (b) and during the second half of the filling phase.

CONCLUSION

A considerable amount of information is necessary to describe the overall instantaneous flow field behind mechanical valve prostheses in mitral and aortic positions. The results gathered in this paper are far from to be complete. However, the quality of the flow visualizations and flow mapping by ultrasound velocimetry, as well as the ability of the numerical model to predict the main features of the flow patterns are very encouraging. A follow-up of these findings, currently in progress, is a detailed comparison between the three-dimensional flow maps of five mechanical valves in aortic position.

REFERENCES

F. Cassot, D. Morvan, P. Issartier, and R. Pelissier, 1985, New versatile physical model fitting the systemic circulation accurately, Med. & Biol. Eng. & Comput., 23:511

F. Cassot, D. Morvan, G. Tonietto, 1988, Unsteady flow through prosthetic heart valves : an integral approach coupled with a vortex method. in: "Boundary Elements X", Vol. 2, C. Brebbia, ed., C.M.P., Southampton and Springer-Verlag, Heidelberg

D. Farahifar, F. Cassot, H. Bodard, and R. Pelissier, 1985, Velocity profiles in the wake of two prosthetic heart valves using a new cardiovascular simulator, J. Biomechanics, 18(10):789

D. Farahifar, H. Bodard, F. Cassot, R. Pelissier, 1987, Velocity field of a Björk-Shiley valve prosthesis : influence of the disc orientation, Cardiovascular Research, 21:90

J.H. Gerrard, 1966, The mechanics of the formation region of vortices behind bluff bodies, J. Fluid Mech., 25:401

J.M. Hasemkam, D. Westphal, H. Reul, J. Gormsen, M. Giersiepen, H. Stodkilde-Jorgensen and P. Kildeberg Paulsen, 1987, Three-dimensional visualization of axial velocity profiles downstream of six different mechanical aortic valve prostheses, measured with a hot-film anemometer in a steady flow model, J. Biomechanics, 20(4): 353

D. Morvan, F. Cassot, A. Friggi, R. Rieu, and R. Pelissier, 1984, In vitro study of intraventricular flow, Life Support Systems, 2:209

FLOWS IN RIGID AND ARTERIAL GRAFT BIFURCATION MODELS

R. Rieu, R. Pelissier, and V. Deplano

Institut de Mécanique des Fluides - UM34 CNRS
1 rue Honnorat   13003 Marseille- France

I.INTRODUCTION

   Reconstructive vascular surgery has become ordinary practice in vascular surgery. As surgical techniques have matured and as graft materials have improved, arterial grafting has become a frequently performed and generally successful operation. Despite the broad experience and the high degree of success, many problems still remain; in particular, the mechanism of formation of the atherosclerotic lesions and the behaviour of the vascular flow after graft implantation.
   In order to develop a better understanding of flow characteristics in bifurcations, experimental investigations were carried out on the development of flow in rigid and elastic models. From the results obtained by these investigations, we have shown in previous papers (Siouffi et al.,1984; Rieu et al.,1985; Rieu et al.,1989): (1) the importance of the three dimensional effects in the daughter branches of the bifurcation, (2) the role played by the flow unsteadiness, (3) the damping effect of the wall elasticity. The purpose of this paper is to discussed the main features of these flow characteristics and to obtain attractive informations for the surgeons in order to find out the best way to implant vascular graft. Characteristics of the flow in symmetric bifurcations in plexiglass with rectangular cross section and in a vascular bifurcated Dacron graft are presented.

II. FLOW IN A BIFURCATION OF RECTANGULAR CROSS SECTION

   The bifurcation model has a rectangular cross-section with a mother branch (section $4ab=4\times2cm^2$; length $l=50cm$) and two symmetric daughter branches (section $4a'b=1.6\times2cm^2$, length $l=50cm$); thus the area ratio of the daugthers and inlet channel is 0.8 (fig.1). The bifurcation angle is 60° and all the angles (vertex and flow divider) are sharp. The study of the velocity distribution was performed with a pulsed Doppler ultrasonic velocimeter (Peronneau et al.,1973). For a given point in the liquid, it is possible to determine three probe positions on the faces of the tube, so as to obtain for this point three non-coplanar components of the velocity in the direction of the ultrasonic beam. Then one may calculate the longitudinal, transverse, and vertical components (Vx,Vy,Vz).

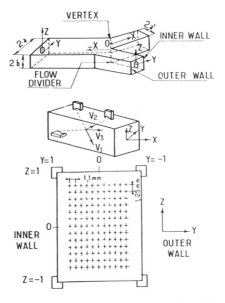

Fig. 1. Velocity measurement method

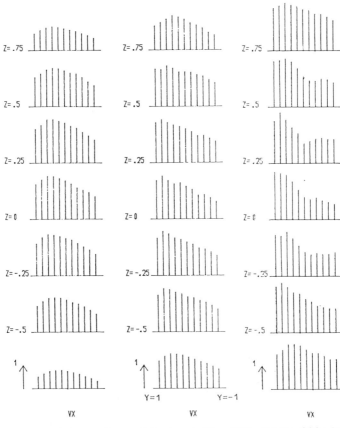

Fig. 2. Reynolds number effects. (a)Re=100; (b)Re=300; (c)Re900

X,Y,Z are the coordinates of a point, non-dimensionalized with respect to the lentgh "l", the half width "a" of the mother branch or "a'" of the daughter branch and the half height "b" respectively. The origin for X is taken on the flow divider. At any given cross-section the origin for Y and Z is the center of the cross-section. The velocity components Vx,Vy,Vz are non-dimensionalized with respect to the mean velocity for steady flows and to the mean velocity at the peak flow rate for pulsatile flows.

II.1 Steady flows

Because of the sudden change in direction, the flow behaviour in the daughter branch is determinate by the opposing influences of the centrifugal force and of the radial pressure gradient. The radial pressure gradient causes a difference of pressure between the apex, where the pressure is maximum, and the vertex; this gradient decreases as X increases. So, downstream of the vertex the pressure gradient exceeds the centrifugal forces. In addition ,due to the split of the flow at the apex, a boundary layer is formed on the inside wall with a maximum axial velocity just outside it (Pedley et al.,1971). Figure 2 shows the Reynolds number effects ($Re=U.Dh/\nu$; $Dh=4ab/(a+b)$,hydraulic diameter; U,mean velocity; ν,kinematic viscosity) on the velocity profile shape. We have illustrated the evolution of the velocity profile. for Re=100,300,900 at the section X=0.084 downstream from the vertex for several vertical values of Z. For Re=100, the velocity profile in the central plane is regular with a maximum situated in the plane Y=0.5 . When the Reynolds number is increased (Re=300),the maximum velocity moves towards the inner wall (Y=0.66) under the conjugate effects of an increase of the centrifugal forces and a decrease of the viscous layer thickness along the inner wall. For Re=900, the maximum velocity increases and is located against the inner wall (Y=0.83) and we observe a change from convex to concave in the shape of the velocity profiles ; this change is due to the increase of the centrifugal forces in the central part of the conduct associated with the development of secondary flows. In the lower and upper planes (Z=-0.75,0.75), the centrifugal forces are less important; the velocity shape become convex and the maximum Vx moves off the inner wall (Y=0.5). With increasing Reynolds number, the amplitude of the longitudinal component Vx near the outer wall in the central plane decreases, and the amplitude of this component in the upper and lower planes increases.

II.1.1 Secondary velocity components

Figure 3 shows the patterns of the transverse Vy and of the vertical Vz velocity components at the section X=0.084 of a daughter branch of the 60° bifurcation for Re=900. The transverse velocity component is negative near the inner wall and positive overall the plane Z=0. These negative transverse velocities correspond to the combined effects of the transverse component of the velocity in the boundary layer on the inner wall and of the radial pressure gradient directed from the inner to the outer wall. Positive transverse velocities are induced by the centrifugal forces that imposed a motion directed from the outer wall to the inner wall. In the planes Z=-0.25 and 0.25 this component remains positive, but its amplitude decreases. In the planes Z=-0.5 and 0.5 the transverse velocity component becomes very small and there is an equilibrium between the negative and positive values. In the upper (Z=0.75) and lower (Z=-0.75) planes this component is negative. These observations are characteristic of a lateral motion directed from the inner to the outer wall in the median plane and from the outer to the inner wall in the upper and lower planes.

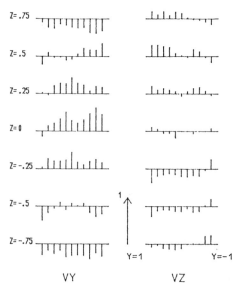

Fig. 3. Secondary velocity components, Re=900

Concerning the vertical velocity component, a first remark is that the amplitude of Vz is smaller than the amplitude of Vy. In the plane Z=0, the amplitude of this component is very small. In the upper plane, Vz is positive near the inner wall and negative near the outer wall and in the lower plane, Vz is negative near the inner wall and positive near the outer wall. In the upper plane, these observations are characteristic of an upward motion near the inner wall and a downward motion near the outer wall.

II.1.2 Vortex structure

The knowledge of the velocity vector at 143 points of each measurement section give some insight on the vortex structure. From the previous radial and vertical velocity distributions, it is possible to formulate a qualitative picture composed of two symmetrical contra-rotating vortex sheets stemmed from the vertex, the plane Z=0 being the plane of symmetry. The projection of the velocity vector onto a plane perpendicular to the daughter branch (Y,Z) aligned in the transverse and vertical directions, allows one to obtain quantitative informations on the vortex structure. Figure 4 shows the evolution of the secondary flow in the section X=0.084 of the 60° bifurcation for several values of Re (100,200,600,900). For Re=100, such velocities records are not representative of closed vortex sheet. Radial velocities Vy are always positive in every Z plane excepted in the viscous layer near the inner wall, and vertical velocities Vz are practically zero in every Z planes; so, at this Reynolds number value, the bifurcation induces only radial flows and centrifugal forces predominate over the pressure gradient in every section. At Re=200, the two vortex sheets are distinct. The two contra-rotating vortices are occuping the entire section, their centers being situated in the plane Y = 0 at the altitudes Z=0.5 and Z=-0.5. When the Reynolds number increases, (Re=600), this zone of radial flow increases and consequently the core of the two vortex sheets are pushed simultaneously towards the outer wall and towards the upper and lower walls. In the same time, the thickness of the viscous layer on the inner wall is reduced and become undetectable by our measurements. This phenomenon is stronger for Re=900.

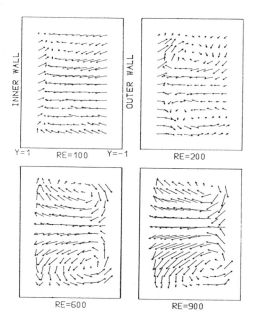

Fig. 4. Reynolds number influence on the vortex structure.

II.1.3 Concluding remarks for steady flows

The main conclusion that ensues from the analysis of the steady flow in a symmetrical bifurcation with rectangular cross-section is the existence of an inward motion in the daughter branch entry. Important three-dimensional effects are present in the daughter branch with secondary velocities as high as 50 % of the longitudinal velocity component. These secondary effects are increasing functions of the Reynolds number.

II.2 Pulsatile flows

Pulsatile flows are characterized by a peak Reynolds number $Rec=Umax.Dh/\nu$, a mean Reynolds number $Rem=Um.Dh/\nu$ and a frequency parameter $\Theta=\Omega.Dh^2/16\nu$ (Umax and Um are respectively the maximum and the mean value during the period of the mean velocity U in the mother branch, Ω is the angular frequency based on the overall single period). In order to demonstrate the effects of unsteadiness, figure 5 shows the longitudinal velocity profiles for a steady flow, a simple pulsatile flow and a physiological type flow in a 60° bifurcation. In all three cases the value of the instantaneous Reynolds number corresponding to the peak flow rate is equal. It can be observed that the very blunt velocity profiles obtained in the mother branch for physiological type flows lead to a small outward shift of the maximum velocity in the daughter branch entry. The small outward maximum velocity shift, observed for physiological type flows, can probably be associated with the high frequencies encountered in the systolic ejection. The explanation is surely that the flow in the core is approximately irrotational so that higher velocities would be expected nearer the centre of curvature of the primary stream-lines. The size of the irrotational core and also, the outward maximum velocity shift is an increasing function of the frequency parameter.

The vortex structure for unsteady flow is qualitatively similar to the structure observed for steady flow. However, the frequency and the amplitude of the oscillation are determining factors in the

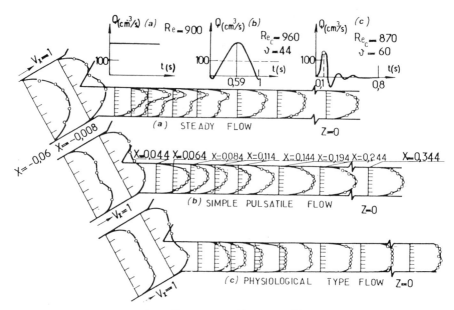

Fig. 5. Unsteadiness effects.

vortex sheet. Figure 6 shows the evolution of the vortex sheet in a section downstream the apex (X=0.084) of a 60° bifurcation at several instants during the cycle for a simple pulsatile flow (Rec=900,Rem=450,Θ=70). The first remark is that the limits of the vortices are less distinct than the steady flow ones in the same section. At the minimum flow rate (instant 16), no ordered structure is established. During acceleration, instants 1 to 7, a structure similar to the steady flow one appears and take up practically the entire section; in this phase, the secondary velocities are small and so, the intensity of the vortices is small. During deceleration, instants 8 to 13, due to the increase of the secondary velocities, the intensity of the two contra-rotating vortices increases. In the central plane near the outer wall, the increase of the transverse velocity profiles is associated with a decrease of the longitudinal velocity profiles. These profiles are very similar to those in steady flow, with a maximum near the inner wall. Along this wall, negative transverse velocities appear, corresponding to an increase of the boundary layer thickness. Consequently, the center of the vortices is pushed away towards the half plane near the outer wall. Towards the end of the pulse, (instants 12,13), the intensity of the vortices is maximum, with higher vertical velocities on the outer wall and higher transverse velocities in the central plane.

III. FLOW IN A VASCULAR BIFURCATED DACRON GRAFT

A point of interest for surgeons in our modelisation is to give thems attractive and useful informations to find out the best way to implant the prostheses. Geometric and dynamic parameters are determinant in the success and longevity of the graft (Rieu et al.,1987). In particular for bifurcated prostheses, anastomosis between the host vessel and the graft, internal geometric shape of the bifurcation and graft resistance play an important role.
We have simulated with our test bench, physiological type flow conditions in a vascular bifurcated graft and have determined for two geometric conditions of implantation the velocity field in the

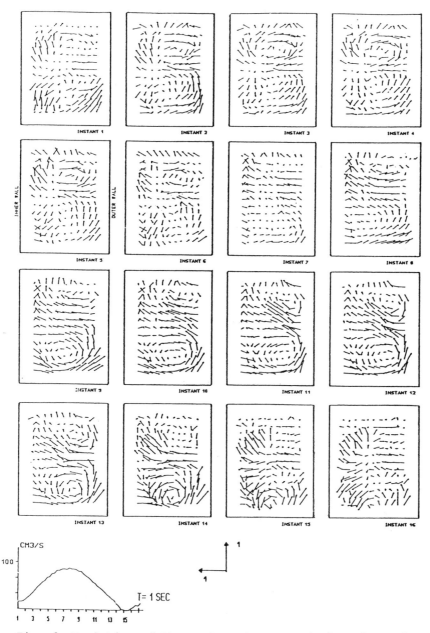

Fig. 6. Evolution of the vortex structure during the cycle

bifurcation. The model used is a low porosity Dacron grafting. The impermeability is obtained using a fibrinogen adhesive and the circulating fluid is a macromolecular solution (viscosity equal to the blood one). The first configuration (A) is a classical implantation with a bifurcation angle of fifty degrees. The second one (B) is used in case of strong aneurism. For this implantation, the surgeon must closed the aneurismal bag around the branches of the bifurcation; so the branches are parallel on thirty millimeters, and then make an angle of thirty degrees. We have perform bidimensional ultrasonic velocity measurements in several horizontal cross sections of the model and simultaneous pressure-diameter variation measurements (Rieu et al.,1987) in the mother branch near the anastomosis and in a daughter branch.

III.1 Pressure-diameter behaviour (Fig. 7)

At the entrance of the mother branch, just behind the anastomosis, we can observe an interesting phenomenon between pressure and diameter variation measurements. Indeed, in the first part of the systolic acceleration, on the contrary of the results obtained with arteries, we observe an increasing pressure and a decreasing diameter, for a very small time; then diameter follow pressure variation. This behaviour is explained by the fact that when the flow acceleration begins, we have first an extension of the prosthesis with a diameter decrease. Indeed the upstream anastomosis is made with a rigid tube. In the daughter branch, this phenomenon disappear and we can observe a classical behaviour of the wall with an increase of the diameter associated to an increase of the pressure.

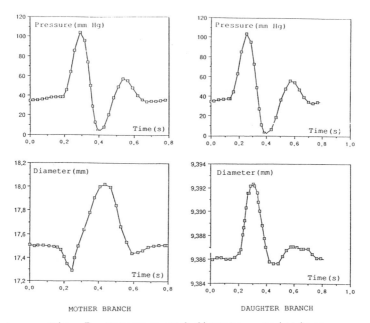

MOTHER BRANCH   DAUGHTER BRANCH

Fig. 7. Pressure and diameter variations

III.2 Velocity vector field

Concerning the velocity vector field in the prostheses, we observe during flow acceleration a similar flow behaviour in the two configurations with a maximum velocity near the inner wall of the bifurcation. During flow deceleration (figure 8),we denote with configuration B, in the parallel branches, high three-dimensional rotational motions corresponding to important vortex structures. In particular, a second vortex zone appears immediatly downstream the flow divider; its rotationnal motion being in opposition with the vortex sheets in the thirty degrees angle branchings.

IV. CONCLUSION

The main conclusions to emerge from these experimental investigations are :
1) The flow downstream of the junction is always three-dimensional.

2) The secondary velocities for steady flows are very important (up
to 50 % of the longitudinal component); for unsteady flows, they
are smaller than for steady flows and we observe a decrease of
these secondary velocities when the frequency parameter increases.
3) Measurements in vascular prosthesis show that the diameter
pressure behaviour at the entrance of the mother branch, by the
formation of a periodic converging tube, can explain the common
stenosis formation in this site and it seems that, by its
hemodynamic implications, configuration A is better than
configuration B.

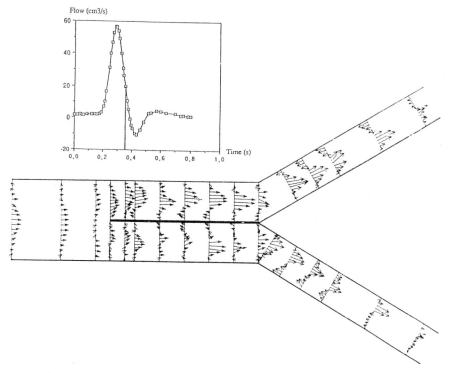

Fig. 8. Velocity vector field in the graft (configuration B)

REFERENCES

Pedley, T.J., Schroter, R.C., Sudlow, M.F., 1971, Flow and pressure
    drop in repeatedly branching tubes. J.Fluid Mech.,46, 365.

Peronneau, P., Hinglais, J., Xhaard, M., Delouche, P., Philippo,
    J., 1973, Cardiovascular Applications of Ultrasound.
    Proceedings International Symposium Janssen Pharmaceutica,
    Beerse, Belgium.(Edited by Reneman R.S., North Holland,
    Amsterdam).

Rieu, R., Bano, S., Pelissier, R., Bergeron, P., Reggi,
    M.,Courbier, R., 1987, Influence of the haemodynamic
    parameters on the repartition of the flow between an
    artery and its graft. International Angiology , Vol.6,
    N°2, pp.147-152.

Rieu, R., Friggi, A., Farahifar, D., Cassot, F., 1987,
Détermination in-vitro de la relation pression-diamètre et
des profils de vitesse par des techniques ultrasonores.
Applications in-vivo. Journal de Physiologie, Vol.82,2

Rieu, R., Friggi, A., Pelissier, R., 1985, Velocity distribution
along an elastic model of an human arterial tree.
J.Biomechanics,18,9,703-715.

Rieu, R., Pelissier, R., Farahifar, D., 1989, An experimental
investigation of flow characteristics in bifurcation models.
European Journal of Mechanics, B/Fluids, vol.8B, n°1.

Siouffi, M., Pelissier, R., Farahifar, D., Rieu, R., 1984, The
effect of unsteadiness on the flow through stenoses and
bifurcations. J. Biomechanics, 17, 5.

A PREDICTIVE SCHEME FOR FLOW IN ARTERIAL BIFURCATIONS :

COMPARISON WITH LABORATORY MEASUREMENTS

M. W. Collins and X. Y. Xu

Thermo-Fluids Engineering Research Centre
The City University
London, U.K.

ABSTRACT

This is an initial study of overall prediction exercise to simulate blood flow through three-dimensional arterial bifurcations, ASTEC code is used with finite element grid definition and finite difference solution methods. Results are compared with laboratory measurements of Ku and Liepsch for T-junctions. Comparison is excellent for two-dimensional steady flow tests, and very good for three-dimensional pulsatile flows.

INTRODUCTION

It is a well known fact that atherosclerosis and thrombosis occur predominantly in arterial bends and bifurcations. Although the exact mechanism is not yet well understood, more recent studies (Zarins et al., 1983; Ku et al., 1985) have confirmed Caro's observation (Caro et al.,1971) that atherosclerotic lesions develop more frequently in regions with low shear stresses and with recirculation. Therefore, detailed insight into the flow phenomena occurring in bifurcations possibly contributes to a better understanding of the mechanism underlying the formation of atherosclerotic plaques and thrombi.

Extensive studies have been performed on flow in arterial bifurcations and branches. O'Brien et al. (1977) made a prediction of unsteady laminar flow through a two-dimensional T-junction based on a Vorticity-Stream function formulation. Detailed measurements and numerical calculations of steady flow in a plane 90 degree bifurcation were presented by Liepsch et al. (1982). The extension to the study of pulsatile flow was reported by Khodadadi et al.(1988). To investigate the effects of non-Newtonian behaviour and wall flexibility, Ku and Liepsch (1986) performed LDA measurements on flow at a three-dimensional T-bifurcation. An excellent review of previous research into the details of flow behaviour in model bifurcations has been made by Liepsch (1986).

The present study is an initial step of the overall prediction exercise which aims to simulate arterial bifurcation flows under real physiological conditions. For convenience of comparison with published measurements, the idealised T-bifurcation models are adopted and the validation of the numerical results is provided. All calculations presented were performed using ASTEC (developed at UKAEA), a fluid flow code applying finite volume solution methods to a finite element mesh (Lonsdale, 1988). As assessed by Xu and Collins (1989), the unique combination of great geometrical flexibility and high efficiency makes ASTEC a well-suited code for the current application.

NUMERICAL METHODS

The numerical predictions are based upon the system of time-dependent Navier-Stokes and continuity equations for viscous, incompressible Newtonian fluid flow in a three-dimensional geometry with rigid walls. These equations can be written in the integral form

$$\int \vec{u} \cdot d\vec{A} = 0 \tag{1}$$

$$\rho \frac{\partial}{\partial t} \int \vec{u} \, dV = -\rho \int \vec{u}\vec{u} \cdot d\vec{A} - \int \nabla P dV + \mu \int \nabla \vec{u} \cdot d\vec{A} \tag{2}$$

Where $\vec{u} = (u,v,w)^T$ is fluid velocity, $d\vec{A}$ and dV represent elements of control area and control volume respectively, and P is pressure. Together with appropriate initial and boundary conditions, equations (1) and (2) are uniquely solvable for velocity components and pressure.

The procedure for solving the discrete equations is iterative and based on the SIMPLE method as inferred by Patankar and Spalding (1972). A preconditioned conjugate gradient algorithm is used to calculate the pressure corrections. To minimise the false diffusion errors associated with the advection term, a vector upwinding scheme is employed. The time differencing is fully implicit, so that there is no Courant stability restriction to the timestep.

To make solution of large three-dimensional problems sufficiently fast, a high degree of vectorization is essential to obtain best performance from present-day supercomputers. ASTEC has been tailored for the supercomputer CRAY in the sense that the programme is written so that the CRAY compiler CFT 77 will produce a highly vectorised code.

RESULTS AND DISCUSSION

Steady Flow in a Two-dimensional T-bifurcation

Particularly with a new code such as ASTEC careful validation exercises should be carried out. For these, the experimental data of Ku and Liepsch were felt to be ideal. To ensure that ASTEC is suitable for the current application, a simple test was carried out on a two-dimensional laminar flow in a T-bifurcation for Reynolds numbers ranging from 250 to 1100. Figure 1 shows the bifurcation geometry and element division used in the calculation. Since ASTEC is a three-dimensional code, the two-dimensional problem was solved on the three-dimensional mesh of one element thick with flow restricted between two symmetric planes. The mesh consists of 840 elements and 1870 nodes. With this mesh a typical calculation for a Reynolds number of 496 and a flow rate ratio $\dot{V}_3/\dot{V}_1$ of 0.44 required 165.8 seconds on a CRAY X-MP/28.

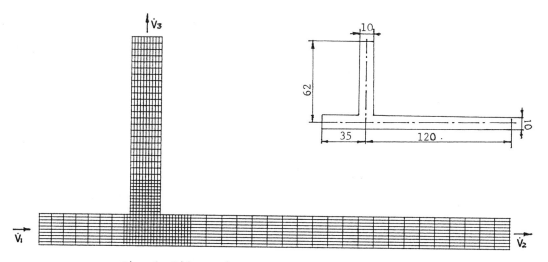

Fig. 1. Bifurcation geometry and mesh division.

In this simulation the density of the fluid is 1000Kg/m^3, and the dynamic viscosity is 1.02×10^{-3} Kg/(ms). At the inlet plane a fully developed axial velocity profile and zero normal velocities are specified. Since the two outlets are well downstream, a fully developed axial velocity profile with specified flow rate has been assumed at one outlet, and zero pressure is assumed at the other. No-slip conditions are imposed on the rigid walls.

Figures 2 and 3 present the velocity vectors in the main and branching tubes respectively, for a Reynolds number of 496 and a flow rate ratio of 0.44. It is seen that the velocity profiles are parabolic at the inlet and outlet regions. Before the bifurcation, the velocity profiles are almost unchanged until about 10mm upstream from the branching. Behind the bifurcation in the main tube, the velocity profiles are skewed towards the upper wall. A reverse flow region of about 20mm in length exists at the bottom wall. The flow farther downstream of the bifurcation becomes fully developed. In the branching tube, the reverse flow region along the upstream wall is more pronounced than in the main tube. The region of reverse flow is about 43mm in height which corresponds to 4.3 times the diameter of the tube.

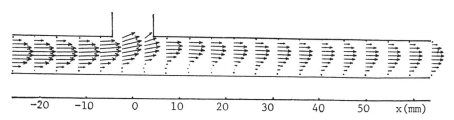

Fig. 2. Velocity vectors in the main tube (Re=496, $\dot{V}_3/\dot{V}_1$=0.44).

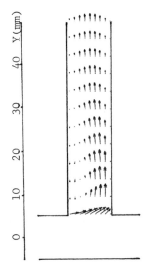

Fig. 3. Velocity vectors in the branching tube (Re=496, $\dot{V}_3/\dot{V}_1$=0.44).

To examine the accuracy of the calculations, a quantitative comparison of the calculated and measured results has been performed. In Figure 4 the calculated axial velocity profiles are compared with LDA measurements obtained by Liepsch et al. (1982). It is observed that there is a very good agreement between the calculations and the measurements; especially in the main tube the axial velocity profiles are predicted quite accurately. Some differences are found in the branching tube. This may be due to the three-dimensional effects of the tube used in the experiments. The differences are, in fact, rather less than with numerical predictions also presented by Liepsch et al.

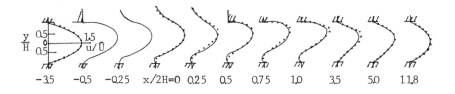

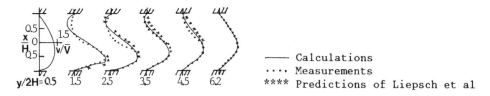

——— Calculations
···· Measurements
**** Predictions of Liepsch et al

Fig. 4. Comparison between calculations and the LDA measurements of Liepsch et al.(1982). (Re=496, $\dot{V}_3/\dot{V}_1$=0.44)

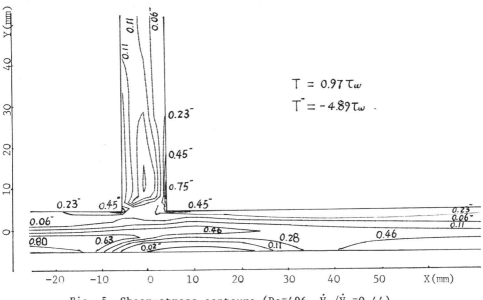

$$T = 0.97\,\tau_w$$
$$T^- = -4.89\,\tau_w$$

Fig. 5. Shear-stress contours (Re=496, $\dot{V}_3/\dot{V}_1$=0.44).

The shear-stress contours for a Reynolds number of 496 and a flow rate ratio $\dot{V}_3/\dot{V}_1$ of 0.44 are shown in Figure 5. For convenience of comparison, the shear-stress contours are plotted as a percentage of their maximum and minimum values respectively. These values are given in terms of the fully developed wall shear-stress at the inlet. Figure 5 shows that there are two zones of low shear-stress formed, one is opposite the branching near the bottom wall of the main tube, with another along the upstream wall of the branching tube. In the region of the divider wall of the branching tube, a high shear-stress zone is formed. The same happens along the upper wall of the main tube around the corner area. Again, the agreement between the predicted shear-stress contours and those presented by Liepsch et al.(1982) is very good.

Steady Flow in a Three-dimensional T-bifurcation

A three-dimensional prediction was performed on steady flow through a T-bifurcation with circular cross-section at an average upstream Reynolds number of 250 and a flow division ratio of 50:50. In this bifurcation, the diameter of the main tube is 6mm and the diameter of the side branch is 3mm. The element division for the model bifurcation is illustrated in Figure 6. Since the flow is symmetric, only half of the bifurcation was considered. The mesh used in the calculation consists of 5360 elements and 6526 nodes. A 70% aqueous glycerine solution with a density of 1180Kg/m^3 and a viscosity of 0.013Kg/(ms) was used in the simulation. The boundary conditions were the same as those applied to the two-dimensional case.

The calculation had a 'cold' start, which means that the initial guess of all variables was zero. When the convergence criterion for the maximum change of velocity components of two successive iterations was less than 1.0×10^{-4} , a converged solution was obtained in 335 seconds on a CRAY X-MP/28.

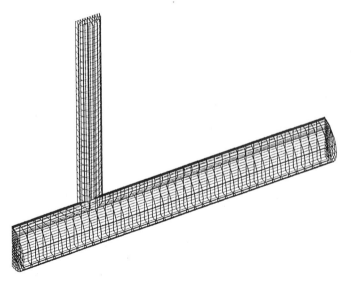

Fig. 6. Finite element mesh used in the calculation.

Figure 7 presents the axial velocity profiles in the symmetry plane. At the inlet and outlet regions, the axial velocity profiles hardly differ from parabolic ones. After the flow divider, maximum axial velocities are shifted towards the divider wall, which is primarily caused by the flow branching. A region with negative axial velocities could be seen at the bottom wall, with negative velocities reaching approximately 2cm/sec.

Figure 9 demonstrates the secondary flow at the cross-sections given in Figure 8. It is observed that secondary flow at the entrance of the main tube (A) is completely directed from the outer wall towards the branching side. At the flow divider site in the main tube (B), secondary flow is almost entirely directed towards the divider wall, where the highest secondary velocities are found. Downstream in the main tube (C) secondary flow shows high resemblance to a Dean type vortex. Near the symmetry plane secondary velocities are directed towards the divider wall and near the side wall they point circumferentially back towards the non-divider wall. At both sites in the branch tube (D and E), secondary velocities are directed towards the divider wall near the symmetry plane and circumferentially back near the side wall.

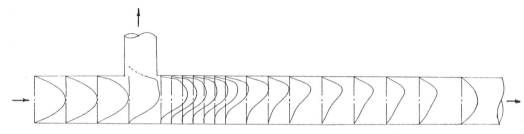

Fig. 7. Axial velocity profiles in the symmetry plane.

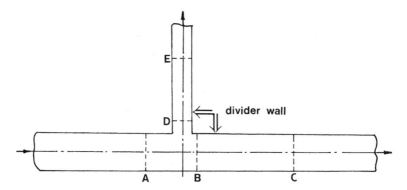

Fig. 8. Cross-sections where secondary flow are presented.

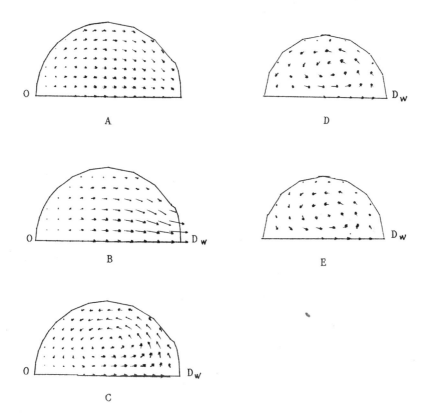

Fig. 9. Secondary flow in the main and the branching tube.
(0 : outer wall, D_W: divider wall)

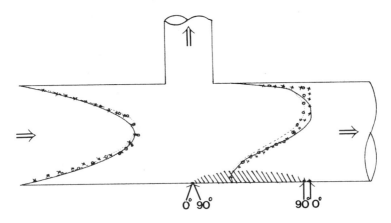

Fig. 10. Calculated (——,....) and measured (×××,∘∘∘∘) axial velocity
profiles at 0 and 90 degrees in the symmetry plane 20mm
upstream and 2.5mm downstream from the branch.   shows the
limits of predicted reverse flow region for phase 0° and
90°.

Pulsatile Flow in a Three-dimensional T-bifurcation

Following the steady flow prediction presented in the previous
section, a preliminary prediction of pulsatile flow in the same three-
dimensional T-bifurcation was performed. The flow conditions reported in
the experimental study of Ku and Liepsch (1986) were reproduced in this
prediction, i.e. an average main tube Reynolds number of 250, an average
upstream pressure gradient of 4000pa/m, a pulsatile pressure amplitude of
1000pa/m, and a Womersley parameter of 2.3 for the sine wave input.
Therefore, with 0° of phase representing the peak pressure, the pressure
gradient at the inlet plane can be given by

$$-\frac{\partial P}{\partial x} = 4000 + 1000 \cos \omega t \qquad (3)$$

By imposing the inlet pressure gradient as one of the boundary conditions,
the calculation was carried out from 0° of phase with the converged
solution of the steady flow case as its initial guess. Thirty-six time steps
in a cycle were used. The calculation for a quarter of a cycle took 15
minutes on a CRAY X-MP/28.

Figure 10 illustrates the velocity profiles at 0 and 90 degrees in
the symmetry plane 20mm upstream and 2.5mm downstream from the branch.
Similar to the steady velocity profiles, the pulsatile profiles in the
main tube 2.5mm downstream of the bifurcation are skewed towards the flow
divider wall. A reverse flow region could always be seen at the bottom
wall of the main tube throughout the cycle, although the extent of this
region varies within the cycle. A most pleasing effect is the prediction
in the right sense of the slight experimental increase and reduction
going from 0° to 90° phase of the velocity profiles upstream and
downstream of the bifurcation. These predictive results, then, are
consistent with the laboratory measurements of Ku and Liepsch (1986).

CONCLUSION

From the results presented above, the following conclusions   can
be drawn :
    1. A predictive scheme for flow in two- and three-dimensional
bifurcations is presented. Calculations for steady flow in a plane T-
bifurcation, both steady and pulsatile flow in a three-dimensional T-
bifurcation have been performed.
    2. The computed results for steady flow in a plane T-bifurcation have
been validated. Comparison of the calculations and the published LDA
measurement is very satisfactory.
    3. The presented predictive scheme has been proved to be efficient
and reliable on three-dimensional problems.
    4. It is intended to complete the pulsatile flow exercise for the
entire 360° cycle, and to investigate the effects of non-Newtonian
viscosity and flexible wall. Then it will be possible to progress to in vivo
bifurcation flow prediction.

ACKNOWLEDGMENTS

We gratefully acknowledge the support of a British Heart Foundation
Research Grant for this programme of work.

REFERENCES

Caro, C. G., Fitz-Gerald, J. M., and Schroter, R. C., 1971, Atheroma and
    arterial wall shear, Proc. Roy. Soc. London B, 177:109.
Khodadadi, J. M., Vlachos, N. S., Liepsch, D., and Moravec, S., 1988,
    LDA measurements and numerical prediction of pulsatile laminar flow
    in a plane 90-degree bifurcation, J. Biomechanics, 15:7:473.
Ku, D., Giddens, D., Zarins, C., 1985, Pulsatile flow and atherosclerosis
    in the human carotid bifurcation. Arteriosclerosis, 5:293.
Ku, D., and Liepsch, D., 1986, The effects of non-Newtonian viscoelasticity
    and wall elasticity on flow at a 90° bifurcation, Biorheology, 23:359.
Liepsch, D., 1986, Review article: Flow in tubes and arteries - a comparison,
    Biorheology, 23:395.
Liepsch, D., Moravec, S., Rastogi, A. K., and Vlachos, N. S., 1982,
    Measurement and calculations of laminar flow in a ninety degree
    bifurcation , J. Biomechanics, 15:No.7:473.
Lonsdale, R. D., 1988, An algorithm for solving thermalhydraulic equations
    in complex geometries: the ASTEC code. in: UKAEA Report.
O'Brien, V., Ehrlich, L. W., 1977, Simulation of unsteady flow at renal
    branches. J. Biomechanics, 10:623.
Patankar, S. V., Spalding, D. B., 1972, A calculation procedure for heat,
    mass and momentum transfer in three-dimensional parabolicflows,
    Int'l J. Heat Mass Trasfer, 15:1787.
Xu, X. Y., and Collins, M. W., 1989, Assessment of the problem of numerical
    simulation of blood flow through three-dimensional bifurcations, in:
    Proc. Int'l. Symposium on Biofluid Mechanics and Biorheology, 671.
Zarins, C. K., Giddens, D. P., and Bharadvaj, B. K., 1983, Carotid
    bifurcations atherosclerosis: Quantitative correlation of plaque
    localization with flow velocity profiles and wall shear stress
    Circ.Res., 53:502.

MODELLING THE PULMONARY CIRCULATION IN HEALTH AND DISEASE

P. E. Hydon†, T. Higenbottam‡ and T. J. Pedley†

† Department of Applied Mathematics and Theoretical Physics
University of Cambridge, Silver Street
Cambridge CB3 9EW, UK

‡ Department of Respiratory Physiology
Papworth Hospital, Cambridge, UK

ABSTRACT

This paper describes work in progress on the pulmonary circulation in health and in pulmonary hypertension. Isolated pig lungs have been perfused and ventilated under steady conditions, and pressure-flow characteristics measured. The purpose of these experiments is to assess the gross hæmodynamic effects of mechanical blockage and of some vasoactive compounds. Pulmonary hypertension, which may occur as a result of various initiating factors, results in occlusion and degeneration of blood vessels. The disease is progressive, but in the early stages, vasodilators such as prostacyclin may drastically reduce the resistance of the pulmonary circulation. In the latter part of the disease, the circulation loses much of its compliance and the right ventricle has to work very hard to supply blood to the lungs.

The pulmonary circulation may be modelled as two networks of branching tubes, connected by capillary sheets. The compliance of the tubes may be varied, to account for vessel tone and the effects of vasoactive drugs. Parts of the structure may be blocked, to simulate thrombi, emboli or degenerative occlusion of vessels. Hydrostatic effects are significant in the larger vessels, and the manner of branching is important.

The model is to be implemented on a computer, with the eventual aim of providing a diagnostic test for the assessment of patients suffering from pulmonary hypertension.

KEYWORDS: Pulmonary Circulation/ Steady Flow/ Pulmonary Hypertension/ Isolated Lungs/ Œdema/ Hydrostatic Effects/ Branching Network

INTRODUCTION

The pulmonary circulation consists of two branching networks (arterial and venous) connected by a sheet-like array of capillaries. The arterial network is made up of short, low resistance tubes which convey blood from the right ventricle to the capillaries. Each of these tubes ends with a bifurcation, where the daughter tubes are of one order lower than the parent (in the Strahler system of ordering: Singhal *et al.* (1973)). In addition, a parent vessel may have several smaller side branches, which lie upstream of the bifurcation, and are approximately at right angles to the parent. The arterial walls are viscoelastic, but the imaginary part of the complex modulus is small compared with the real (purely elastic) part. Therefore, the assumption will be made that the walls are elastic, although not necessarily Hookean. Arteries are not passive tubes, but respond actively to various stimuli. For example, hypoxia induces vasoconstriction; vasodilation may occur as a consequence of an increase of endothelium derived relaxing factors (EDRF) or an infusion of prostacyclin. A further important function of the arterial network is to sieve the circulating blood, removing emboli that have originated in systemic veins. This means that some parts of the network may be blocked.

The capillaries are small thin-walled tubes, which line the alveolar walls, forming the site for gas exchange. These can be modelled as a fluid sheet bounded by two thin membranes, which are connected by posts of tissue in a hexagonal array (Fung,1984). The capillaries are fully open if the downstream blood pressure (P_{ven}) exceeds the surrounding pressure (P_{Alv}). If P_{Alv} is less than the upstream blood pressure (P_{art}), but exceeds P_{ven}, the capillary sheet will be partially collapsed, with a waterfall flow condition (Permutt and Riley (1963)). If P_{art} is less than P_{Alv}, the capillary sheet will be completely collapsed; in this case it is necessary to supply energy to reopen the sheet.

The veins are in a similar network to the arteries but, unlike the major arteries, they do not follow the bronchi closely. In man, the four largest veins empty directly into the left atrium. The venous walls are thinner than those of arteries of comparable size, and may be subject to collapse. (This is contentious; Fung (1984) has argued that intra-parenchymal veins are tethered by the surrounding tissue, and do not collapse.) Venous walls have little muscle, and may be thought of as passive elastic tubes.

Pulmonary hypertension has a diversity of initiating factors, but once it is established, the disease progresses in a number of degenerative stages (Heath and Edwards (1953)). Initially, there is intimal thickening of up to 90 % of the small arteries, which results in narrowing or total occlusion of the affected vessels. This increases the resistance of the lung, and an adequate blood supply can be maintained only by increasing the pulmonary artery pressure. This is generated by the right ventricle, which becomes grossly hypertrophied. After the initial stage, the larger pulmonary arteries thicken, and eventually become very stiff. It is not uncommon to encounter atheroma in the largest pulmonary arteries of patients with pulmonary hypertension. The duration of the disease is variable; in patients with longstanding disease, thin-walled anastomoses develop, which provide a collateral circulation to the capillaries, bypassing the blocked vessels. These anastomoses are not found in patients who die shortly after diagnosis. A typical cause of death is right heart failure, although the disease is incapacitating to such an extent that turning over in bed may cause syncope. In the early stages of pulmonary hypertension vasodilators may provide considerable relief, by lowering the resistance of those vessels which are patent. This treatment is not effective after the vessels have stiffened, although vasoconstriction may still be induced by appropriate stimuli.

The only cure currently available for pulmonary hypertension is heart-lung transplantation. Papworth Hospital is a major centre for heart-lung transplantation in the UK. Therefore we are in a position to collect lung function data from patients prior to transplantation and to carry out *in vitro* experiments on the same lungs immediately after removal. We are also able to compare hypertensive lungs with healthy lungs. The aim of this project is to create a valid fluid dynamical model of the healthy pulmonary circulation and to imitate pulmonary hypertension in the model, by blocking vessels and reducing vascular distensibility. The model should be validated by comparison with measurements from patients and experiments, and it is hoped that it will lead to a tool for the assessment of a patient's condition, using information gained from routine diagnostic tests.

This paper reports the progress made so far in our initial attempts to model the problem, and discusses our first experimental observations.

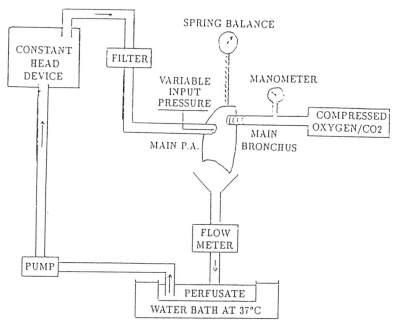

Fig. 1 The constant-head perfusion rig.

EXPERIMENTS

The steady pressure-flow behaviour of isolated pig lungs was studied, using a constant head perfusion circuit. Pigs and humans have lungs of similar size, enabling us to develop a technique suitable for use with excised human lungs. Physiological flow is unsteady, and *in vivo* pressure and flowrate data are available, but it is not possible (for obvious reasons) to determine the steady characteristics *in vivo*. The results of steady experiments will provide further information about the lungs, and will provide an additional test of any model that is proposed.

Procedure. Pig lungs were obtained from a local abattoir. Immediately after removal from the animal, the lungs were intubated, inflated with $95\%O_2/5\%CO_2$ to a pressure of 20mm Hg and then allowed to deflate, after which the trachea was clamped. The main pulmonary artery was intubated and the pulmonary circulation was washed out with Ringers solution, until no blood was visible in the solution returning from the pulmonary veins. In addition, a bolus of prostacyclin was injected, to dilate the vessels and facilitate flow at low washing-out pressures. The lungs were then placed in cold Ringers solution and transported in a cool-box. (The above procedure is a modification of a method of lung preservation used in the procurement of donated organs for transplantation (Hakim *et al.* (1987).) After removal from the cool-box, the lungs were again inflated, and visually inspected for signs of ischæmia, œdema, or other damage. Provided that a lung was not obviously damaged, it was placed in the perfusion rig (Figure 1) and the pulmonary artery supplying the lower lobe was intubated. The lobe was perfused with Gelofusine (a plasma substitute: Hausmann Laboratories Inc., Switzerland) at varying steady pulmonary artery and alveolar pressures. The left atrium was opened to the atmosphere: liquid leaving the pulmonary veins was collected and the flowrate was measured before being returned to the constant head device. The lung's weight was continually monitored, and the condition of the lung was periodically checked.

Results. In all but one pig lung, œdema became pronounced after a short time (approximately 30 minutes). The effect of this œdema was to increase the overall resistance of the lung, as can be seen in Figure 2a). The lung was also tested for its response to phenylephrine and acetylcholine. The response to the latter was generally small, leading us to the conclusion that a substantial amount of endothelial damage had occurred. The ischæmic time for an animal killed in the abattoir is approximately 20 minutes, as the skin is removed by scalding

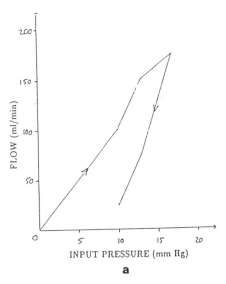

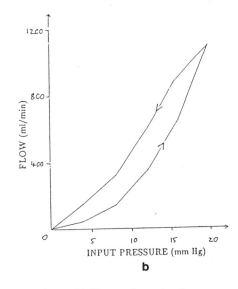

| a | b |

Fig. 2 Pressure-flow curves: a) Pig lung developing œdema; b) Human hypertensive lung, with no œdema.

before the carcass is opened. We feel that this has led to the endothelial damage. We are now able to obtain lungs within 5 minutes of death; these are in much better condition.

We have also perfused a small number of human lungs. They have not developed any noticeable œdema, and have a marked response to acetylcholine. Flow in a pulmonary hypertensive lung was considerably less than in the normal lung. Hysteresis was observed in the pressure-flow curve (see Figure 2b)). The limited supply of human lungs means that we do not yet have a statistically significant body of data, but initial results are encouraging.

MODELLING

Fung's model. Initially, we examined the model of steady flow in the human lung described by Fung (1984). This model assumes Poiseuille flow in each artery and vein, neglecting the effects in the entrance region of the vessel. Continuity of pressure and conservation of mass is assumed at each bifurcation, but the complicated flow patterns at the bifurcation are assumed to have a negligible effect on the pressure distribution in the lung. The capillaries are treated as a sheet bounded by two membranes, which are connected by an array of tissue posts. This model has been successfully used for cat lungs, where the hydrostatic pressure difference across the lung is small. Experiments have produced a full set of elasticity data for the vessels of the cat lung (Zhuang *et al.* (1983)), and the predictions of the model have been experimentally verified by Krishnan *et al.* (1986).

Within the range of physiological pressures, the pressure-radius relation for a vein or artery is of the form

$$a(x) = a_o(1 + \beta\,p(x))\tag{1}$$

where

$a(x)$ is the radius of the vessel at a distance x from the upstream end;

a_o is the radius at zero transmural pressure;

$p(x)$ is the transmural pressure at x;

β is the relative compliance of the vessel.

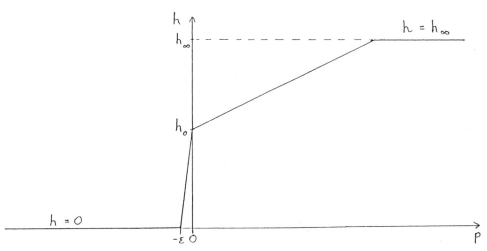

Fig. 3 Pressure-thickness relationship for a pulmonary capillary sheet.

When this is combined with the Poiseuille equation

$$-\frac{dp}{dx} = \frac{8\,\mu\,Q}{\pi\,a^4(x)} \tag{2}$$

(μ is the blood viscosity; Q is the volume flowrate), integration over a vessel of length L leads to the result

$$(1 + \beta\,p(0))^5 - (1 + \beta\,p(L))^5 = \frac{40\,\beta\,L\,\mu\,Q}{\pi\,a_o^4} \tag{3}$$

The pressure-thickness relation for a capillary sheet is piecewise linear, allowing the sheet to collapse (see Figure 3), *i.e.* the thickness $h(x)$ is given by the equation

$$h(x) = h_o\,(1 + \beta(p)\,p(x)) \tag{4}$$

where β will take one of three possible values, depending upon the transmural pressure. Fung has shown that the situation may be more complicated when p is close to zero: patches of a capillary sheet may be collapsed, whilst nearby capillaries may be patent. We will assume that transmural capillary pressures lie in the range $0 < p(x) < p_m$, in which case β may be treated as a non-zero constant. This assumption is not unrealistic in pulmonary hypertension, where the entire circulationis working at grossly elevated pressures. Combining the linear pressure-thickness equation with the pressure-flow relation for Stokes flow in a sheet leads to

$$(1 + \beta\,p_a)^4 - (1 + \beta\,p_v)^4 = G\,\mu\,Q \tag{5}$$

where p_a, p_v are the transmural pressures upstream and downstream (respectively) of the sheet, and where G is a function of the geometry of the sheet and of β.

Hydrostatic effects. In the human, there is a hydrostatic pressure difference of some 20 mm Hg between the top and bottom of the lung, so it important to adjust the model to incorporate hydrostatic effects. Consider flow in a tube aligned at an angle α to the vertical. For human pulmonary arteries, the transverse hydrostatic variation is very small, so we assume that the flow satisfies

$$-\frac{dp}{dx} - \rho\,g\,\cos\alpha = \frac{8\,\mu\,Q}{\pi\,a^4(x)} \tag{6}$$

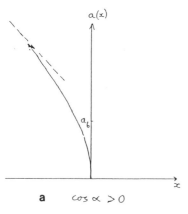
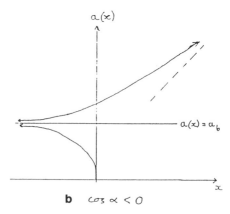

Fig. 4 Pressure drop along a vessel inclined to the vertical: a) Flow is opposed by gravity; b) Flow is assisted by gravity.

together with the pressure-radius relation (1). (ρ is the blood density; g is gravitational acceleration.)

The solution is sketched in Figure 4. When flow is being opposed by gravity (Figure 4a)), the pressure drop is accentuated. When there is a gain in hydrostatic pressure along the vessel (Figure 4b)), there are three types of solution:

   i) If the hydrostatic effects dominate, the vessel will increase in radius with distance downstream

   ii) If the viscous pressure drop dominates, the vessel will decrease in radius with distance downstream

   iii) if viscous and hydrostatic effects precisely balance, the vessel will remain cylindrical.

The balance occurs when the vessel radius is

$$a_b = \left(\frac{8\mu Q}{\pi \rho g |\cos \alpha|} \right)^{\frac{1}{4}} \tag{7}$$

At physiological flowrates, hydrostatic effects are negligible only in the smallest vessels and in those vessels which are almost horizontal.

Predictions. The model has been tested on the cat lung, using the elastic data from Zhuang *et al.* (1986). Cat lungs have 22 (Strahler) orders of vessels. Results were computed for several regular network configurations, at varying pressures in the main artery, left atrium and alveoli. The pressure drop in a lung with symmetrically bifurcating arteries (and veins) is sketched in Figure 5. Approximately 80% of the pressure drop occurs in vessels of radius less than 50μm. A similar pressure profile is obtained in a symmetrically trifurcating lung. Blockages in the model lung appear to affect only nearby vessels, *i.e.* the parent and grandparent (and their offspring) of the blocked artery.

FOR FUTURE STUDY

For any mathematical model of the human pulmonary circulationto be valid, it is necessary that the elastic properties of the vessels should be known. Greenfield and Griggs (1963)

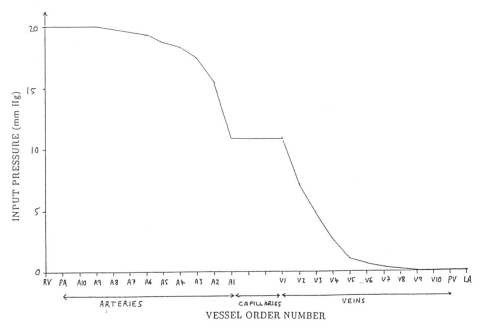

Fig. 5 The pressure drop in each vessel of the model pulmonary circulation.

have tested the large pulmonary arteries, and Yen *et al.* (1988) have examined the smallest pulmonary vessels, but the elasticity of the remaining vessels is not yet known. Quantitative measurements of the elasticity of vessels in hypertensive lungs have yet to be made. We also intend to continue with the steady flow experiments in order to build up a substantial body of information about the behaviour of the lung under these conditions. We will refine the mathematical model of a single tube to incorporate pressure loss in the entrance region and the effects of bifurcations. It is easy to compute pressure and flow in a model lung with homogeneous properties, but the simplicity is lost when blockages are introduced. We are in the process of developing an efficient algorithm which will resolve this difficulty. The unsteady flow problem will be examined using a wave propagation analysis, and the results will be compared with data obtained *in vivo*. It is hoped that the effects of vessel blockage and stiffening will be detectable using this approach.

CONCLUSIONS

Pulmonary hypertension is accompanied by a drastic increase in the resistance of the lung to blood flow. This paper reports the initial stages of an attempt to model this change with the tools of fluid dynamics. In addition, we are carrying out steady flow experiments on isolated lungs to obtain data for comparison with the model.

DISCUSSION

Dr. Greenwald What is the effect on the pressure-flow curves of the lung being removed from the thorax, given that the airway compliance of an isolated lung is known to be different to that of a lung in the thorax?

P. Hydon Fung (1984) states that the transmural pressure in the intraparenchymal extra-alveolar vessels is the difference between the blood pressure and the pleural pressure. Hence the chief effect of removing the lung from the thorax is to alter the transmural pressures, and thus the resistance of the lung. It is easy to take account of this effect when comparing model predictions with experimental data.

Professor Kitney In your capillary bifurcation modelling, have you considered the use of fractal theory?

P. Hydon The morphometric measurements made by Singhal *et al.* (1973) show a remarkable uniformity in the ratios of numbers of arteries of successive orders, for the eight smallest orders. This seems to indicate that the small arteries may branch in a self-similar manner, suggesting that ideas from fractal theory could be useful. Nevertheless, the data is statistical, so there could be spatial inhomogeneity in the manner of branching. The larger vessels do not have the same branching ratio, and it is necessary to incorporate measurements of their anatomical locations in the model.

REFERENCES

Fung, Y. C., 1984, "Biodynamics: Circulation," Springer-Verlag, New York.

Greenfield, J. C., Jr., and Griggs, D. M., Jr., 1963, Relation between pressure and diameter in main pulmonary artery of man, J. Appl. Physiology, 18:557.

Hakim, M., Higenbottam, T., English, T. A. H., and Wallwork, J., 1987, Distance procurement and preservation of heart-lung homografts, Transplantation Proceedings, 19:3535.

Heath, D., and Edwards, J. E., 1953, The pathology of hypertensive pulmonary vascular disease, Circulation, 18:533.

Krishnan, A., Linehan, J. H., Rickaby, D. A., and Dawson, C. A., 1986, Cat lung hemodynamics; comparisons of experimental results and model predictions, J. Appl. Physiology, 61:2023.

Permutt, S., and Riley, R. L., 1963, Hemodynamics of collapsible vessels with tone: the vascular waterfall, J. Appl. Physiology, 18:924.

Singhal, S., Henderson, R., Horsfield, K., Harding, K., and Cumming, G., 1973, Morphometry of the human pulmonary arterial tree, Circulation Res., 33:190.

Wiener, F., Morkin, E., Skalak, R., and Fishman, A. P., 1966, Wave propagation in the pulmonary ciculation, Circulation Research, 19:834.

Yen, R. T., and Sobin, S. S., 1988, Elasticity of arterioles and venules in postmortem human lungs, J. Appl. Physiology, 64:611.

Zhuang, F. Y., Fung, Y. C., and Yen, R. C., 1983, Analysis of blood flow in cat's lung with detailed anatomical and elasticity data, J. Appl. Physiology, 55:1341.

NUMERICAL AND EXPERIMENTAL INVESTIGATION

OF LUNG BIFURCATION FLOWS

P. Corieri[1], C. Benocci[2], M. Paiva[3] and M. Riethmuller[2]

[1] von Karman Institute (Rhode St Genèse) and Université Libre de Bruxelles (Brussels)
[2] von Karman Institute (Rhode St Genèse)
[3] Institut de Recherche Interdisciplinaire,Université Libre de Bruxelles (Brussels)

von Karman Institute, 72 ch de Waterloo, 1640 Rhode St Genèse, Belgium

ABSTRACT

The aim of this contribution is to describe a study of pulmonary flows performed using both experimental and computational methods.

The lung consists of a network of bifurcating tubes through which air flows from the trachea through 17 generations of airways where convective processes dominate to the 6 further generations (alveolar zone) where gas exchange occurs and where transport is dominated by molecular diffusion. The part of the lung studied here is the one between generations 12 and 17, just before the alveolar zone,and is characterized by small dimensions and small Reynolds numbers. The main characteristics of this pulmonary flow are: incompressibility, three dimensionality, unsteadiness, laminarity.

Because of its complexity, the problem considered was simplified to consider a single bifurcation. Measurements of velocity were made in a glass model of a bifurcation using Laser Doppler Velocimetry. The geometry of the scale model was chosen to respect geometrical similarity with the physiological data, while dynamics similarity was obtained by using glycerine and varying the velocity at the entry to have the same Reynolds number. Flow visualisations and results for the velocity field are presented for steady state measurements.

A study of this flow is also being made using a computational method. An existing code was adapted to simulate two dimensional, incompressible, steady flow in a symmetric bifurcation. The code employs finite differences in a curvilinear coordinate system. Results of the first stage of this numerical simulation will be presented and compared with existing data in the literature.

KEYWORDS : lung/ airduct bifurcation/ low Reynolds number/ 3-D experimentation / 2-D numerical modelisation/ Laser Doppler Velocimetry.

INTRODUCTION

The main function of the lung is to insure gas exchange between air and blood. The lung configuration is usually described by a network of 23 generations of tubes (Weibel (1963)) through which air flows in a oscillatory regime. The network is constituted by tubes that branch as a succession of bifurcations.

Anatomical data (Horsfield et al.,1968) provide information concerning the diameters, lengths and angles from which the average velocities in the different parts of the lung can be computed. Those data are reported in figure 1. Two different regions are differentiated by their gas transport process: the first one is the convective zone constituted by the first sixteen generations, and the second one is the alveolar zone through which gas exchange occurs.

Along the pulmonary tree the characteristics of the flow change significantly; this is caused by:

• variation of velocities, diameters, lengths and angles of the tubes.

• change in the flow transport process.

From these one may deduce the characteristics of the flow through the network :

• the air may be assumed as an incompressible fluid because of the small velocities involved.

• the Reynolds number varies between 0.02 and 2500 for a flowrate at the mouth equal to 0.5 l/s showing that the flow properties vary from turbulence to laminarity and even Stokes flow. Because of the complexity of the geometry it is difficult to establish the transition zone in the lung where the flow changes from turbulence to laminarity.

• the complexity of the geometry makes the flow three dimensional; this effect will be stronger for higher Reynolds numbers.

• the breath is an unsteady quasi sinusoidal mechanism. Pedley (1977) has justified that in a first approximation of the problem the flow may be supposed as steady.

Another unsteady effect is due to the moving wall of the lung. This is assumed by Pedley (1977) as a second order effect on the flow mechanism.

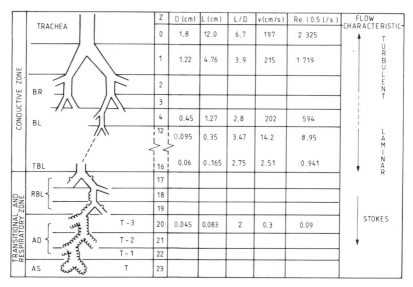

Fig. 1   Pulmonary background

At this step it is interesting to compare air flow in lung bifurcations, and in blood flow vessels bifurcations. The main differences are : the blood is a non-newtonian fluid, blood flow is pulsed. The similarities are : same range of Reynolds numbers, incompressible flow, same type of geometry, moving wall. Since in several modelisations of blood flow, unsteadiness and non-newtonian effects are neglected, the technique used for both flow case are similar and this is to be kept in mind for comparison of the techniques used and the results obtained.

The aim of this study is to characterize the flow in the air ducts preceding the alveolar zone i.e. generations 12 to 17. In this zone of the lung, the Reynolds number varies between 10 and 1 (cfr. figure 1), the diameters between 0.1 and 0.06 mm and the length to diameters ratio of the tubes between 3.5 and 2.8. The Peclet number that characterizes the ratio of the convection to the diffusion transport is equal to 0.7 at generation 16 from which the transition between convective and diffusive flow.

Because of the small dimensions involved, in vivo measurements in the lung are not feasible. To approach this problem we have combined an experimental and a numerical modelisation. The next section of the paper will describe the technique chosen in both experimental and numerical approaches but a complete comparison between results obtained is not yet possible. Because of the complexity of the problem several assumptions have to be made. The assumptions are : incompressibility of the fluid, steady state and fixed wall.

The experimental technique chosen is Laser Doppler Velocimetry, and the numerical technique uses a discretisation of the Navier-Stokes equations by a finite difference method. The two next sections will describe in details the methods used and the results obtained at this step of the study.

EXPERIMENTAL APPROACH

Material and methods

Laser velocimetry measurements are performed in a scaled-up glass model of a bifurcation. The increase of the characteristic dimensions is compensated by the use of glycerol in order to keep the Reynolds similarity. The model, built to a scale of approximately 40, respects the ratio of diameters between parent and daughter branches and the angles corresponding to the anatomical data for the generations 12 to 17. It is assumed that the 3 branches are in the same plane. The lengths of the branches are chosen long enough to obtain a fully developed flow at the inlet and outlet. The fluid used for the experiments is pure glycerol, a highly viscous fluid that has the other advantage to have a refractive index matching the glass.

The bifurcation is placed in a closed loop as shown on figure 2. The characteristics of the flow system are:

• the system is fixed; the position of the measurement points will be adjusted by moving the optical system.

• the bifurcation model is submerged in a plexiglass box filled with glycerol to avoid diffraction and dioptrical effects when laser beams enter the tube.

• the model is placed in a loop and a continuous pump of " Moineau " type circulates the glycerol.

• the direction of the flow corresponds to inspiration.

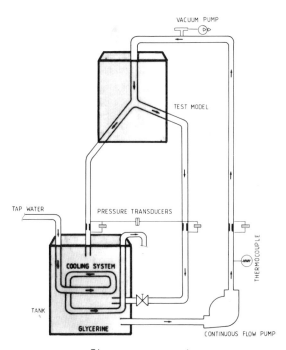

Fig. 2   Flow system

145

Measurements of the parameters of the loop consist of (figure 2):

• three flowmeters that measure the flow rates in the tubes.

• a copper-constantan thermocouple that measures the temperature. A serial cooling system is installed in the fluid loop to maintain the temperature constant during a measurement procedure in order to keep viscosity, refractive index and flow rate constant.

• a Validyne pressure transducer is placed between the outlet of the daughter tube to ensure equality of pressures at the outlets. This system verifies the conditions that we have chosen to impose: newtonian fluid, incompressibility, steady state, model of bifurcation as close as possible to the one of the lung and boundary conditions consisting of a fully established flow upstream of the bifurcation and equality of pressures at the outlets of the 2 daughter tubes.

The velocity is measured by Laser Doppler Velocimetry (LDV) for different positions of the LDV probe volume. The Laser Doppler set up consists of two parts represented on the figure 3 : the first part is the optical system, and the second part is the data processing system. The technique used is detailed by Corieri and Riethmuller, 1989.

The optical system consists of a conventional backscatter laser system (fig 3). The position of the probe volume is adjusted in three orthogonal directions by moving a motorised table. Particles used to seed the glycerol are TSI particles (model 10087) recognized for their good reflection property.

The processing system is particular, because the velocities to be measured are very low. The first low pass filter of our standard period counter is set to 10 kHz which was higher than the frequency that we had to measure. This is the reason why in addition to the common period counter we also use a Spectrum analyser to measure the lower Doppler frequency that is proportional to the velocity.

In this experimentation a large number of parameters have to be controlled, adjusted and measured. Because all these operations are very time consuming, the measurement process has been automatised using a PC computer.

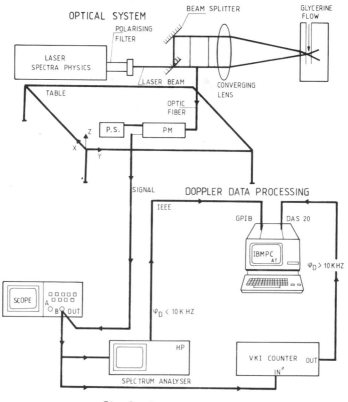

Fig. 3 Laser system

Experimental results

Before performing the measurements several flow visualisations were made to obtain a qualitative idea of the flow at low Reynolds number. The method of visualisation consists of producing a Laser sheet by passing a Laser beam through a cylindrical lens. A long time exposure photograph allows us to follow the streamlines. In this case the particles are little air bubbles confined in the glycerol. Result of the visualisations is presented on picture 1. The position of the sheet with respect to the tube is presented on drawing 1. The picture corresponds to a time exposure of 0.5 sec and the Reynolds number is equal to 1.
The Reynolds number is defined as follow :

$$RE = \frac{v.d}{\nu} \begin{cases} v = & average\ velocity \\ d = & diameter \\ \nu = & \text{viscosity of the glycerol} \end{cases}$$

The picture shows the laminarity of the flow and the stagnation zone at the junction of the bifurcation. Since the Reynolds number is small and the wall smooth, no recirculation and significant separation zone appear on the picture. In agreement with this flow visualisation, the figure 4 gives an overview of a non dimensionalised axial velocity profile for a Reynolds number of 1. As well as in the visualisation the velocity profiles show laminar flow, without any separation zone. The measurements show that the daughter velocity profiles are fully established close to the bifurcation. This result is interesting because, in this area of the lung the length to diameter ratio is around 3; thus we may conclude that there is no influence from one bifurcation on the following ones. This should be confirmed by further studies.

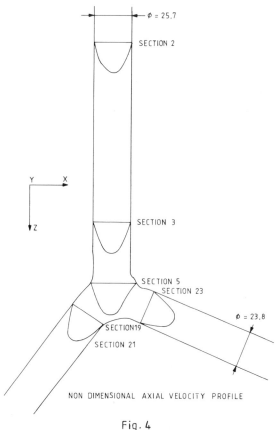

Fig. 4

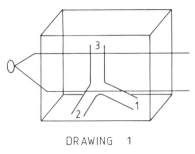

DRAWING 1

Picture 1

NUMERICAL STUDY

Theoretical formulation of the problem

Important features of pulmonary flows are their geometrical complexity and the high number of parameters affecting their behaviour. The recent development of numerical techniques and increasingly powerful computers offer an effective tool to solve numerically the pulmonary bifurcation problem in two and three dimensions.

The equations representing the physical problem are the incompressible Navier-Stokes(N-S) equations which are in non dimensional form,

the Conservation of Momentum(1):

$$\frac{\partial u_i}{\partial t} + \nabla p + \nabla.(u_j u_i) - \frac{1}{RE}\nabla^2 u_j = 0 \tag{1}$$

and the Conservation of Mass (2):

$$\nabla.u_i = 0 \tag{2}$$

where u_i is the velocity in the coordinate direction i, t the time, p the pressure to density ratio and Re the Reynolds number based on the average velocity and the diameter.

The above equations have to be discretized and solved over a grid of discrete mesh points, whose locations are selected as a function of the method of solution and the geometry of the problem. For complex geometries, like the present one, appropriate meshes are generated by "ad hoc" codes or "grid generators".

Methods of solutions

An existing code, solving the two dimensional incompressible N-S equations has been used to simulate the flow in a two dimensional symmetrical bifurcation. Details concerning this code are reported by Ruddick and Benocci,1989, and only the main outlines will be given below.

The code solves the N-S equations written in the non orthogonal curvilinear form, making it possible to use curvilinear coordinate systems (s,n) in order to fit exactly the physical boundaries of geometrically complex domains. An example of curvilinear mesh for the present case is shown in Fig.5.Equations 1 and 2 are discretized in space with centered finite differences. Equation 1 is advanced in time by using an implicit time marching technique, while mass conservation is enforced indirectly, by computing a pressure field such that the corresponding velocities satisfy equation 2.

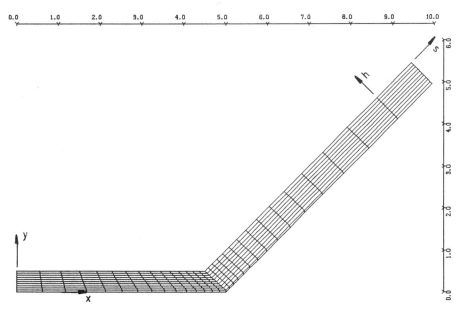

Fig. 5   Mesh (Physical domain)

The discrete equivalent of equation 1 , with an implicit marching in time can be written as:

$$\frac{u_i^{n+1} - u_i^n}{\triangle t} = S(u_j^{n+1}) + R(u_j^{n+1}) - \frac{\partial p^{n+1}}{\partial x_i} \tag{3}$$

where x_i is the generic space direction and $S(u)$ and $R(u)$ are discrete operators containing all derivatives of u with respect to space; namely $S(u)$contains the mixed $(\partial\ /\partial s\partial n\ )$ terms coming out of the curvilinear formulation of the viscous terms and $R(u)$ contains the convective and the non-mixed viscous terms .

Each time step is split in a Predictor and a Corrector step. In the Predictor step equation 4 for the intermediate variable u^* is solved:

$$u_i^* = u_i^n + \triangle t[R(u_j^*) + S(u_j^n)] - \frac{\partial p^n}{\partial x_i} \tag{4}$$

where p^n is the pressure field at the old time level. To determine u^* a fractional step method, is adopted i.e. an intermediate time step is used for each coordinate direction; in such a way that only tridiagonal systems of linear equations have to be solved (the mixed derivatives operator $S(u)$ has to be solved explicitly because it does not fit a tridiagonal scheme).

The solution of 4 will satisfy the conservation of momentum, but not, in general the conservation of mass. The latter is imposed modifying the value of the pressure and correcting as:
Corrector step

$$u_i^{n+1} = u_i^* + \triangle t[\frac{\partial}{\partial x_i}(p^n - p^{n+1})] \tag{5}$$

Deriving 5 with respect to x_i and imposing the incompressibility condition $(\nabla.u_i = 0)$ on the new solution u (n+1) a Poisson equation for the pressure can be obtained, namely:

$$\nabla^2(p^{n+1} - p^n) = \frac{\nabla.u_i^*}{\triangle t} \tag{6}$$

149

The new pressure field ensures mass conservation in the resolution of the N-S equations. Then by solving 6 and introducing p^{n+1} so obtained in 5, the final solution u^{n+1} is obtained and the time step is completed.

The boundary conditions imposed are discussed in the next section describing the computation for a bifurcation.

The program used has already been tested for flow problems which have the advantage of containing some of the flow features that we may encounter in pulmonary flows.

The first test case was the prediction of flow in a straight channel, with an uniform velocity profile imposed at the inlet: in such a case the velocity profile has to evolve gradually to a parabolic shape, and the aim of the test was to verify that the evolution predicted by the code (establishment length for the parabolic profile) was in good agreement with the theory. Secondly, the flow in a curved channel where recirculating flow regions were expected has been computed. In this case the aim of the test was to verify the capability of the code to predict correctly the existence and shape of separated flow regions. Finally, the flow over a backward facing step was simulated to verify the capability of the code to deal with flows including sharp corners and fixed separation points. These three test cases have shown (Ruddick and Benocci,1989) a good agreement with theoretical and experimental solutions.

Numerical results

Computations have been made for low Reynolds numbers in a symmetrical bifurcation with sharp corners at the connection wall. The geometry adopted is similar to the one used in Bramley and Dennis's paper, 1984, in order to compare results.

The applied boundary conditions are :

• at the inlet, the two velocity components are specified. A half parabolic profile is fixed at the entrance of the tube.

• at the solid wall, a no slip boundary condition (u=v=0) is required

• at the outlet, boundary conditions are a problem with incompressible flow. After several numerical experiments, an extrapolation condition giving

$$\frac{\partial^2 u}{\partial s^2} = 0 \qquad (7)$$

$$\frac{\partial^2 v}{\partial s^2} = 0 \qquad (8)$$

has been adopted, where s is the mesh direction normal to the outlet boundary.

• at the "fluid" symmetry wall, a symmetrical boundary condition is applied.

$$\frac{\partial u}{\partial y} = 0 \qquad (9)$$

$$v = 0 \qquad (10)$$

Following Teman (1979) the approximate boundary condition of zero pressure derivative in the direction normal to the boundary is imposed everywhere.

Tests were performed for Reynolds numbers of 50,100 and 200. In order to avoid confusion, the same Reynolds number definition than Bramley (1984) has been adopted :

$$RE = \frac{RU_{ave}}{\nu} \qquad (11)$$

Where

    R : Radius of the parent tube

    U_{ave} : Average velocity in the parent tube

    ν : Kinematic viscosity

150

The geometry adopted is a half bifurcation (fig 5), the ratio parent over daughter tube is equal to $\frac{1}{\sqrt{2}}$. The branching angle is equal to 45 degrees. The length of the parent tube is equal to 5 times the diameter and the length of the daughter tube varies as function of the Reynolds number computed.

Tests were performed with different meshes. Table 1 gives the number of meshes and length of daughter tube as a function of the Reynolds number. Figure 5 represents the mesh corresponding to the case A in table 1.

Table 1

Re	Parent tube Grid				Daughter tube Grid				Case
	D	Length	s	h	D	Length	s	h	A
50	1	5D	15	11	2	7D	16	11	B
			15	16			26	16	
			15	21			46	21	C
			15	26			56	26	D
100	1	5D	15	26	2	7D	56	26	E
			15	41	2	14D	111	41	F
200	1	5D	15	41	2	14D	111	41	G

The grid is generated in order to obtain:
• higher resolution near the wall and around the bifurcation since in these areas the velocity gradients are largest.
• mesh lines orthogonal to the symmetry wall to facilitate the application of symmetrical boundary condition.

The grid is generated by using a multi-directional scheme technique described by N.P Weatherill (1989).

In order to evaluate the quality of the results, the first step is to make sure they are independent of the mesh. This is achieved by the comparison of the upper wall vorticity of the daughter tube for several meshes.

Figure 6 represents the results for a Reynolds number equal to 50. Size of the computational domain and number of mesh points for the cases A,B,C,D are described in the table 1. It can be observed that the differences in results for different meshes decrease with a quadratic low, which is consistent with the discretization used, and that the solution for case C and D can be taken as "mesh free". The computed vorticity in the last part of the tube, where the flow is fully established, converges to 3.92 for the more refined meshes. The theoretical value being equal to 4, taking into account that the velocities close to the wall are the most affected by discretization errors, this is a satisfactory result. Figure 7 compares the vorticity at the upper wall for Reynolds numbers of 50,100 and 200. No recirculation zone exist for the Reynolds number of 50, while a small one is present for Reynolds number 100, which is in agreement with findings of Bramley et al., 1984. Lengths of velocity profile establishment are comparable to the ones computed by Bramley(1984).

Figure 8,9 show streamlines for Reynolds numbers of 50 and 200. The recirculation regions are observed for Re 200.

The tests presented show that the code used give satisfactory results for a 2D symmetrical bifurcation. The second step of the work will consist in the extension of the code in order to model a 3D bifurcation with curved wall. The main difficulties will be the amount of CPU time required by 3D calculation and the discretization of a 3D bifurcation system which will require a sophisticated 3D grid generator. Because the code is implemented in curvilinear coordinates the 3D bifurcation model could be close to a in vivo lung bifurcation.

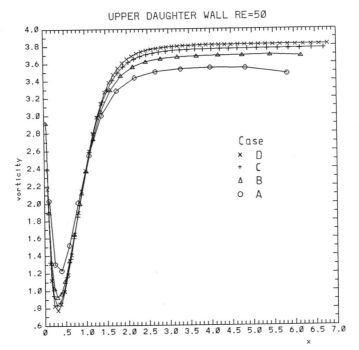

Fig.6 Vorticity

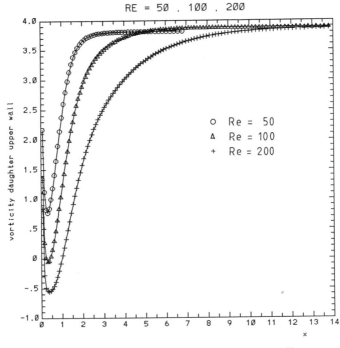

Fig.7 Vorticity upper daughter wall

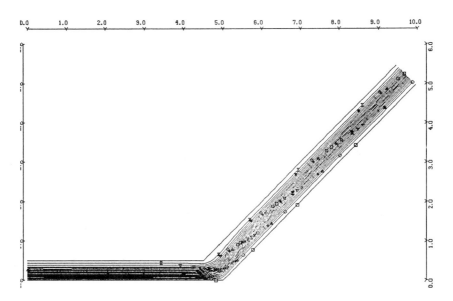

Fig. 8   Streamlines pattern RE= 50

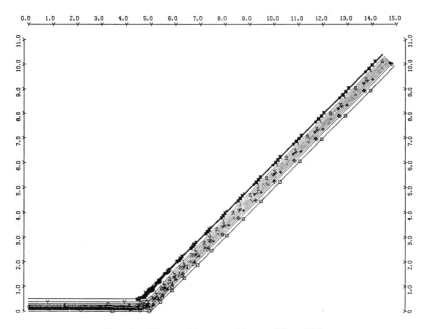

Fig. 9   Streamlines pattern RE = 200

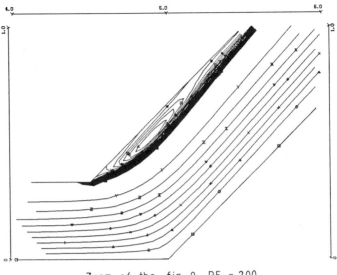

Zoom of the fig. 9 RE = 200

CONCLUSIONS

This paper has described experimental and numerical modelling of flow in lung bifurcation at low Reynolds numbers.

Laser Doppler velocimetry has been shown to be extremely valuable to perform measurements at low velocities in the complex configuration of the lung bifurcation model. Velocity measurement has shown that the flow is fully established close to the bifurcation. Further measurements should confirm that at low Reynolds numbers there is no interaction between successive bifurcations.

A 2D incompressible solver in curvilinear coordinates has been applied to the case of a symmetrical bifurcation. Good agreement with studies performed by Bramley has been shown. The next step of this work will consist of modelling the 3D bifurcation with curved wall. The main difficulties will be the amount of CPU time required and the discretisation of the 3D bifurcation in a curvilinear system.

REFERENCES

Benocci, C., Ruddick, K.G., 1989, Solution of the steady state incompressible Navier-Stokes equations at high Reynolds numbers. VKI Technical note 170.

Bramley, J.S., Dennis, S.C.R., 1984,The numerical solution of a two-dimensional flow in a branching channel., Computers & fluids, Vol. 12, No. 4, pp. 339-335.

Corieri, P., Riethmuller, M.L., 1989,Laser Doppler Velocimetry and computer automation to measure low velocities in a pulmonary model", Proceedings of the ICIASF 1989 record, International congress on instrumentation in aerospace simulation facilities, Gottingen, W-Germany.

Horsfield, K., Cumming, G., 1968, Morphology of the bronchial tree in man., Journal.Appl.Physiol.24 : 384-390 .

Pedley, T.J., 1977,Gas Flow and Mixing in the Airways. Bioengeneering Aspect of the Lung, M. DEKKER INC.

Ruddick, K.G., 1988, A Navier-Stokes solver for incompressible flow on a curvilinear, non-orthogonal, staggered mesh. VKI project report 20.

Ruddick, K.G., Benocci, C., 1989, An implicit Navier-Stokes Solver optimised for a vector-parallel computer. VKI Technical note 169.

Teman, R., 1979, Navier Stokes Equations, Revised Edn., North Holland.

Weatherhill, N.P., 1989, Mesh generation in CFD, von Karman Institute Lecture series 4.

Weibel,E.R., 1963, Morphometry of the human lung. New York, Academic Press .

REGULATION OF FLOW IN VASCULAR NETWORKS BY EDRF

T M Griffith and D H Edwards*

Departments of Diagnostic Radiology and *Cardiology
University of Wales College of Medicine
Heath Park, Cardiff, CF4 4XN, UK

The endothelial lining of arteries is anatomically well sited to convey physiological information about local blood flow to underlying smooth muscle cells, and by releasing endothelium-derived relaxing factor (EDRF) in response to shear stress (Pohl et al., 1986; Rubanyi et al., 1986) mediates flow-dependent dilatation of both conduit (Holtz et al., 1983; Melkumyants et al., 1989) and resistance arteries (Griffith et al., 1987; Griffith and Edwards, 1990). Flow-dependent dilatation provides a way of coordinating responses in different parts of the arterial tree: a metabolically-induced reduction in distal resistance, for example, can produce a secondary fall in the resistance of more proximal "feed vessels" simply as a consequence of the accompanying increase in flow (Holtz et al., 1983).

We have used X-ray microangiographic techniques to study the relationship between flow and EDRF activity in resistance arteries ranging from ca. 70 to 1000 µm diameter (Fig 1) in isolated rabbit ear preparations perfused with physiological buffer under conditions of non-pulsatile flow (Griffith et al., 1987, 1988, 1989). EDRF is now believed to be nitric oxide (Palmer et al., 1987) and is avidly scavenged by haemoglobin in buffer-perfused preparations without complicating effects on either prostacyclin production or endothelium-dependent hyperpolarisation (Martin et al., 1985; White and Martin, 1989; Nishiye et al., 1989). EDRF activity is also inhibited by N^G-monomethyl-L-arginine (L-NMMA) which competitively blocks its synthesis from L-arginine (Rees et al., 1989). In some experiments constriction was induced by 0.1 µM 5HT in order to mimic the in vivo situation in which resistance arteries possess significant tone. 5HT does not actively stimulate EDRF release in the rabbit ear (Griffith et al., 1988).

EDRF LIMITS PRESSURE GRADIENTS

In resting (ie pharmacologically-unconstricted) preparations the diameters (d) of the different generations of arteries can be related to the input flow rate into the central ear artery (Q) by expressions of the form $Q=ad^4+b$, although only in the presence of EDRF activity (Griffith et al., 1987). This implies that the pressure gradients (dP/dL) necessary to maintain increased flow are limited by EDRF activity at high flow rates. In Poiseuille flow, $dP/dL \propto Q/d^4$, so that at high flow rates (ie when $b \ll ad^4$, and therefore $Q/d^4 \rightarrow a$), dP/dL will asymptotically

Biomechanical Transport Processes, Edited by F. Mosora et al.
Plenum Press, New York, 1990

155

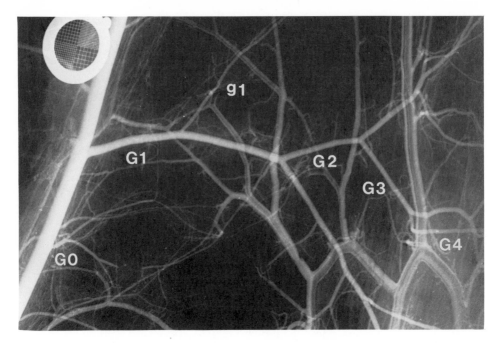

Fig 1 Representative microangiogram of successive branch generations (G) in the vascular network of an isolated rabbit ear preparation. The central ear artery is labelled as G0 and subsequent generations as G1, G2, G3 and G4. Small branches arising from G1 are labelled g1. Note streaming of contrast medium in venous arcades. The calibration grid has an external diameter of 3 mm and the grid wires have diameters of 10 μm.

approach a constant value. There is evidence that endothelial cells are similarly able to stabilise pressure gradients in conduit vessels (Melkumyants *et al.,* 1987). This effect will serve to limit the energy losses incurred by large increases in flow, and may be of particular importance in resistance vessels.

EDRF MAY PREVENT VASCULAR "STEAL"

EDRF activity helps to preserve constancy of relative arterial diameters ("geometrical similarity") in pharmacologically-constricted rabbit ear preparations at different flow rates (Griffith *et al.,* 1987). This is illustrated in Fig 2 in which the ratios of the diameters of the daughter arteries (asymmetry ratios, $\alpha = d_2/d_1$) at certain bifurcation types are shown to vary with flow rate in 0.1 μM 5-HT constricted preparations after, but not before, inhibition of EDRF activity. As hydraulic resistance depends on $1/(diameter)^4$, geometrical similarity implies constancy in the relative spatial distribution of flow, so that a physiological role of EDRF activity may be to prevent vascular "steal". Indeed, abolition of basal EDRF activity (by the irreversible inhibitor gossypol) leads to mismatches between metabolism and flow and consequently to regional hypoxia in rabbit skeletal muscle (Busse and Pohl, 1989). Geometrical similarity is maintained in resting preparations both in the presence and the absence of EDRF activity, but is lost in preparations highly constricted by 1 μM 5HT even in the presence of EDRF activity (Griffith *et al.,* 1987). This suggests that there may be an "optimal" balance between competing vasodilator and vasoconstrictor influences.

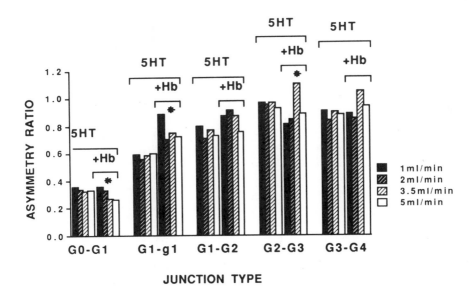

Fig 2   Plots of asymmetry ratios ($\alpha=d_2/d_1$, the ratio of daughter artery diameters) at the different bifurcation types as a function of flow rate in preparations constricted by 0.1 µM 5HT. α was independent of flow rate in the absence, but not the presence of haemoglobin (+Hb) at the G0-G1, G1-g1 and G2-G3 junctions (*p<0.05, non-parametric Friedman two-way analysis of variance by ranks). This is consistent with maintenance of "geometrical similarity" by EDRF activity in pharmacologically constricted preparations, so that the equation $d_1^x + d_2^x = d_0^x$ (where x is the "junction exponent") will have the same solution for x all flow rates (see text).

EDRF AND THE OPTIMALITY OF BRANCHING GEOMETRY

Murray proposed in 1926 that local arterial diameters (d) and local flow rate (q) were related in such a way as to minimise the sum of viscous power losses ($\propto \frac{q^2}{d^4}$) and the energetic "cost" of intravascular volume ($\propto d^2$). When "total" energy losses, $\frac{aq^2}{d^4} + bd^2$ (a and b are constants) are minimised, the expression $q=kd^3$ (where $k = \sqrt{\frac{b}{2a}}$) can be shown to give the "optimal" relationship between flow and diameter in a segment of artery (Murray, 1926). If this cubic law applies to each of the three vessels at a bifurcation, then, from the equation of continuity $q_0=q_1+q_2$, it follows that $d_0^3=d_1^3+d_2^3$ (where the subscripts 0 denote parent, 1 major daughter and 2 minor daughter arteries respectively). In order to determine if EDRF activity influences the "optimality" of arterial branching, the influence of EDRF activity on the junction exponent, x, which is obtained from the equation $d_0^x=d_1^x+d_2^x$, was investigated at four flow rates at the five topographically distinct types of bifurcation shown in Fig 1.

In resting preparations, either in the absence or presence of EDRF activity, and also in preparations constricted by 0.1 µM 5HT alone, the frequency distributions of x at the different flow rates possessed modes close to the Murray optimum of 3 (Figs 3 and 4). Median values of x were, however, slightly larger

157

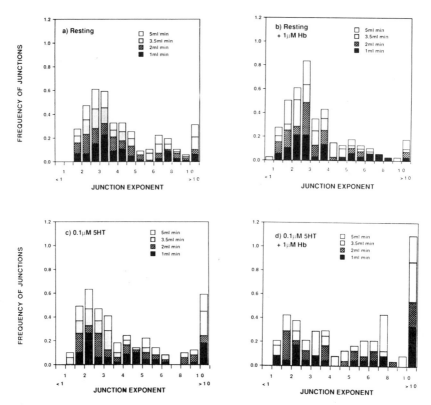

Fig 3   Stacked frequency distributions of junction exponents at each of the four
flow rates employed. (a) before (n=63 junctions) and (b) after (n=38
junctions) inhibition of EDRF activity by 1 μM haemoglobin (Hb) in resting
(ie pharmacologically-unconstricted) preparations, (c) before (n=49
junctions) and (d) after (n=24 junctions) inhibition of EDRF activity by 1 μM
haemoglobin in preparations constricted by 0.1 μM 5HT. In (a), (b) and (c)
the modes and medians of the distributions were close to 3 at each flow
rate but in (d) no clearly defined mode common to the four flow rates
could be identified and the medians were increased to ca. 5 (see also Fig
4).

than 3 because of skewness in the frequency distribution. In the presence of
haemoglobin, in contrast, the frequency distribution of x exhibited no consistent
mode and the median values of x were approximately 5 in preparations
constricted by 0.1 μM 5HT (Figs 3 and 4). These findings suggest that EDRF
activity contributes to optimality of branching according to the principles
suggested by Murray when there is significant vasomotor tone (as is the case *in
vivo*).

An alternative method of analysing branching geometry is to derive
"optimal" branching angles from mathematical models which minimise the "total
cost", $\sum_{i=0}^{2} c_i l_i$, of a vascular bifurcation, where l_i denotes the length of its
component vessels and c_i is a notional "cost" per unit length or so-called "cost
function" (Murray, 1926; Zamir, 1976). Woldenberg and Horsfield (1986)
showed that in four models, in which the cost functions are chosen to depend on

Table 1 Cost Functions for the Four Minimisation Principles

MODEL TYPE	SURFACE	VOLUME	DRAG	POWER LOSS
COST/UNIT LENGTH (c)	$\propto d$	$\propto d^2$	$\propto q/d^2$	$\propto q^2/d^4$

(note that surface = drag, and volume = power loss if $q=ad^3$ as proposed by Murray)

surface area, volume, drag due to shear stress or viscous power losses respectively (Table 1), the angle between the two daughter arteries at a bifurcation (ψ) is related to the junction exponent x and asymmetry ratio α in such a way that plots of x against ψ are insensitive to the exact value of α. Plots of median values of x (averaged with respect to flow rate) against median values of ψ for resting preparations (either in the presence or the absence of haemoglobin) and for preparations constricted by 0.1 µM 5HT alone, were close to the intersection of curves reflecting optimality in terms of minimum volume and minimum power losses (Fig 4). This was not the case, however, for the corresponding plot of x and ψ in 5HT-constricted preparations in the presence of haemoglobin (Fig 4). As the curves of the minimum power loss and the minimum volume model cross at x=3, these observations provide additional support for the idea that EDRF maintains optimality in accordance with Murray's hypothesis when constrictor tone is elevated. The finding that median values of the junction exponent x remained close to 3, except under those conditions where geometrical similarity was lost, is consistent with the fact that the equation $d_0^x=d_1^x+d_2^x$ will have the same solution for x at all flow rates if there is constancy of diameter and asymmetry ratios (Fig 2).

"Optimality" according to Murray's equation has two main physiological implications. The arterial side of the mammalian circulation is a low volume/high pressure system. Flow is determined by pressure divided by resistance so that to produce large increases as are necessary, for example, during exercise when there may be a 20-fold rise above resting conditions, it is biologically more advantageous to reduce resistance though powerful local vasodilator mechanisms (including flow-dependent release of EDRF), than increase pressure to levels which could precipitate left ventricular failure (Harris, 1983). From a physical point of view, large and rapid changes in resistance would be facilitated by a "minimum volume" circulation, as small changes in arterial/arteriolar diameters would then have a relatively large haemodynamic effect. A minimum power loss system would carry the obvious advantage of limiting cardiac work.

EDRF ACTIVITY OPPOSES MYOGENIC AUTOREGULATION OF FLOW

The influence of basal EDRF release on the ability of the rabbit ear to autoregulate (ie) to maintain "constant" blood flow when perfusion pressure is varied was investigated by constructing pressure-flow (P-Q) and diameter-flow (d-Q) relationships under conditions of both "controlled-flow" and "controlled-pressure" perfusion (Griffith and Edwards, 1990). This was achieved either by regulating flow rate with pump speed, or by keeping perfusion pressure constant through a feedback loop which compared the signal from the pressure transducer with a control voltage. Experimentally, flow was increased in discrete steps and "controlled-flow" data acquired when perfusion pressure was again steady (generally after ca. 10 mins). "Controlled-pressure" data was subsequently obtained by "switching in" the feedback loop to hold perfusion pressure constant.

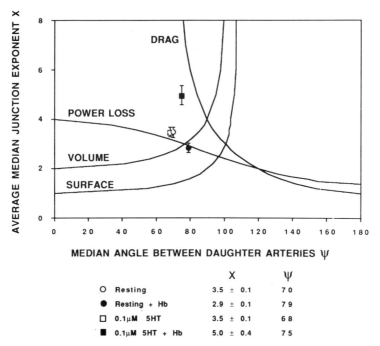

		X	ψ
○	Resting	3.5 ± 0.1	7 0
●	Resting + Hb	2.9 ± 0.1	7 9
□	0.1μM 5HT	3.5 ± 0.1	6 8
■	0.1μM 5HT + Hb	5.0 ± 0.4	7 5

Fig 4 Plots of the median junction exponent (mean ± SEM over the range of flow rates 1-5 ml/min) against corresponding median values of the angle between the two daughter arteries, ψ. Numerical values are shown in inset. The continuous curves represent the ideal relationship between x and ψ for the four minimisation principles (see Woldenberg and Horsfield, 1986), and are given by the equations

$$\cos \psi = \frac{(1+\alpha^x)^{2/x} - \alpha^2 - 1}{2\alpha} \qquad \text{(minimum surface)},$$

$$\cos \psi = \frac{(1+\alpha^x)^{4/x} - \alpha^4 - 1}{2\alpha^2} \qquad \text{(minimum volume)},$$

$$\cos \psi = \frac{(1+\alpha^x)^{2-4/x} - \alpha^{2x-4} - 1}{2\alpha^{x-2}} \qquad \text{(minimum drag)},$$

$$\cos \psi = \frac{(1+\alpha^x)^{4-8/x} - \alpha^{4x-8} - 1}{2\alpha^{2x-4}} \qquad \text{(mimimum power loss)}.$$

The experimental data points lie close to the intersection of the optimal minimum volume and minimum power loss curves except in 5HT-constricted preparations in the absence of EDRF activity (+Hb). α was chosen as 0.6, which corresponds to the G1-g1 bifurcation, although the shape and position of the curves do not change significantly when α is varied.

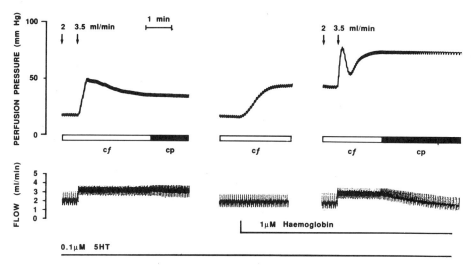

Fig 5 Changes in perfusion pressure in an isolated ear preparation preconstricted by 0.1 µM 5HT. The rapid rise induced by increasing flow from 2 to 3.5 ml/min was followed by a slower decline ("stress relaxation") in the absence (left) of haemoglobin. In the presence of haemoglobin however (right), a slowly-developing secondary rise in pressure ("myogenic response") was also observed. Changing the perfusion mode from "controlled-flow" (cf) to "controlled-pressure" (cp) had little effect on flow in the absence of haemoglobin (left) but resulted in a slow decline in its presence (right). L-NMMA behaved in a fashion analogous to that of haemoglobin (not illustrated). The phasic nature of the flow trace reflects the electronic signal controlling the pump, actual flow was damped.

In the presence of 1 µM haemoglobin or 100 µM L-NMMA but not in their absence or in the presence of 100 µM D-NMMA (which is the biologically inactive enantiomer of L-NMMA), rapid increments in flow rate induced a secondary "myogenic" constrictor response (Fig 5). Flow rate generally remained unchanged when the perfusion mode was changed from controlled-flow to controlled-pressure in the absence of inhibitors of EDRF activity, but declined in their presence (Fig 5). This may be explained by the fact that constriction of perfused arteries against a constant transmural pressure is not isotonic: circumferential wall stress first increases during the initial phase of contraction, but then declines as diameter continues to decrease (Speden, 1985). This makes it possible for constriction to induce further constriction in controlled-pressure perfusion mode. As a consequence of this effect, controlled-pressure, but not controlled-flow, pressure-flow (P-Q) relationships became sigmoidal in shape in the presence of either haemoglobin or L-NMMA, so that there was then a range of perfusion pressures over which flow rate remained relatively constant and the preparations were able to "autoregulate" (Fig 6). The corresponding conductance-flow (G-Q) relationships and diameter-flow (d-Q) relationships obtained directly by microangiography both exhibited regions with a negative slope which coincided the range of flow rates (1.5-2.5 ml/min) over which the preparations "autoregulated" (Fig 6). This is consistent with the existence of a "flow-dependent" constrictor response which is able to mediate autoregulation but is normally overridden by basal EDRF activity. Interestingly, pharmacological concentrations of acetylcholine, an agent which potently stimulates EDRF release (Furchgott, 1983; Griffith et al., 1984), have been

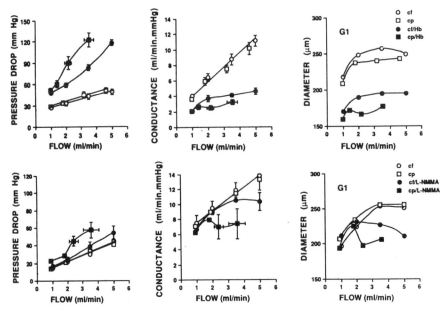

Fig 6 Left panels: Pressure-flow (P-Q) relationships of 0.1 μM 5HT-constricted rabbit ear preparations in the absence (open symbols) and presence (closed symbols) of haemoglobin (upper, n=11) and 100 μM L-NMMA (lower, n=5) under conditions of controlled-flow (cf, circles) and controlled-pressure (cp, squares) perfusion. In the presence of the EDRF inhibitors flow rates were lower in controlled-pressure than controlled-flow mode at equivalent pressures, and the sigmoidal shape of the P-Q relationship in controlled-pressure mode indicates suppression of autoregulation by basal EDRF activity (over flow rates between ca. 1.5-2.5 ml/min).

Centre panels: In the presence of EDRF activity, conductance-flow (G-Q) relationships derived from the P-Q data were straight lines common to both controlled-flow and controlled-pressure perfusion. In the absence of EDRF activity ie in the presence of haemoglobin (upper) or L-NMMA (lower), the shape of the G-Q relationships differed between the two perfusion modes, exhibiting a negative slope over the flow range 1.5-2.5 ml/min with controlled-pressure perfusion. As this range of flow rates corresponds to the sigmoidal part of the P-Q relationships, flow-dependent constriction appears to be the mechanism underlying autoregulation.

Right panels: The existence of flow-dependent constriction was experimentally confirmed by microangiography. Diameter-flow (d-Q) relationships exhibited regions of negative slope in all arterial branch generations (illustrated here for G1) over the flow range 1.5-2.5 ml/min in controlled-pressure mode in the presence of either haemoglobin or L-NMMA. Diameters were similar in both perfusion modes in the presence of EDRF activity: in its absence they were smaller with controlled-pressure than controlled-flow perfusion at all flow rates. SEM averaged ca 10% of mean diameter (not shown for clarity).

D-NMMA, the biologically inactive enantiomer of L-NMMA, exerted no significant influence on the P-Q, G-Q and d-Q relationships (not shown).

known for some time to abolish the ability of the canine kidney to autoregulate (Gross *et al.,* 1976).

Although haemoglobin and L-NMMA both induced vasoconstriction in all arterial branch generations (eg G1, Fig 6), diameters still increased monotonically with flow in controlled-flow perfusion mode. This is consistent with the observation that flow-dependent dilatation is reportedly only partially abolished by endothelial removal in isolated rabbit ear arteries (Bevan *et al.,* 1988). Under certain experimental conditions, however, "flow-dependent" dilatation could simply represent passive distension in response to increases in pressure rather than an active, flow-dependent EDRF-mediated phenomenon. In controlled-pressure perfusion mode, the unmasking of frank flow-dependent constriction by haemoglobin and L-NMMA indicates that flow-dependent dilatation is strictly EDRF dependent over flow rates between 1.5 and 2.5 ml/min.

CONCLUSIONS

The data are consistent with the idea that EDRF can mediate active flow-dependent dilatation and attenuates autoregulation by opposing a flow-dependent constrictor response. This latter phenomenon could be either a pressure- or a flow- sensitive phenomenon and, indeed, there is evidence for the existence of both mechanisms (Bevan and Joyce, 1988). Interestingly, so-called "myogenic" responses are endothelium-independent in some artery types (Bevan and Joyce, 1988), but endothelium-dependent in others (Harder, 1987; Vanhoutte, 1987). In spite of the potential complexity introduced by such opposing vasodilator and vasoconstrictor mechanisms, overall network "design" appears to be governed by the twin optimality principles of minimum volume and minimum power losses as proposed by Murray, and EDRF appears to contribute to this optimality when there is significant vasomotor tone.

ACKNOWLEDGEMENTS

The work was supported by the British Heart Foundation. The authors are grateful to R Newcome for statistical and W Simons for secretarial assistance.

REFERENCES

Bevan, J.A. and Joyce, E.H., 1988, Flow-dependent contraction observed in a myograph-mounted resistance artery, Blood Vessels., 25:261-264.

Bevan, J.A., Joyce, E.J. and Wellman, G.C., 1988, Flow-dependent dilation in a resistance artery still occurs after endothelium removal., Circ. Res., 63:980-985.

Busse, R. and Pohl, U., 1989, Role of endothelial cells in the control of vascular tone, in: "Vascular Dynamics", Nato ASI Series, Plenum Press 1989, pp.161-175.

Furchgott, R.F., 1983, Role of endothelium in responses of vascular smooth muscle, Circ. Res., 53:557-573.

Griffith, T.M. and Edwards, D.H., 1989, Myogenic autoregulation of flow may be inversely related to EDRF activity, Am. J. Physiol., in press.

Griffith, T.M., Edwards, D.H., Lewis, M.J., Newby, A.C. and Henderson, A.H., 1984, The nature of endothelium-derived relaxant factor, Nature., 308:645-647.

Griffith, T.M., Edwards, D.H., Davies, R.Ll., Harrison, T.J. and Evans, K.T., 1987, EDRF coordinates the behaviour of vascular resistance vessels, Nature., 329:442-445.

Griffith, T.M., Edwards, D.H., Davies, R.Ll., Harrison, T.J. and Evans, K.T., 1988, Endothelium-derived relaxing factor (EDRF) and resistance vessels in an intact vascular bed: a microangiographic study of the rabbit isolated ear, Br. J. Pharmacol. 93:654-662.

Griffith, T.M., Edwards, D.H., Davies, R.Ll. and Henderson, A.H., 1989, The role of EDRF in flow distribution: a microangiographic study of the rabbit isolated ear, Microvascular Res., 37:162-177.

Gross, R., Kirchheim, H. and Brandstetter, K., Basal vascular tone in the kidney: evaluation from the static pressure-flow relationship under normal autoregulation and at maximal dilatation in the dog, Circ. Res., 38:525-531.

Harder, D.R., 1987, Pressure-induced myogenic activation of cat cerebral arteries is dependent on intact endothelium, Circ. Res., 60:102-107.

Harris, P., 1983, Evolution and the cardiac patient: origins of the blood pressure, Cardiovasc. Res., 17:373-378.

Holtz, J., Giesler, M. and Bassenge, E., 1983, Two dilatory mechanisms of anti-anginal drugs on epicardial coronary arteries *in vivo*: indirect, flow-dependent, endothelium mediated dilation and direct smooth muscle relaxation, Z. Kardiol., 72:98-106.

Martin, W., Villani, G.M., Jothianandan, D. and Furchgott, R.F., 1985, Selective blockade of endothelium-dependent and glyceryl trinitrate-induced relaxation by haemoglobin and by methylene blue in the rabbit aorta, J. Pharmacol. Exp. Ther., 232:708-716.

Melkumyants, A., Balashov, S.A. and Khayutin, V.M., 1989, Endothelium dependent control of arterial diameter by blood viscosity, Cardiovasc. Res., 23:741-747.

Melkumyants, A.M., Balashov, S.A., Veselova, E.S. and Khayutin, V.M., 1987, Continuous control of the lumen of feline conduit arteries by blood flow rate, Cardiovasc. Res., 21:863-870.

Murray, C.D., 1926, The physiological principle of minimum work applied to the angle of branching of arteries, J. Gen. Physiol., 9:835-841.

Nishiye, E., Nakao, K., Itoh, T. and Kuriyama, H., 1989, Factors inducing endothelium-dependent relaxation in the guinea pig basilar artery as estimated from the actions of haemoglobin, Br. J. Pharmacol., 96:645-655.

Pohl, U., Busse, R., Kuon, E. and Bassenge, E., 1986, Pulsatile perfusion stimulates the release of endothelial autocoids, J. Appl. Cardiol., 1:215-235.

Rees, D.D., Palmer, R.M.J., Hodson, H.F. and Moncada, S., 1989, A specific inhibitor of nitric oxide formation from L-arginine attenuates endothelium-dependent relaxation, Br. J. Pharmacol., 96:418-424.

Rubanyi, G.M., Romero, J.C. and Vanhoutte, P.M., 1986, Flow-induced release of endothelium-derived relaxing factor, Am. J. Physiol., 250:H1145-H1149.

Speden, R.N., 1985, The use of excised pressurised blood vessels to study the physiology of vascular smooth muscle, Experentia, 41:1026-1028.

White, D.G. and Martin, W., 1989, Differential control and calcium-dependence of production of endothelium-derived relaxing factor and prostacyclin by pig aortic endothelial cells, Br. J. Pharmacol., 97:683-690.

Woldenberg, M.J. and Horsfield, K., 1986, Relation of branching angles to optimality for four cost principles, J. Theor. Biol., 122:187-204.

Zamir, M, 1976, Optimality principles in arterial branching, J. Theor. Biol., 62:227-251.

FLOW STUDIES IN A MODEL OF THE ABDOMINAL AORTA

D. Liepsch, A. Poll, and P.D. Stein*

Fachhochschule und Technische Universität München
Lothstraße 34, D-8000 München 2, W.Germany
*Henry Ford Heart and Vascular Institute, Detroit, Mich. 48202

ABSTRACT

Biofluidmechanic effects play a role in the formation of atherosclerotic plaques, especially near branch points and bifurcations.
As an example, we chose a rigid polyester resin and an elastic silicone
rubber model of the abdominal aorta. The model was prepared from a cast of
a 27 year old woman who died of trauma.
The flow in the model was visualized using a birefringent solution
(vanadiumpentoxide). Zones of flow separation, disturbed flow regions,
coherent structures can be seen clearly. This method can be used for steady
and pulsatile flow, in contrast to the dye method.
Laser-Doppler velocity measurements were carried out using a one component
LDA system. 3D velocity measurements were done simultaneously in a simplified 90°-T-junction.
All the biofluid mechanic factors were studied separately in the models;
factors such as steady and unsteady flow, elasticity of the wall, geometry,
flow rate ratio, and non-Newtonian flow behavior of blood in secondary
flow regions.
At high branch-to-trunk flow ratios (> 0.3), flow separation zones were
observed in the aorta downstream of the renal arteries.
In pulsatile flow, flow separation zones were found at nearly all branch-
to-trunk ratios. With increasing flow (higher Reynold-numbers) and in a
range of normal pulse waves, flow separation almost disappears.
The elasticity of the wall reduces the flow separation effects, in comparison to a rigid wall.
Also non-Newtonian flow behavior can not be neglected. The flow always
formes coherent structures at bifurcations. A totally different velocity
distribution was found compared to a Newtonian fluid with the same viscosity.

Key Words: Biofluid Mechanics, flow visualization, photoelasticity,
           Laser-DopplerAnemometry, Elastic silicon rubber models,
           abdominal aorta.

INTRODUCTION

Atherosclerotic plaques are mostly found in large arteries at bends and bifurcations, where the flow separates from the wall (Ross, 1976; Liepsch, 1986 and 1990). Several studies have demonstrated this already in different simplified and true-to-scale models (Stehbens, 1975; Rodkiewicz, 1981; Ku et al., 1983; Karino et al., 1983). It is assumed that hemodynamic factors play an important role with chemical reactions in forming atherosclerotic plaques and thrombosis (Liepsch, 1989 and 1990; Schmid-Schönbein et al. 1984). Such biofluid mechanic factors are: geometry, pulsatility (unsteady flow), elasticity of the wall and the blood cell membranes, non-Newtonian flow behavior of blood (blood viscosity). We studied these factors separately in rigid and elastic 90°-T-junctions, simulating the left descending coronary artery of the septum branch (Liepsch, 1986). Some results in a rigid elastic model of the abdominal aorta with the renal arteries will be shown in this paper.

MODELS AND METHODS

The used models were a simplified rigid plexiglass and an elastic silicon rubber model. The inner diameter for the abdominal aorta was 10 ± 0.05 mm and for the two renal arteries 5.1 ± 0.05 mm. The model geometry is shown in Figure 1. We also used a true-to-scale 1:1 polyester resin model (Figure 2) and an elastic model with the same geometry. Furthermore we worked also with a true-to-scale plexiglass model (Sabbah et al., 1984) of an abdominal aorta prepared from a cast of a 27 year old woman who died of trauma.
The methods we used were the following:
Flow was visualized by using dyes and also with a photoelastic apparatus and a birefringent solution.
Only single streampaths can be visualized using dyes, whereas with the birefringent solution and the photoelastic apparatus the whole flow field can be visualized also under pulsatile flow. This method was described in detail already (Liepsch, 1974 and 1975; Liepsch et al., 1989). Having localized the length of the flow separation regions, the velocity distribution was measured with an one and two component Laser-Doppler-Anemometer (LDA). We want to report only on the one component measurements in this paper. The principals of the LDA and the experimental set up is already described (Liepsch, 1986; Liepsch et al. 1989).
The model was mounted on an x-y-z-moving table, so that the velocity distribution could be measured over the whole cross section.
An vanadiumpentoxide solution was used for the flow visualization and the photoelastic apparatus. The velocity measurements were done with an aqueous glycerol solution with a dynamic viscosity of Ro = 10 mPas and a density eta = 1155 kg/m^3 and two different non-Newtonian polyacrylamide solutions. One with a representative viscosity of Ro = 4.9 mPas and another with a glycerol Separan (AP 30/45) Dow Chemical mixture with a representative dynamic viscosity Ro = 15.5 mPas and a density eta = 1150 kg/m^3. The last solution has a similar refraction index as the silicone rubber wall of the model (n = 1.413). The elastic model was embedded in an aqueous glycerol solution with the same refraction index as the model wall, so that the beams of the laser were not reflected by the movement of the wall for pulsatile flow. The velocities were measured perpendicular to the plain of the branches in this report. A steady pressure of about 100 mm mercury creates an expansion of the tube diameter of about delta = 0.6 mm for a wall thickness of 1 mm. The pulse wave was obtained by using a heart pump (Reul, 1983).

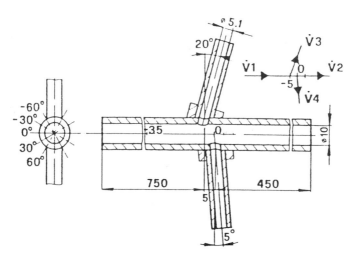

Figure 1 SIMPLIFIED PLEXIGLASS MODEL

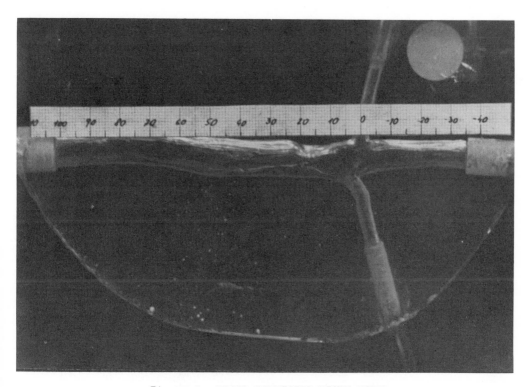

Figure 2 RIGID POLYESTER RESIN MODEL

RESULTS AND DISCUSSION

Figure 3 shows the viscosity versus shear rate for human blood with a hemotocrit of 46 % compared to a polyacrylamid mixture (AP 30/45) to which is added 4 % isopropanol and magnesium chloride to stabilize this solution (Liepsch, 1986). The viscosity is nearly the same as human blood in a shear rate range above 5 1/s.
Whereas the glycerine water solution shows a Newtonian flow behavior, the polyacrylamide mixture shows a shear thinning behavior. We have analyzed this fluid several times and we know that the long chain molecules have a higher elasticity compared to blood. However the flow behavior of such a fluid is much closer to that of blood compared to glycerine water. More details are described in (Liepsch 1974, 1975; Liepsch 1986). From the experiments in a 90°-T-junction we know that the flow behavior of such fluids show large differences in the velocity distribution. This was demonstrated under steady flow conditions with dyes (Liepsch, 1986).
Figure 4 shows the rigid human abdominal aorta of the 27 year old lady at an entrance Reynolds-number 1365 and a branch-to-trunk flow rate ratio of 0.175, that means 17.5 % of the whole flow streams in the branches. Distal to the branches a large separation zone can be seen but no high velocity fluctuations. Whereas at lower Reynolds-numbers and higher branch-to-trunk flow ratio the flow shows larger coherent structures and velocity fluctuations. Further results are reported in (Liepsch et al. 1989). A video tape demonstrates the influence of the pulsatility to the flow separation regions. The amplitude of the pulse wave influences the size of the flow separation. The flow separation region oscillates with the frequency in the same areas observed at steady flow.
Figure 5 shows the velocity distribution for a pulsatile flow in the simplified rigid model. The Reynold-number proximal of the bifurcations was 498. 25 % of the whole flow is directed in each renal artery. The velocity was measured 35 mm proximal to the bifurcation till 15 mm distal of the branches.

The Womersley parameter Alpha = $R \sqrt{omega / nue}$ = 6.52 and the amplitude of the pressure drop was Am 1230 Pa/m. The local velocity profiles were recorded over 14 cycles. The average velocity was always plotted. The figure shows the velocity distribution over a whole pulse cycle at phases omega•t: 0°, 45°, 90°, 135°, 180°, 225°, 170°, 315°, for a sinusoidal wave form. The velocity was recorded over the diameter in the measured plane 0°, 30°, -30°. As already reported the flow shows nearly at every measuring point higher fluctuations for a glycerine water solution compared to the polyacrylamid solution (Liepsch, 1986; Liepsch et al., 1987). Also the flow separation region for glycerine water is larger and is almost reaching till the tube center, whereas for polyacrylamide this region is more located around the wall.
We observed a strong influence of the Womersley parameter, the amplitude of the pressure drop (Pa/m) and the entrance Reynolds-number. We calculated from these measurements the shear stress distribution (Figure 6). It can be seen at 5 mm in the 30° plane that the highest values are not located at the wall. They are found at the border of the separation region and the forward flow. We have observed the highest value with about 5-8 Pa. However it is not shown in this figure. These shear stresses are high enough to change endothelial cell structures and also the membrane of blood cells (Schmid-Schönbein et al., 1984), so that chemical reactions are activated. A counter movement was observed between the forward and backward flow of the main stream and the flow separation region at pulsatile flow.
We also demonstrated that high wall elasticity reduces the backward flow of about 25 % or more changing the diameter of about 5 %. Further experiments in a model of a femoral artery have shown this clearly.
Finally we can say: none of the biofluid mechanic factors can be neglected. The elasticity reduces the reverse flow, the pulsatility creates

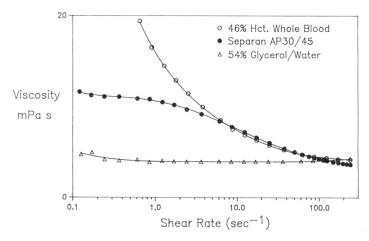

Figure 3   VISCOSITY VERSUS SHEAR RATE FOR HUMAN BLOOD,
SEPARAN AND A 54 % GLYCEROL/WATER SOLUTION

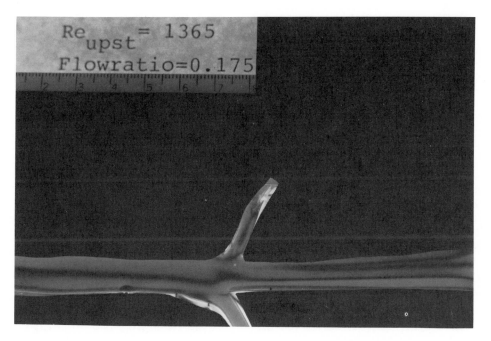

Figure 4   FLOW VISUALIZATION WITH A BIREFRINGENT SOLUTION

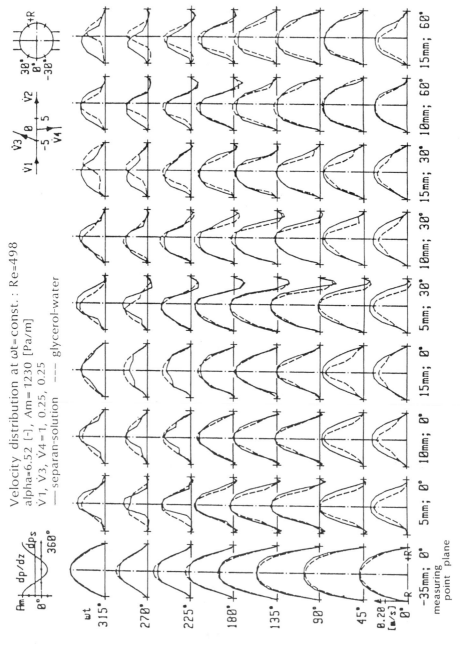

Figure 5 VELOCITY DISTRIBUTION FOR SEVERAL MEASURED POINTS AND PLANES AT PULSATILE FLOW

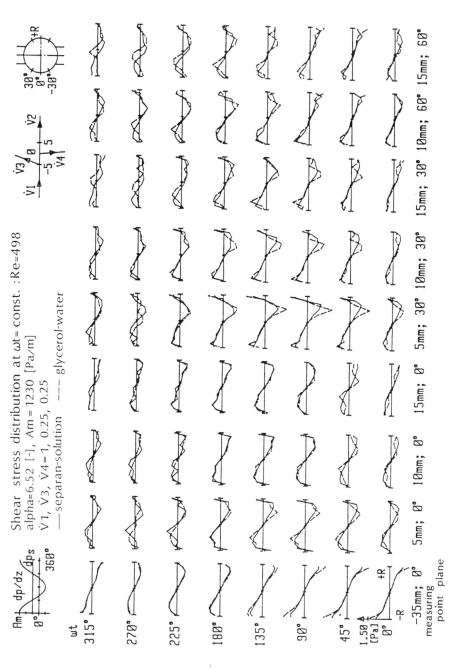

Figure 6 SHEAR STRESS DISTRIBUTION CALCULATED FROM THE MEASURED VELOCITY PROFILES OF FIGURE 5

171

movement; particles are moved out of the flow separation region, and the non-Newtonian flow behavior creates a different flow field in regions where the flow is disturbed. Further detailed studies are necessary in anatomical models for analyzing the flow structures accurately. Therefore we have constructed a 3D-LDA system to measure all three velocity components simultaneously, so that the correct velocity vector can be calculated.

ACKNOWLEDGMENT

The authors would like to thank the Deutsche Forschungsgemeinschaft for supporting this project under contract Li 256-15 and Li 256-19. I would like to thank Joyce McLean and Ingeborg Hopf for preparation of this manuscript.

REFERENCES

Karino, T. and Notomiya, M., 1983, Flow Visualization in Isolated Transparent Natural Blood Vessels. Biorheology Vol. 20: 119-127.

Ku, D.N. and Giddens, D.P., 1983, Pulsatile Flow in a Model Carotid Bifurcation, Atheroclerosis, Vol. 3: 31-39.

Liepsch, D., 1974, 1975, Untersuchung der Strömungsverhältnisse in Verzweigungen von Rohren kleiner Durchmesser (Koronararterien) bei Stromtrennung. Diss TU München und VDI-Berichte 232, 423-441.

Liepsch, D., 1986, Flow Studies in Models of the Human Vascular System. Fortschrittsberichte VDI, Reihe 7: Strömungstechnik Nr. 113, VDI-Verlag.

Liepsch, D., 1986, Flow in Tubes and Arteries - A Comparison, Biorheology, Vol. 23: 395-433.

Liepsch, D., Poll, A., and Moravec, St., 1987, Flow Studies in True-to-scale Models of Human Renal Arteries. in: Role of Blood Flow in Atherogenesis, ed. Yoshida, Tokyo: Springer 1-96.

Liepsch, D., Poll, A., Strigberger, J., Sabbah, H.N. and Stein, P.D., 1989, Flow Visualization Studies in an Mold of the Normal Human Aorta and Renal Arteries. Journal of Biomechanical Engineering Vol. III: 222-227.

Liepsch, D., 1989, Blood Flow in Large Vessels. An Interdisciplinary Conference. Proceedings of the 2nd Internat. Symposium on Biofluid Mechanics and Biorheology. June 25-28, 1989, Munich.

Liepsch, D., 1990, Blood Flow in Large Arteries. Applications to Atherogenesis and Clinical Medicine. Monographs on Atheroclerosis Vol. 15, Karger, Basel.

Reul, H., 1983, Hydraulic Analog Model of the Systemic Circulation-Designed for Fluid Mechanical Studies in the Left Heart and Systemic Arteries. Adv. Cardiovasc. Phys. Vol.: 43-54.

Rodkiewicz, C.M., 1981, Arteries and Arterial Blood Flow. Biological and Physiological Aspects. CISM Courses and Lectures, No. 270, Springer, Wien.

Ross, R., 1986, The Pathogenesis of Atherosclerosis-An Update. New England Journal of Medicine, Vol. 314: 488-500.

Sabbah, H.N., Hawkins, E.T. and Stein, P.D., 1984, Flow Separation in the Renal Arteries, Atheroclerosis, Vol. 4: 28-33.

Schmid-Schönbein, H. and Naumann, A., 1984, Fluiddynamische, zellphysiologische und biochemische Aspekte der Atherogenese unter Strömungseinflüssen. Rheinisch-Westfälische Akademie der Wissenschaften, Vortrag N 331, Westdeutscher Verlag.

Stehbens, W.E., 1975, Flow in Glass Models of Arterial Bifurcations and Berry Aneurysm at Low Renolds-Numbers. Quantitative Journal of Experimental Physiology, Vol. 60: 181-192.

THREE DIMENSIONAL SHEAR STRESS DISTRIBUTION AROUND SMALL

ATHEROSCLEROTIC PLAQUES WITH STEADY AND UNSTEADY FLOW

Takami Yamaguchi, Atushi Nakano, and Sotaro Hanai

Department of Vascular Physiology
National Cardiovascular Center Research Institute
Osaka 565, Japan

ABSTRACT

Results of a numerical model study of the flow field and the wall shear stress distribution patterns in the vicinity of early atherosclerotic plaques were examined using a computational fluid mechanics method. It was shown from the computation that unsteadiness significantly affected the wall shear stress distribution in the neighborhood of minimal deformations of the arterial wall. It was also shown that upstream condition did not severely influence the downstream flow structure. In other words, local flow structure was mainly determined by the local geometry. Complexity of flow field invoked by the irregularity of vascular walls was stressed and the necessity of fluid mechanical studies of the blood flow under three dimensional unsteady flow conditions was pointed out. The computational fluid dynamical method incorporating supercomputers was shown to be very useful to understand the fine structure of those flow fields.

KEYWORDS

atherosclerosis, early plaque, blood flow, macromolecular transport, shear stress, computational fluid mechanics

INTRODUCTION

Many studies indicated that the so-called low wall shear stress regions may be specific arterial locations prone to atherosclerosis in human arteries (Caro et al. 1971, Yoshida et al., 1988). These regions were usually called so because low wall shear stress was expected from two dimensional steady flow studies assuming normal intact arteries without any lesions. However, once atherosclerosis is initiated,   it is

Biomechanical Transport Processes, Edited by F. Mosora *et al.*
Plenum Press, New York, 1990

very likely that the wall shear stress distribution pattern changes markedly. This seems particularly important when atherosclerotic wall irregularities develop and subsequently alter arterial geometry as a whole. It is thought that the three dimensional unsteady nature of the blood flow become more important under these conditions.

We have been trying to estimate the flow field, pressure, and wall shear stress distribution patterns in regions surrounding realistic three dimensional atherosclerosis plaque models using a three dimensional computational fluid dynamical method (Yamaguchi and Hanai, 1989). In the present study, some results of a numerical model study of the flow field and the wall shear stress distribution patterns around early atherosclerotic plaques are discussed.

METHOD

Models. The model is based on the well-known shape of the real early atherosclerotic plaque. Early atherosclerotic plaque is known to grow in semilunar shape in cross-section unilaterally in the lumens of arteries. Along the longitudinal axis, the atherosclerotic plaque can be considered smoothly rising from normal intimal surface.

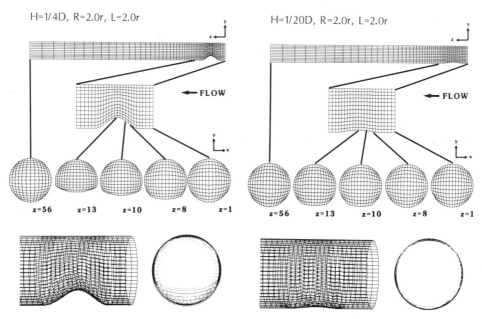

Fig.1 Model plaques and their corresponding computational meshes. See text for explanations. Direction of the flow is from right to left.

We used three models of early atherosclerotic plaque with different heights, namely 1/4D, 1/8D, and 1/20D, residing on the luminal surface of a circular pipe. The 1/4D and 1/20D models are shown in Fig.1. Their cross section was an eccentric circle of which radius was 2.0 times that of the pipe. At the longitudinal section, the luminal surface of the model plaque had a curve of the cosine bell of which streamwise base length was 2.0 times the radius of the pipe. Since we were interested mainly in the flows in the vicinity and downstream of the model plaque, the model plaques were positioned close to the inlet of the model artery with their peak point at three diameters downstream from the inlet. All three types of models of atherosclerotic plaque had an identical streamwise base length and the phase angle of the cosine bell.

Computational mesh defined for the models are also illustrated in Fig.1. Throughout this study, dimension of the mesh was 17 in horizontal or x direction, 13 in vertical or y direction, and 56 in streamwise or z direction, thus giving a total of 12,376 nodes in the computational mesh. No symmetry was assumed in the definition of the mesh.

In the following figures and their explanations, by the word "bottom", the wall containing the model plaque is designated. "Left" means the left hand side wall, and "Right" means the right hand side wall as viewed from upstream. "Top" is the opposite wall to the wall containing the model plaque.

Computation. The three dimensional Navier-Stokes equations for incompressible Newtonian fluid flow as shown below, were solved.
Mass conservation;

$$\frac{\partial u_i}{\partial x_i} = 0 \tag{1}$$

momentum transport;

$$\frac{\partial u_i}{\partial t} + \frac{\partial u_j u_i}{\partial x_j} = -\frac{1}{\rho}\frac{\partial p}{\partial x_i} + \nu \frac{\partial^2 u_i}{\partial x_j \partial x_j} \tag{2}$$

where, u_i is velocity and x_i, x_j is the cartesian coordinate, p is pressure, ρ is density, and ν is kinematic viscosity. The Reynolds number is defined as following.

$$Re = \frac{LU_0}{\nu} \tag{3}$$

where L is the diameter of the artery model, and U_0 is the average inflow velocity. In the unsteady flow, the frequency parameter α defined as following was used.

$$\alpha = \frac{d}{2}\sqrt{\frac{\omega}{\nu}} \tag{4}$$

where ω is an angular frequency ($\omega = 2 \cdot \pi \cdot f$, f is frequency in Hz).

The boundary conditions for the momentum equations were: steady and unsteady fully developed inflow, non slip rigid walls and zero pressure at the outlet cross section. In the present study, the frequency parameter α was approximately 25, with time course of central velocity shown in Fig.2.

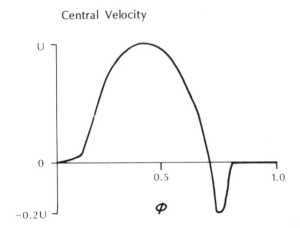

Central Velocity

Fig.2. Velocity waveform used in unsteady calculations. U is the maximum central velocity, and **f** is the phase of the waveform.

In the unsteady case, calculation was carried out for at least 2 cycles and the first cycle result was used as the initial conditions of the second cycle. The computer used was an Alliant FX-40 system with 96 MBytes of main memory. The computation code used was the SCRYU system by Software Cradle Ltd. This system performed a finite volume integration of the Navier-Stokes equation by so-called SIMPLE algorithm with the body-fitted coordinate (BFC) system. Magnitude of the wall shear stress was obtained from the computed flow field and mapped by color graphics.

RESULTS

The secondary flow velocity vector maps at several selected cross sections in the model of 1/4D height under an unsteady flow condition are shown in Fig.3. Even at early phases, f=0.1 and 0.2, when virtually no z-direction flows in inlet cross-section existed, remaining vortices from the previous cycle were found. When the inlet flow velocity increased, at phases of =0.3 and after, remarkable secondary flow was invoked. At phases f=0.4 to 0.6 forward flow velocity kept its maximum. During these phases, secondary flow velocity patterns seemed to be rather stable, that is, the whole field resembles that of the corresponding steady flow. From the phase of f=0.6, the forward flow velocity started to decrease, and around f=0.7-0.8, the inlet flow reversed. It can be seen from the vector maps, secondary flow velocity

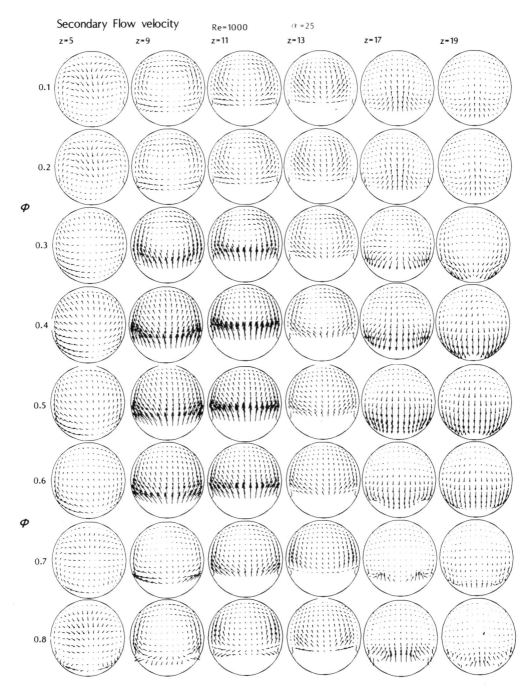

Fig.3 Secondary velocity vector maps of some selected cross sections in the vicinity of the model plaque (1/4D Height, Re=1000, α =25.) φ is phase of the waveform, shown in Fig.2.

patterns were markedly different from the previous phases, particularly near the downstream side surface of the model plaque.

Corresponding wall shear stress patterns under the same condition are shown in Fig.4. In Figs.4 and 5, the direction of the main flow is from right to left. Each strip comprising the figures represents a color map of spatial distribution of the shear stress on a luminal surface of the model at a phase. Intervals between phases were 1/10 of the whole cycle. Colors represent magnitude of the wall shear stress, and the average wall shear stress value of Poiseuille flow are referred in the middle of the color-bar shown in the right hand side of the figure. By the word "Poiseuille" in the color bar, an average wall shear stress value of fully developed steady flow in a straight pipe, with the same peak velocity as the peak central velocity is designated.

A characteristic wall shear stress distribution pattern was shown in the vicinity and downstream of the model plaque. The low wall shear stress pattern should be considered as resulting from a complex system of vortices even under steady flow conditions, though it is not shown here. The high shear stress zones also showed a complex pattern as illustrated. The highest shear stress points were on the lateral edges of the plaque at $\varphi = 0.4$. There were a few high shear stress spots around the model plaque, particularly on the lateral and top walls.

Because the calculation illustrated here was for the second cycle, a wall shear stress distribution pattern probably due to large vortices was found to remain even before the main flow started to increase, just as observed in velocity vector maps in Fig.3. During systole, the wall shear stress distribution resembled that of steady flow. As was pointed out in the foregoing, a characteristic low wall shear stress structure downstream of the model plaque could be observed. After the main flow reached its peak, in the decelerating phase, large low shear regions appeared on the side walls in addition. At the later phases of the cycle, when the main flow decreased and reversed, complicated wall shear stress distribution patterns appeared. Complexity of the wall shear stress distribution was remarkable especially at the phases of $\varphi = 0.7$ to 1.0. There were high shear stress areas on the bottom wall just downstream of the plaque at those reversed flow phases. The low wall shear stress area, which was clearly found in steady calculations, was completely washed out during the diastole, and a relatively high shear area could be observed where the shear stress was considered to be low under steady flow conditions. In other words, those so-called low wall shear stress regions found under steady flow conditions could be the sites where the shear stress changed most during the whole cardiac cycle under unsteady flow conditions.

When the model atherosclerotic plaque was very low in its profile, the pattern of low wall shear stress area under steady flow condition did not show significant spatial variation (not shown here). Although it still showed a characteristic twin vortices pattern, and the lateral high shear stress zone was resident under steady flow conditions, an extent of affected area around a small plaque was limited in general. However, dramatic changes were found when the main flow

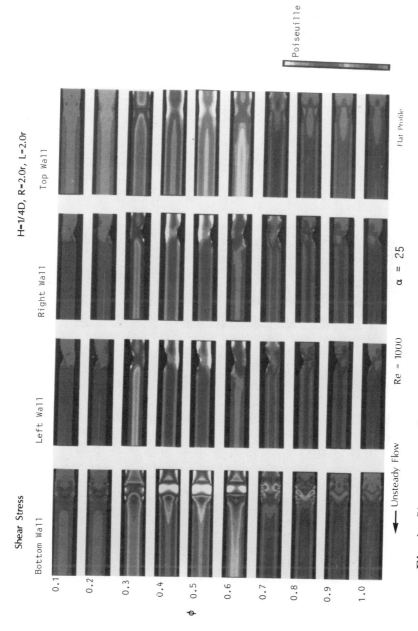

Fig. 4. Shear stress distributions in the vicinity of the model of an early atherosclerotic plaque. Height of the model: 1/4 D, Reynolds number=1000, frequency parameter α = 25.

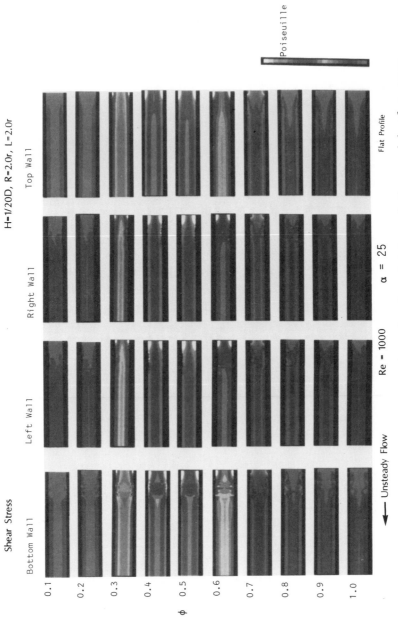

Fig. 5. Shear stress distributions in the vicinity of the model of a very early atherosclerotic plaque. Height of the model: 1/20 D, Reynolds number=1000, frequency parameter α = 25.

became unsteady. Wall shear stress distribution in the vicinity of a model of very early atherosclerotic plaque is shown in Fig.5. Its height was 1/20th of the main vessel diameter. Other factors were the same as aforementioned.

As shown in Fig.5, the wall shear stress distribution pattern, in the systolic phase, and even more in the diastolic phase, shows an extremely complicated pattern in which high shear area and low shear area reside next to each other, changing drastically in time. This spatial distribution patterns resembles the complex shapes of fatty streaks or sudanophillic lesions in young human aorta. From the steady flow calculations, it was extremely difficult to imagine that such kind of complex wall shear stress distribution patterns were produced by an unsteady flow around such nearly flat lesions. Another remarkable phenomenon, commonly found in these unsteady flow fields, is that unsymmetrical upstream flow did not influence the flow field downstream of such a small deformation. Even when the inlet flow condition showed remarkable distortion in spatial distribution, the flow around and downstream of the model plaque became symmetrical. In other words, the flow was rectified by the symmetrical geometry of the flow field.

DISCUSSION AND CONCLUDING REMARKS

Our results so far suggested that complex and profound changes in wall shear stress distribution could be induced by the appearance of an early atherosclerotic lesion particularly when the flow is unsteady.

Human precursory or early atherosclerotic lesions are known to develop at particular sites of the arterial wall. It is not known what happens as the very first event in atherogenesis. This could be a complicated problem in which a number of biological and physical phenomena including the blood flow may contribute. Once a very small distortion of the wall occurs, it is highly conceivable that regions surrounding the earliest plaque will be occupied by complex bound vortices as shown in the present study. Since human atherosclerosis needs a very long period to grow, the developed lesions should be considered as a net result of mutually interrelating phenomena, such as physical influence of the blood flow, biological reactions of the vessel wall, and biochemical behavior of blood constituents. In this context, it should be kept in mind that the blood flow may completely change during the course of the development of the atherosclerotic plaques.

From the observations presented here, the following findings should be pointed out; Firstly, unsteadiness affected very much the wall shear stress distribution when the arterial wall has minimal deformation, particularly in diastole. Secondly, upstream conditions did not influence the flow structure downstream. In other words, local flow structure is mainly determined by local geometry. In conclusion, complexity of vortex system around any kind of irregularity of vascular walls, either natural or pathological, must be stressed when any effects of blood flow on the vascular wall is discussed. It was shown in the present study that the fluid mechanics of the blood flow should be studied under three dimensional unsteady flow conditions. In order to carry out these studies, the computational fluid dynamical method incorporating supercomputers would be of immense help to understand the fine nature of the structure of the flow field.

ACKNOWLEDGMENT

A part of this study was supported by the Grant-in-Aid from the Ministry of Education, Japan, No.01571243, and was also supported by a grant from the Japan Foundation for Health Sciences, 1988.

REFERENCES

C.G.Caro, J.M.Fitz-Gerald, and R.C.Schroter, 1971, Atheroma and arterial wall shear - Observation, correlation and proposal of a shear dependent mass transfer mechanism for atherogenesis, Proc. Roy.Soc.Lond., B177 :109-159.

Y.Yoshida, T.Yamaguchi, C.G.Caro, S.Glagov, and R.M.Nerem (Eds.), 1988, "Role of Blood Flow in Atherogenesis", Springer-Verlag, Tokyo.

T.Yamaguchi and S.Hanai,1989, Atherosclerotic plaque and the three dimensional distribution of wall shear stress - A model and numerical study. in "The Proceedings of the 2nd International Symposium on Biofluid Mechanics and Biorheology," Munich pp733-744.

HEMORHEOLOGY AND MICROCIRCULATION

SYNERGETICS OF O2-TRANSPORT IN THE MAMMALIAN MICROCIRCULATION:

COOPERATIVITY OF MOLECULAR, CELLULAR, KINEMATIC AND HEMODYNAMIC FACTORS

IN "RECTIFIED BLOOD FLOW"

Holger Schmid-Schönbein

Department of Physiology
RWTH-Klinikum
D-5100 Aachen, W. Germany

ABSTRACT

The transport efficacy of the mammalian cardiovascular system for oxygen (energetic expenditure in Joule/Mol O_2-transferred) is unusually high both in comparison to that prevailing in hemoglobin-plasma mixtures and in comparable suspensions of avian (nucleated) red blood cells. This can be reduced to the "rectification process" of the "fluid-drop-like" mammalian erythrocytes induced by the shear conditions prevailing in rapidly perfused microvessels. All aspects of the rectification process (orientation, deformation, membrane tanktreading and passive axial drift) are interpreted as "synergetic processes", based on the principle of energy minimization of multiphase flow in Stokes flow. The latter is associated with steep gradients of driving pressure (acting along the vessel axis) and of shear stresses (acting across the vessel). The causes of red cell rectification (fluidity of RBC cytosole and of RBC membrane as a fluid-tetra-laminate), as well as some of the functional consequences are outlined under aspects of local and global energy minimization principles and are interpreted as the spontaneous generation of an ordered state ("dissipative structure" in the sense of PRIGOGINE).

KEYWORDS

CAPILLARY FLOW - CHAOS THEORY - DISSIPATIVE STRUCTURE - ERYTHROCYTE
"DEFORMABILITY" - MEMBRANE TANKTREADING -OXYGEN TRANSPORT - RED CELL
"RECTIFICATION" - SYNERGETICS

INTRODUCTION: SYNERGETICS OR JOINT EFFORTS OF BIOCHEMICAL AND BIOPHYSICAL
FACTORS IN CARDIOVASCULAR O_2-TRANSPORT

The mammalian blood is a multiphase fluid exhibiting marked deviations from Newtonian behaviour. In vivo, its composition and thence its flow characteristics are highly variable (review see Schmid-Schönbein, 1988), depending on local chemical (genetic and metabolic), fluid-mechanic and thence hemodynamic variables (Schmid-Schönbein, 1990a). Hemorheological research, in having uncovered various previously hidden s y n e r g i s m s (and their breakdown in diesease (Schmid-Schönbein, 1990a)) has paved the way for more comprehensive biomedical concepts The experimental work in this field can be greatly aided by theoretical cross-fertilization from "chaos theory" (Prigogine, 1986;Libchaber, 1982) and synergetics (Haken, 1978).

In biology as in fluid-dynamics the concepts of synergetic have opened our eyes for so-called "dissipative structures", i.e. temporally and spatially ordered, cooperative multiphase systems resulting from the (more or less stationary) flow of energy and matter. Many forms of biological "order" in multiphase systems are now recognized as self-sustaining, autopoetic and cooperative structures; they can generally be explained as result of spontaneously occuring dynamic processes in circumscript systems kept from physical and chemical equilibrium. The apparently paradox term "dissipative structure" describes the "morphogenetic" influences of complex kinematic processes associated with phase transitions, which only become "manifest" if a continuous flow of energy and matter creates a "boosted steady state" (Schmid-Schönbein, 1990a). PRIGOGINE's pioneering work has opened the road to a broader thermodynamic understanding of the well known interdependence of function and structure on all levels of biological activities.

Lorenz's (Lorenz, 1963) approach to understanding the synergetics of u n s t a b l e aerodynamic motions (as they are found in metereology) has generated criteria for factors promoting the kinematics and dynamics of many s t a b l e fluid-mechanic phenomena including those found in hemodynamics. As shown in Schmid-Schönbein (1990a), it is helpful to consider biofluid-dynamic phenomena under the aspects of "attractor processes", i.e. as the consequences of mechanical, chemical and electrostatic interactions of highly different energy flux. In this context, however, it must be kept in mind, that the circulation of blood not only serves transport purposes (supply of nutrients, mediators and oxygen and removal of metabolic waste products) but entirely different within the realm of host-defense and hemostatic processes (where creeping flow is mandatory as discussed in Schmid-Schönbein, 1990a). As will be shown subsequently, the high flow situations associated with maximum oxygen consumption ("functional hyperemia) prove to be characterized by high stability, which (from a rheological standpoint) as an "energized" motion far from thermo- and fluiddynamic equilibrium.

ENERGETICS OF BLOOD FLOW AS A DEFORMATION PROCESS IN MICRO-TUBE NETWORKS

As shown by Schmid-Schönbein (1990b) the motion of the blood in the vascular bed is a continuous deformation process of a composite fluid, involving frictional work most of which, however, is performed in the exchange vessels. As a rough estimate of the true extent of this work, one can calculate the energy associated with this deformation process as the product of flow rate (volume deformed in unit time) and the pressure difference. If we assume a flow rate of 5 L/min in 60 beats or 83.3. cm^3/sec) $8.3 \cdot 10^{-5}$ m^3/s) in a human, the power required for this process (neglecting accelerational work and work in the minor circulation) is

$$E = \Delta P \times \dot{Q} \quad (N/m^2 \cdot m^3/sec)$$

$$= 1,33 \cdot 8,33 \ (Nm \ s^{-1}) \quad = 1,104 \ Nm/s$$

which at a heart rate of 1/sec corresponds to a work of 1,104 Joule.

Table 1. Comparison of transport efficacy of human whole body (at rest and under maximum physical exercise) and that for the theoretical case of oxygen transport in plasma

	Normal human body Rest	Normal human body $\dot{V}_{O_{2max}}$ -condition	Theoretical case convective transport of plasma 10 L/min at 100 mgHG
C.O. ΔP Work+ (sec)	$8.3 \cdot 10^{-5}$ m^3/s $1.33 \cdot 10^4$ N/m^2 1.1 Nm	$3.3 \cdot 10^{-4}$ m^3/s $1.33 \cdot 10^4$ N/m^2 4.4 Nm	$16.6 \cdot 10^{-4}$ m^3/s $1.33 \cdot 10^4$ N/m^2 2.2 Nm
O_2-Transport (sec)	5 ml/s $= 2.2 \cdot 10^{-4}$ Mol	100 ml/sec $4.4 \cdot 10^{-3}$ Mol	0.8 ml $3.4 \cdot 10^{-5}$ Mol
Transport Efficiency (J/Mol)	$\dfrac{1.1 \text{ J}}{2.2 \cdot 10^{-4} \text{ Mol}}$ 5 kJ/Mol O_2	$\dfrac{4.4 \text{ J}}{4.4 \cdot 10^{-3} \text{ Mol}}$ 1 kJ/Mol O_2	$\dfrac{2.2 \text{ Nm}}{3.4 \cdot 10^{-5} \text{ Mol}}$ 64 kJ/Mol O_2

+ Work = Power x Time ; Power = C.O x ΔP

    If we assume a net oxygen uptake of 300 ml/min (= 5 ml/s or 2.2 · 10^{-5} Mol/s) we can calculate the "cost" of resting oxygen transport as roughly 5 kJ/Mol. Knowing that at rest there is only 25 % average oxygen extraction, one can attribute 3/4 of this work as the cost of "luxury" perfusion. This becomes evident when one calculated the limiting case of maximum oxygen uptake under conditions of physical exercise, namely 20 L/min in 180 beats per min (Stroke volume 111 ml) and oxygen transport of 3000 ml/min = 60 ml/s. The work performed in 1 sec is equivalent to 4,429 Nm (assuming constant mean arterial pressure), the O_2 transported equals 2.2 · 10^{-4} Mol/s, the "cost" dropping to roughly 2 kJ/Mol (see Table 1).
    The entropy flux associated with the distribution of oxygen from the lungs to the exchanging capillaries can best be estimated if one compares it – in an experiment of thought – to the cost of oxygen transport for the theoretical case of pumping the cell free plasma *) and the maximum possible amount of oxygen, physically dissolved in it: (assuming a BUNSEN-coefficient @$_{o2}$ of 2.4 ml/L · atm = 0.032 ml/L · Torr) this amounts to 1.6 ml/min at cardiac output of 10 L/min and thence to 22.16 Nm for 0.027 ml O_2/sec at a cost of 1800 kJ/Mol.

---

*) As was elaborated by Gaehtgens et al. (1979) and can be seen from fig. 13 in Schmid-Schönbein (1990), there is a linear hematocrit/conductance relationship in entire vascular beds (between about 40 and zero percent red cell volume fraction). The surprisingly flat regression line shows by the presence of red cells in physiological concentrations the whole body conductance is only reduced to about one half of that can be found by extrapolation to zero hematocrit (flow of plasma): in other words the presence of 40 % RBC only increases mechanical work of pumping RBC by a factor 2.

The biochemical basis of enhanced O_2 -transport c a p a c i t y is
obvious (reversible binding of oxygen to haemochromes, here especially that
to hemoglobins) augments the oxygen b i n d i n g process of blood over
that of plasma by roughly a factor of 65 (210 ml/L vs 3.2 ml/L at 100
Torr). However, since the O_2-binding occurs to complex colloidal structure
required for caging Fe^{++} as found in the hemochromes, it is surprising how
small the effect of the high concentrations (160 g/L) of this proteins is
on the frictional energy dissipation and thence e n e r g e t i c s of
blood motion.

Table 2. Comparison of transport efficacy of mammalian and avian red cell
plasma mixtures, of plasma and of plasma-hemoglobin mixtures in individual
nutritive capillaries.

	ΔP $(10^3$ Pa)	Work $(10^{-10}$ J)	$\dot{V}_{O2}$ $(10^{-12}$ Mol/s)	Convective oxygen conductance $(10^{-10}$ Mol/Pa)	Transport Expenditure (J/J)
Mammalian RBC $H_T = 0.3$ $H_T/H_D = 0.7$ $\eta = 1.375 \cdot 10^{-3}$ Pa·s	2.66	12.63	4.06	15	$6.9 \cdot 10^{-4}$
Free hemoglobin in plasma: 0.096 L/L $\eta = 2.5 \cdot 10^{-3}$ Pa · s	4.8	22.8	2.84	5.9	$1.74 \cdot 10^{-3}$
Avian RBC $H_T = 0.3$ $H_T/H_D = 1.0$ $\eta = 5.0 \cdot 10^{-3}$ Pa · s	9.54	45.3	2.93	3.07	$3.4 \cdot 10^{-3}$
Cell free plasma $\eta_0 = 1.25 \cdot 10^{-3}$ Pa $(O_2)_{phys} = 0.133$ Mol/m^3	2.39	12.49	0.07	0.26	$4.42 \cdot 10^{-2}$

There is a substantial "improvement" of transport efficacy in non-
nucleated mammalian RBC over that found in nucleated avian RBC. Again
taking data from Gaehtgens, as recalculated by Schmid-Schönbein (1990) one
can calculate convective oxygen conductance (Mol/Pa) and transport
expenditure by calculating pressure drop and work associated with perfusion
of a 8 μm exchange capillary. In this calculation, the effect of axial
drift of RBC (as expressed by the FAHRAEUS-factor (H_T/H_D, see Schmid-
Schönbein 1988a and Schmid-Schönbein 1990b) was also considered. as can be
seen, the "invention" of the non-nucleated mammalian RBC has, however,
created a "multiphase system" resulting from the "packaging" of hemoglobin
(and oxyhemoglobin) into non-nucleated, highly flexible erythrocytes. This
"trick" conveys unique t r a n s p o r t features to the f l o w i n g
blood resulting in marked amplification of transport efficiency by coherent
behavior of "rectified" erythrocytes (see Table 2).
The efficiency of this transport system reflects itself in the
exceedingly high apparent fluidity of suspensions of mammalian RBC in
microscopic tubes. Here, the relative apparent fluidity is almost 1.0 (see
Schmid-Schönbein 1988a and 1990b) in vessels with diameters (< 10 μm) equal
to or smaller than that of resting RBC. This superfluidity is reminiscent
of that found for electrons in superconductors and for liquid Helium[4] as a
superfluid, since it is based on the functional "segregation" of a normal
fluid (normal helium, plasma) and a superfluid (helium[4], hemoglobin
contained in RBC membranes (Landau and Lifschitz 1969). The present report
is based on purely phenomenological considerations ("hematokinematics",

v.i.) and some net fluid dynamic considerations, for reasons of space restraint, rigorous thermodynamics analysis will have to be omitted but will be presented shortly elsewhere (Schmid-Schönbein, H., Biorheology, manuscript in preparation).

SYNERGETIC CONSIDERATIONS OF RBC-DEFORMATION AS A RESPONSE TO SHEAR STRESSES IN STOKE's FLOW IN NARROW CYLINDRICAL CONDUITS

In ordinary fluids, the convective transport is associated with irreversible deformation of fluid elements, potential energy inducing incoherent motion of molecules (and/or) particles with short range organization (Landau and Lifschitz, 1969). The thermal energy dissipation (in the continuous phase Newtonian fluids) is directly proportional to the r a t e of deformation following from the forces acting and the geometry of the perfused conduit. Many textbooks of biophysics and physiology fail to underline the fundamental fact that in Poiseuille flow the maximum entropy generation is near the wall, whereas the axial fluid laminae move fast but with very little frictional energy dissipation (see fig. 1).

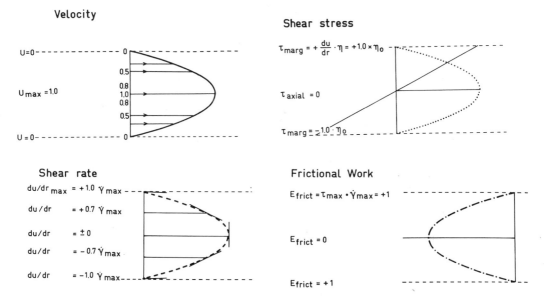

Fig.1. Schematic representation of the distribution of the velocities (u = f (r)), the velocity gradients, the shear stresses and the frictional energy dissipation in ideal Poiseuille flow. Note that there exist a gradient of frictional work i n v e r s e to the gradient of forward displacement: axial parts of the flowing medium are moving in a kind of energy furrow (see Fig. 2).

Blood behaves strikingly non-Newtonian, a fact reflected in the dependence of energy dissipation on the diameter (and thence on the ratio of characteristic cell dimension to tube radius) and on the magnitude of driving forces (and thence on the pressure, shear stress and shear stress distribution) acting along it. In essence, the effect of RBC on the energy dissipation associated with shearing of plasma is highly non-linear, decreasing with increasing s h e a r s t r e s s gradients ($d\tau/dr$) normalized for the "thickness" h of the deformed RBC (v.i.)

Shear Stresses Gradients as "Ordering" Principle

As predicted by the EINSTEIN-STOKES equation, the presence of a (chemically dissolved or mechanically dispersed) phase attenuates the net shearing motion in the continuous phase for any given energy gradient, thus reducing "fluidity" of the dispersion over that of the continuum and/or the

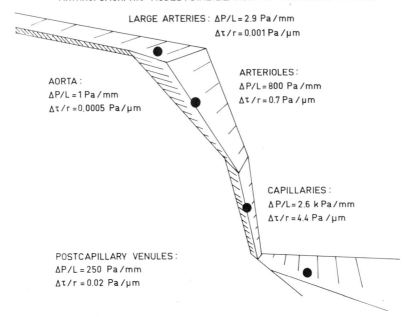

RBC MOTION DOWN THE MICROVASCULAR BED:
ANTHROPOMORPHIC MODEL: STABILIZATION IN "ENERGETIC FURROW"

LARGE ARTERIES: $\Delta P/L = 2.9$ Pa/mm
$\Delta \tau/r = 0.001$ Pa/µm

AORTA:
$\Delta P/L = 1$ Pa/mm
$\Delta \tau/r = 0.0005$ Pa/µm

ARTERIOLES:
$\Delta P/L = 800$ Pa/mm
$\Delta \tau/r = 0.7$ Pa/µm

CAPILLARIES:
$\Delta P/L = 2.6$ kPa/mm
$\Delta \tau/r = 4.4$ Pa/µm

POSTCAPILLARY VENULES:
$\Delta P/L = 250$ Pa/mm
$\Delta \tau/r = 0.02$ Pa/µm

Fig.2. Schematic representation of the pressure gradients (pressure difference for a given length) and the shear stress gradients (fall in shear stress for a given distance from the tube wall) in the different parts of the macro- and microvasculature. The motion of the suspended phase can be compared conceptually to that of a ball rolling down a set of furrows which increasing steepness in both forward direction and lateral direction. Following a principle of energy minimization, the suspended phase is "stabilized" in the center of the "furrows" used here as an analogue to the cardiovascular system.

averaged velocity of fluid motion. The magnitude of this effect is proportional to the volume fraction of the dispersed phase and the shape of the molecules or particles; in multiphase Poiseuille flow, the effect cannot be predicted from the volume fraction of the stagnant fluid if particles are not evenly distributed across the tube axis. In emulsions or suspensions of deformable particles, the shape factor is not a constant.

Furthermore, the viscous drag ($\tau \times A$) acting on the surface of ordinary particles induces orbiting motions (see Goldsmith and Mason, 1967), thus increasing their hydrodynamically effective volume fraction (if they are not spherical in shape). For all these reasons, Poiseuille flow of dilute suspensions, where preferred orientation and/or axial drift (induced by the shear stress gradient v.i.) may in part reduce the viscous energy dissipation in particle suspensions (Goldsmith and Mason, 1967; Jeffrey, 1922). The shear dependent effects described dominate the pattern of fluid element deformation and motion ("hemato-kinematics") of the highly flexible mammalian RBC when subjected to high driving pressures in narrow cylindrical conduits. Even though many details of the deformation process are not fully understood, a phenomenological theory for "ordering" effects by the shear stress generated in RBC-suspensions can now be presented. The presented synergetic interpretation of shear induced phase separation in red cell suspensions in microvessels is merely based on a principle of energy minimization; whether or not a comprehensive microscopic theory evolves from these considerations remains to be seen.

Our analysis has to start from the phenomenon of "red cell membrane tanktreading" as the guiding principle of hematokinematics in all, especially in microscopic tubes. It is associated with continuous viscous energy dissipation and "non-conservative" response of membrane and cytosol. The most surprising phenomenon in this context is the "steadiness" of the deformation despite the fact that the RBC are undergoing a continuous "membrane tanktreading motion". A novel membrane model had to be developed (see contribution Grebe and Schönbein), explaining the continuous, regular and remarkably stationary orientation of RBC in tube flow (Gaehtgens and Schmid-Schönbein, 1982) and viscometric flow. The proposed "fluid quatrolaminate" model of the membrane incorporates all requirements to explain the experimentally established facts about the RBC behavior, which we paraphrase as the combination of "resilience", "compliance" and "mechanical stability" (Schmid-Schönbein and Teitel, manuscript in preparation).

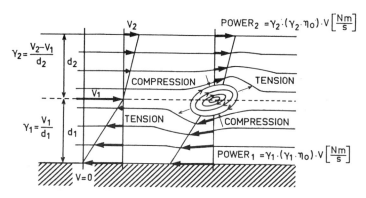

Fig. 3. Schematic depiction of the distribution of the forces, shear stresses, viscous drags and the power associated with the flow of an elongated oriented ellipsoidal mammalian erythrocyte termed "rectified RBC", assuming a non-linear shear field (as it occurs in Poiseuille flow). Note that the compressive and tensile forces, the shear rates (and thence the shear stresses and the viscous drag) are l o w e r at a greater distance from the wall (v = o). For this reason, there is a gradient of power associated with the motion of a "rectified" erythrocyte in Poiseuille flow, likely to displace it away from the wall and preventing its movement towards the wall.

Phenomenologically, orientation is associated with passive axial drift of RBC, thence as the motion of suspensions of highly shear compliant RBC is induced. They "escape" from the marginal areas associated with high frictional energy dissipation to the axial core with low entropy generation. We postulate this to be caused by the potential gradient caused by the steep gradient of shear stresses in rapidly perfused microvessels. A first phase transition occurs in the membrane, causing it to acquire fluid properties and allowing its orbiting motion around the fluid cytosol (much like that of interfaces of ordinary fluid droplets in shear fields). One also observes a transition of a l l cells into stable orientation, i.e. a collective, highly cohaerant particle orientation (Schmid-Schönbein, 1984).

In the whole blood element subjected to flow, self organization into a kind of "dissipative structure" occurs: the shear stresses generated by the mutual gliding of fluid elements induce an order by "enslaving" the motion of the individual "fluid drop like" red blood cells. By this mechanism, a second phase transition occurs, separating the highly fluid, newtonian plasma from the less fluid, yet "rectified" RBC moving in the axial core of high velocity but low velocity gradients (and thence less energy dissipation for any given forward motion).

The shear stresses generated by pressure gradients thus play a dual role:

1) They induce microscopic "order" termed ("rectification") by inducing coherent, long range cooperativity of the elements in the individual RBC with the effect of reducing the entropy flux of a unique transport system.

2) They induce a macroscopic order measurable as a marked increase in velocity (velocity gradient, apparent fluidity) for the given energy dissipation as the ultimate cause of the high transport efficiency (v.s.) during the convection of energy and matter.

As the frictional energy dissipated for any given flow rate is drastically reduced, a larger fraction of the potential energy of the arterial blood is conserved as kinetic energy in both the capillaries and of the venous system. The former energy is utilized to induce vortex flow within and between travelling RBC (Gaehtgens et al. 1980; Schmid-Schönbein, 1990b and Gaehtgens and Schmid-Schönbein, 1982) thus producing convective mixing of both hemoglobin plasma. The kinetic energy of the venous blood is utilized to deform the right ventricle in cardiac diastole (a phenomenon beyond the scope of the present treatize).

Boundary Conditions for "Blood Cell Rectification"

The potential energy of the arterial blood, in combination with the network characteristics of the vascular bed and the specific rheological characteristics of mammalian blood give rise to a set of boundary conditions that favour the establishment of stable cell orientation, marked cell deformation and rapid axial drift of cells resulting in the generation of a lubricating sleeve of low viscosity plasma. The combination of these entirely "passive", yet red cell specific responses will be termed "cell rectification" or "rectified blood flow" and is favoured by the following physical boundary conditions:

1) very steep pressure gradients ($\Delta P/L$) as they exist in the terminal vascular bed. These result in very nigh wall shear stresses, i.e. a high magnitude of the forces orienting highly flexible red blood cells (see fig. 2),

2) in the microscopic vessel (2.5 µm > r > 100 µm), the combination of high shear stress and small radii create a very steep shear stress gradient with deep minima of frictional energy dissipation near the axis, associated with very steep gradients of energy dissipation towards the wall (see fig. 3).

In other words: what was formerly referred to as an "anomaly" now proves to be a pivotal feature of hematokinematics, dominating blood motion through all exchange vessels which - from an energetic point of view - are also those vessels where the mechanical energy generated by the heart is utilized to induce convective motion of O_2-carrying hemochromes. In synergetic terms, the potential energy gradients create a highly stable orientation giving rise to a unique kind of "dissipative structure" of blood optimizing the cooperativity processes necessary to pass blood through a complex system of highly branching cylindrical conduits.

Two assets of fluid dynamics in microvessels thus explain that the dispersed phase are driven by into the area of locally low energy dissipation . Both with regards to conservation of the kinetic energy of the moving particle, as well as to energy minimization of the entire system, cell accumulation near the axis amplifies efficacy of the transfer of energy and matter in the microvasculature.**)

PHENOMENOLOGICAL, KINEMATIC AND DYNAMIC ASPECTS OF AXIAL RBC DRIFT

Qualitatively speaking, the passive drift as a major factor g e n e r a t i n g the rectified flow as an dissipative structure reflects a "secondary motion" of oriented and deformed RBC (undergoing "membrane tanktreading") in response to gradients in mechanical parameters. Passive axial drift of highly deformed and oriented RBC is a transverse motion in a non-linear shear field, well known phenomenon in the rheology of multiphase systems and common to flow of deformable and asymmetric particles, in spheres and in liquid droplets. It has been established that in all suspensions of such particles, motion across streamlines of the undisturbed continuum is possible under a set of conditions, all of which are amply fulfilled in blood flow in capillaries. It is also known that the specific properties of the RBC-membrane (see Grebe and Schmid-Schönbein, this volume).

The "rectifying" conditions have been summarized by LEAL (1980) as follows:
1) Asymmetric geometry of deformed particle relative to undisturbed streamlines.
2) Non-zero Reynolds number, i.e. finite yet small effects of kinetic energy.
3) Non-newtonian, i.e. shear stress dependent flow behaviour.
4) Position relative to either the wall or the undisturbed flow.
According to LEAL, a cumulative effect of all these conditons for the r a t e of transverse motion has to be assumed, while the asymmetric geometry of the deformed particle appears to direct the transverse motion towards the tube axis. A more detailed analysis of the kinematics and dynamics of this functionally important phaenomenon is difficult, since to date experimental investigations are mostly performed under conditions inducing rapid drift and strong deformation. On the other hand, theoretical studies have as yet only been possible for small contributions of the above conditions, mostly studied in isolation. Nevertheless, both theoretical and

---

**) The fluid-dynamic consequences of the coherent motion minimize the consequences of the presence of RBC, which increase apparent viscosity (as well as the pressure gradient and energy dissipation for a given flow rate) by only about 10 % over that of cell free plasma. Despite of this, the motion is associated with eddy formation in the plasma. In the smallest capillaries (r = 2.5 μm) cells are stationarily deformed and oriented (as more or less regular rotationally symmetric paraboloids) assuming a minimum radius of about 1.5 μm, thus leaving a lubricating layer of about 1 μm and a ring vortex of about 2 μm in the plasma between cells (Schmid-Schönbein, 1990b). Thus, isentropic RBC motion is associated with a kind of microturbulence, so called "bolus flow" (Gaehtgens, 1980; Burton, 1969) a mixing process of the continuous phase due to vortex flow behind cells moving in the exchange capillaries.

experimental results strongly suggest a cumulative effect on the time required to assume stable orientation, the pivotal determinant of the drift direction.

Fig. 4 depict some potential mechanisms that might cooperate displacing in the elongated cell in stationary orientation from the region of high shear near the wall to the region of low shear near the vessel axis. The dynamic situation is highly asymmetric due to the close proximity

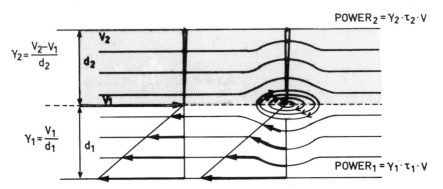

Fig.4. Schematic representation of a rectified red blood cell moving at the interface between the plasma (section 1) and the densile packed erythrocyte core. As the viscosity of the core increases, the velocity gradients for any given forward velocity d e c r e a s e s, thus further diminishing the power associated with the forward displacement of the erythrocytes in the cell rich core. Thence, the difference in forces, viscous drags, shear stresses and power is further increased, stabilizing the position of the individual red cells near the cell rich axial plug.

of a rigid wall. Near the wall there is a reduction of the volume fraction of the dispersed phase already at rest ("VAND-effect", see Bayliss (1965)). The situation of a continuously falling (yet presently unknown) velocity gradient, thence the uneven distribution of shear stress ($\tau = \dot{\gamma} \cdot \eta_0$), of viscous drag on the surface of the cell ($\tau \cdot A_{eff}$) of the power associated with motion ($\tau \cdot \gamma \cdot V$) and the inverse distribution of velocities is schematically depicted. During rapid cross stream migration from the marginal area (with low velocity and high shear stresses) to the axial core (rapid motion and low shear stresses) the cells obviously move "down" gradients of shear stresses ($\tau = \dot{\gamma} \cdot \eta_0$) and of viscous drags on the surface of ($\vartheta = \tau \cdot A_{eff}$), where τ = tangential force and A_{eff} = effective surface area. It must also be remembered that the tank treading RBC membrane is exposed to tensile (due to tangential stresses) and compressive (due to normal stresses) forces in alternate quadrants as they move in a shear field (Goldsmith and Mason, 1967).

Whatever the ultimate cause of the drift, the asymmetry of forces is the most likely mechanism of action. It is interesting to note that the geometric features of the microvascular networks and the resulting fluiddynamic conditions create a situation where during the passage of blood elements from larger to smaller arterioles and eventually the capillaries both the pressure gradients along the vessel segments ($\Delta P/L$) and the shear stress gradients ($\Delta \tau/r$) become progressively steeper. In other words, the potential gradients become gradually steeper, and thence the action of the (v.s.) become more and more dominant.

REFERENCES

Bayliss, L.E., 1965, The flow of suspensions of red blood cells in capillary tubes. Changes in "cell free" marginal sheath with changes in the shearing stress. J.Physiol.Lond. 179:1-25.

Burton, A.C., 1969, The mechanics of red cell in relation to its carrier function, in: Ciba Symposion: Circulatory and Respiratory Mass Transport. G.E.W. Wolstenholme and J. Knight, eds., Churchill, London, pp. 67-84.

Ebeling, W., 1989, Chaos-Ordnung-Information. Selbstorganisation in Natur und Technik, Harry Deutsch, Frankfurt.

Gaehtgens, P., Kreutz, F. and Albrecht, K.H., 1979, Optimal hematocrit of canine skeletal muscle during rhythmic isotonic exercise, Eur.J.Appl.Physiol. 41:27-39.

Gaehtgens, P., Schmidt, F., Will, G., 1981a, Comparative rheology of nucleated and non-nucleated red blood cells. I. Microrheology of avian erythrocytes during capillary flow, Pflügers Arch. 390:278-290.

Gaehtgens, P., Schmidt, F., Will, G., 1981b, Comparative rheology of nucleated and non-nucleated red blood cells. II. Rheological properties of avian red cell suspensions in narrow capillaries. Pflügers Arch. 390:291-304.

Gaehtgens, P. and Schmid-Schönbein, H., 1982, Mechanisms of dynamic flow adaptation of mammalian erythrocytes. Naturwissenschaften 69:294-297.

Goldsmith, H.L. and Mason, S.G., 1967, The microrheology of dispersions, in: Rheology, Theory and Applications, Vol. IV., F.R.Eirich, ed., Academic Press, New York, pp. 86-249.

Haken, H., 1978, Synergetics. An Introduction - nonequilibrium phase transitions and self-organization in physics, chemistry and biology, Springer, Berlin, Heidelberg, New York.

Jeffrey, G.B., 1922, The motion of ellipsoidal particles immersed in a viscous fluid. Proc.Roy.Soc. (London) Ser. A 102:162-179.

Landau, L.F. and Lifschitz, E.M., 1969, Fluid Mechanics, Pergamon, Oxford.

Leal, L.G., 1980, Particle motions in a viscous fluid, Ann.Rev.Fluid Mech. 12:435-476.

Libchaber, A., 1982, Experimental study of hydrodynamic instabilities: Rayleigh-Benard-Experiment (helium in a small box), in: Nonlinear Phenomena at Phase Transitions and Instabilities, H. Riste, ed., Plenum, New York, p. 259ff.

Lorenz, E., 1963, Deterministic non-periodic flow, J.Atmosph.Sci. 20:130-141.

Perkkiö, J., A model for non-symmetric flow of suspensions through cylindrical channels. Z.Angew.Math.Mech., in press.

Prigogine, I., Stengers, I., 1986, Dialog mit der Natur. Neue Wege naturwissenschaftlichen Denkens, 5. Aufl., Piper, München.

Schmid-Schönbein, H., 1981, Perspectives in hemorheology 1980: an interdisciplinary science returning to the center of medicine, in: The Rheology of Blood, Blood Vessels, and Associated Tissues, D.R. Gross, N.H.C. Hwang, eds., NATO Advanced Study Institute Series: Appl.Sci, pp. 1-21.

Schmid-Schönbein, H., Gaehtgens, P., Fischer, Th., Stöhr-Liesen, M., 1984, Biology of red cells: non-nucleated erythrocytes as fluid-drop-like cell fragments. Int.J.Microcirc.:Clin.Exp. 3:161-196.

Schmid-Schönbein, H., 1988, Fluid dynamics and hemorheology in vivo: the interactions of hemodynamic parameters and hemorheological "properties" in determining the flow behavior of blood in microvascular networks, in: Clinical Blood Rheology, Vol. I, G.D.O. Lowe ed., CRC Press, Boca Raton, FL, pp. 129-219.

Schmid-Schönbein, H., 1988, Thrombose als ein Vorgang in "strömendem Blut": Wechselwirkung fluiddynamischer, rheologischer und enzymologischer Ereignisse beim Ablauf von Thrombozytenaggregation und Fibrinpolymerisation, Hämostasiologie 8:149-172.

Schmid-Schönbein, H., 1990, Blood rheology and oxygen conductance from the alveoli to the mitochondria, in: Dugs and the Delivery of Oxygen to Tissues, J.S. Fleming ed., CRC Press, Boca Raton, Fl., pp. 15-86.

Schmid-Schönbein, H., 1990, Synergetic order and chaotic malfunctions of the circulatory systems in "multiorgan failure": breakdown of cooperativity of hemodynamic functions as cause acute microvascular pathologies, in: Uptdate in Intensive Care and Emergency Medicine, Vol. 11, Springer, Heidelberg, in press.

Schmid-Schönbein, H., 1990, Synergetics of fluid-dynamic and biochemical catastrophe reactions in coronary artery thrombosis, in: Unstable Angina, Bleifeld, Braunwald, Hamm, eds., Springer, Heidelberg, in press.

Schmid-Schönbein, H., 1990, Biology and rheology of the "acute phase reaction" in chronic degenerative diseases, Clin. Hemorheology, in press.

WALL SHEAR RATE IN ARTERIOLES: LEAST ESTIMATES FROM IN VIVO RECORDED VELOCITY PROFILES

Robert S. Reneman, Theo Arts, Dick W. Slaaf and Geert Jan Tangelder

Departments of Physiology and Biophysics, Cardiovascular Research Institute Maastricht, University of Limburg Maastricht, The Netherlands

ABSTRACT

In this chapter velocity profiles, as recorded in mesenteric arterioles of the rabbit with the use of fluorescently labeled platelets as natural flow markers, are described. These velocity profiles are flattened parabolas with maximal and mean velocity ratios varying between 1.39 and 1.54 (median; 1.50). The wall shear rate values estimated from these velocity profiles, considering a two phase model, range from 472 to 4712 s^{-1} (median: 1700 s^{-1}) for centerline red blood cell velocities varying between 1.3 and 14.4 $mm.s^{-1}$. These wall shear rate values are at least 1.46 to 3.94 (median 2.12) times higher than those expected on the basis of a parabolic velocity profile, but with the same volume flow in the vessel.

Keywords: Velocity profiles - wall shear rate - arterioles - intravital microscopy, mesentery - rabbit.

INTRODUCTION

Wall shear rate is an important parameter in describing the interaction between a blood component and the vessel wall (Goldsmith and Turitto, 1986; Turitto, 1982). For example, in vitro it has been shown that the adhesion of blood platelets to subendothelial

Biomechanical Transport Processes, Edited by F. Mosora *et al.*
Plenum Press, New York, 1990

structures is determined by wall shear rate (Turitto and Baumgartner, 1979), because both transport of platelets toward the wall and the chemical reactions involved in the binding depend on this parameter. Most of the wall shear rate values in microvessels presently available were calculated based on assumption of a parabolic velocity profile (Turitto, 1982). In arterioles, however, the velocity profiles differ significantly from a parabola; they are more blunt in both systole and diastole which has its consequences for the assessment of wall shear rate (Tangelder et al, 1986). Therefore, wall shear rates should be calculated on the basis of actual velocity profiles rather than assumed ones. In this chapter the least estimates of actual wall shear rate calculated from platelet velocity profiles, as determined in vivo in rabbit mesenteric arterioles, will be discussed.

ASSESSMENT OF VELOCITY PROFILES IN ARTERIOLES

To assess velocity profiles in vivo fluorescently labeled blood platelets were used as natural flow markers. The in vivo labeling of platelets (Tangelder et al, 1982a), their localization within a thin optical section (Tangelder et al, 1982b) and the determination of platelet velocity profiles (Tangelder et al, 1986) have been described in detail before. In anesthetized rabbits, circulating blood platelets were labeled by an intravenous injection of the fluorescent dye acridine red (absorption and emission peaks at 525 and 625 nm, respectively). The fluorescent platelets flowing in mesenteric arterioles (17-32 μm diam) were visualized by intravital fluorescence video microscopy with a Leitz x100 salt-water immersion objective (numerical aperture 1.20), a final magnification at the front plane of the TV camera of x200 and incident flashed illumination. Pairs of flashes were given with a short, preset time interval between the two flashes, yielding in one video image two images of the same platelet displaced over a certain distance for the given time interval (Fig. 1). The first flash was given in the blanking period of the TV camera.

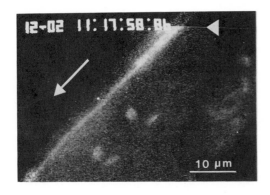

Fig. 1. Fluorescently labeled platelets flowing in an arteriole, as observed with dual-flash illumination, yielding in one TV picture two images of the same platelet displaced over a certain distance during the time interval between the two flashes. The first flash is given in the blanking period of the TV camera. Arrowhead indicates the moment of the second flash (4 msec after the first one). Arrow indicates the direction of flow. After Tangelder et al, 1986. With permission of the American Heart Association.

In each experiment, the time interval between the first and the second flash (range 1-5 ms) was selected in such a way that the two concomitant images of a platelet showed no or only little overlap. Within a measuring period (30-60 s), all flash pairs were recorded at a selected moment in the cardiac cycle (± 20 msec) by triggering the light flashes by the R wave of the electrocardiogram (ECG) and by using a preset delay (Tangelder et al, 1986). To be able to determine velocity profiles in both systole and diastole, instantaneous red blood cell velocity was assessed photometrically using a prism grating system (Slaaf et al, 1981).

Video recordings were analyzed frame by frame. To determine the velocity profile, the geometric centers of the images of a platelet were identified by eye and used to measure with vernier calipers 1) the displacement of the platelet in the preset time interval, yielding its velocity, and 2) its relative radial position in the vessel, defined as the mean of the radial positions of the two images relative to the vessel radius. The vessel walls were also labeled with the dye (Fig. 1). Only reasonably sharp images were used. This means that, with the high numerical aperture of the objective lens and the total optical magnification used, the data were obtained from an optical section around the median plane of the vessel with a depth of ~5 μm or less (Tangelder et al, 1982b). Because the geometric center of a platelet image was used for the measurements, no data points could be obtained closer to the wall than ~0.5 μm, because of the size (Frojmovic and Panjwani, 1976) and orientation (Teirlinck et al, 1984) of the platelets. A platelet-velocity profile obtained in this way is presented in Fig. 2.

The shape of velocity profiles in arterioles

The experimental platelet-velocity profiles can be adequately described with the following equation modified after Roevros (1974):

$$V(r) = V_{max}(1 - | a(r/R + b|^K), \quad a > 0, \tag{1}$$

where V(r) is the velocity at radial position r, the vertical stripes denote absolute values, V_{max} is the maximal velocity in the vessel, R is the radius of the vessel, a is a scale factor allowing a non-zero intercept of the fit with the vessel wall and b is a parameter correcting for a shift of the top of the profile away from the vessel center.

If the scale factor a is smaller than 1, the intercept of the fit with the vessel wall will be positive and will increase towards the value of V_{max} when a approaches zero. If K = 2, a parabolic velocity destribution is obtained. An increase in K yields a progressively flatter profile. The ratio of V_{max} and the mean velocity of the profile (V_{mean}) can also be used as an index of the degree of blunting of a profile, being 2 in the case of a parabolic profile and 1 in the case of complete plug flow, i.e. all layers of fluid are travelling at the same speed.

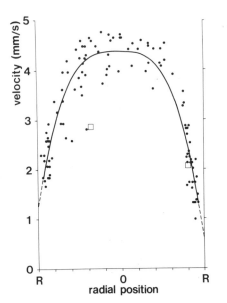

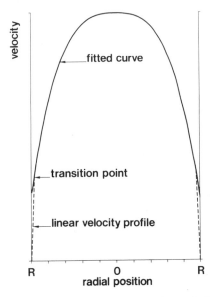

Fig. 2. Velocity profile of blood platelets flowing in an arteriole with a diameter of 23 μm. Each dot represents the velocity of one platelet. The open squares indicate the velocities of the two leukocytes which could be observed during determination of the platelet velocity profile. In addition, the best fit is shown as obtained with equation (1). Ordinates at R indicate position of vessel walls. R = vessel radius, O = center of the vessel. The dashed part of the fit is the extrapolated part. After Tangelder et al, 1988. With permission of The American Physiological Society.

Fig. 3. Schematic representation of the curves used to describe the velocity distribution in the arterioles and to calculate the least estimates of wall shear rate in these vessels. At the wall velocity is zero (no-slip condition). In thin layer of fluid near the wall, which has a lower apparent viscosity than fluid in the remainder of the vessel, the mean velocity gradient is determined, resulting in a linear velocity profile (dashed lines). In the remainder of the vessel velocity profile is described by fitted platelet velocity profile (solid line). After Tangelder et al, 1988. With permission of The American Physiological Society.

As previously described (Tangelder et al, 1986), the ratio of V_{max} and V_{mean} can be approximated, with the use of the following equation:

$$V_{max}/V_{mean} = (K + 2)/(K + 2 - 2(a)^K) \qquad (2)$$

The best fits are obtained when the position of the vessel walls is ignored ($a \neq 1$), leading to a positive intercept (see Table I for the values of a) with the vessel wall (Fig. 2). The latter is irrealistic because flow velocity at the vessel wall must be zero due to no-slip conditions (Caro et al, 1978). This indicates that the velocity profiles in arterioles have to be described with a two phase model and a steep velocity gradient near the wall may be assumed (Fig. 3). By means of this way of analysis K values varying between 2.3 and 4.0, and V_{max} and V_{mean} ratios ranging from 1.39 to 1.54 are found (Table I). These findings show that the velocity profiles in arterioles can be considered to be flattened parabolas. The finding that at each relative radial position within a vessel the residuals scattered around zero, with no systematic over or understimation of velocities, indicates that the velocity profiles are symmetric.

WALL SHEAR RATE IN VIVO

By describing the velocity profiles, as recorded in arterioles, by the best fit through the measuring points with the use of equation (1) and a linear intrapolation from the point closest to the wall at which the velocity could be measured (transition point) to zero flow at the wall (Fig. 3), the least estimate of wall shear rate (WSR) can be determined by means of the following equation (Tangelder et al, 1988):

$$WSR = 2V(x)/D(1-|x|) \qquad (3)$$

where $V(x)$ is the velocity at x, which is the relative radial position of the transition point, and D is the vessel diameter.

The least estimates of average wall shear rate, i.e. the mean of the values obtained at both vessel walls, range from 472 - 4712 s^{-1} (median: 1700 s^{-1}) for centerline red blood cell velocities ranging from 1.3 to 14.4 $min.s^{-1}$ (median; 3.4 $mm.s^{-1}$). These data are derived from 12 velocity profiles in 9 arterioles (Table I). The values of the least estimates of wall shear rate at opposite vessel walls may differ from the mean value for that vessel by 2 to 44%. These differences are caused by 1) unequal thickness of the thin layer of fluid near the wall on opposite sides and 2) asymmetry of the velocity profile with a shift of its peak away from the vessel center line. Although the latter effect is small (<4%), its influence on wall shear rate cannot be neglected.

Table I. Velocity profile parameters and least estimates of wall shear rate as assessed in arterioles of various diameters in vivo.

Diam.	Cardiac Cycle Phase	K	a	Vmax/Vmean	Distance Trans. Point from Vessel Wall	Least Estimates WSR	Least Estimates WSR/WSR-P
(μm)					(μm)	(s^{-1})	
17	D	2.4	0.83	1.41	0.5	1,750	2.65
23	D	3.2	0.93	1.44	0.8	2,114	2.00
23	D	4.0	0.96	1.39	1.3	472	1.46
24	D	2.6	0.90	1.49	0.7	1,942	2.23
24	D	2.7	0.86	1.40	0.5	2,866	3.94
	S	2.3	0.87	1.51	0.8	2,044	2.42
25	D	2.9	0.94	1.52	0.7	626	2.00
	S	2.7	0.93	1.54	0.9	1,224	2.26
27	D	2.9	0.92	1.47	0.7	1,649	2.64
29	D	2.8	0.93	1.52	1.0	4,712	1.81
32	D	3.0	0.95	1.52	1.3	980	1.58
	S	3.0	0.95	1.52	0.8	1,481	2.01
Median		2.9	0.93	1.50	0.8	1,700	2.12

D, diastolic phase; S, systolic phase, i.e., when the velocity in the microvessel reached its minimum and maximum value, respectively. For transition point see Fig. 3; for each experiment the mean of values obtained at right and left vessel wall is presented. See for K and a equation (1) and for Vmax/Vmean equation (2).

The wall shear rate values expected in the case of a parabolic velocity distribution (WSR-P), but with the same volume flow in the vessel, can be derived with the following equation:

$$WSR\text{-}P = 4\ V_{max}\ parabola/D$$
$$= 8\ V_{mean}/D \qquad (4)$$

The extent to which the wall shear rate values, as derived from the actual velocity profiles in vivo, are higher than those calculated assuming a parabolic velocity distribution for the same volume flow is given by the relation:

$$\frac{WSR}{WSR\text{-}P} = 0.25 \cdot V(x)/(1\text{-}|x|) \cdot V_{mean} \qquad (5)$$

At peak red blood cell velocities ranging from 1.3 to 14.4 mm.s^{-1} (median 3.4 mm.s^{-1}) the estimated wall shear rate values are at least 1.46 to 3.94 (median: 2.12) times higher than those calculated on the basis of a parabolic velocity profile.

CONCLUDING DISCUSSION

The data presented in this chapter indicate that in arterioles wall shear rate cannot be determined based on the assumption that the velocity profile is parabolic. The velocity profiles, as recorded in mesenteric arterioles with the use of fluorescently labeled platelets as natural flow markers, are flattened parabolas.

The wall shear rate data presented are a least estimate because velocities close to the wall cannot be determined with the technique used to assess velocity profiles. The position closest to the wall where platelet velocity can be determined is about 0.5 μm. Therefore, it cannot be excluded that the velocity gradient near the wall is steeper than estimated from the actual velocity profiles.

Previous studies in our laboratory have shown that near the wall platelets tend to align with the flow stream (Teirlinck et al, 1984), exposing their largest surface area to the endothelial cells, and that the platelet concentration is significantly higher near the wall than in the center of the bloodstream (Tangelder et al, 1985). These observations, combined with the low platelet velocity near the wall, indicate that the conditions are very favorable for platelet-endothelium interactions under pathological circumstances.

Acknowledgements

The authors are indebted to Jos Heemskerk, Karin van Brussel and Rosy Hanssen for their help in preparing the manuscript.

REFERENCES

Caro, C.G., Pedley, T.J., Schroter, R.C., and Seed, W.A., 1978, The Mechanics of the Circulation, Oxford, Oxford University Press.

Frojmovic, M.M., and Panjwani, R., 1976, Geometry of normal mammalian platelets by quantitative microscopic studies. Biophys. J. 16:1071-1089.

Goldsmith, H.L. and Turitto, V.T.,1986, Rheological aspects of thrombosis and haemostasis: basic principles and applications, Thromb. Haemostas. 55:415-435.

Roevros, J.M.J.G., 1974, Analogue processing of CW Doppler flowmeter signals to determine average frequency shift momentaneously without the use of a wave analyser, Cardiovascular Applications of Ultrasound, edited by R.S. Reneman, Amsterdam, North Holland, Publishing Company, pp 43-54.

Slaaf, D.W., Rood, J.P.S.M., Tangelder, G.J., Jeurens, T.J.M., Alewijnse, R., Reneman, R.S., and Arts, T., 1981, A bidirectional optical (BDO) three-stage prism grating system for on-line measurement of red blood cell velocity in microvessels, Microvasc. Res. 22:110-122.

Tangelder, G.J., Slaaf, D.W., Arts, T., and Reneman, R.S., 1988, Wall shear rate in arterioles in vivo: least estimates from platelet velocity profiles, Am. J. Physiol. 254:H1059-H1064.

Tangelder, G.J., Slaaf, D.W., Muijtjens, A.M.M., Arts, T., oude Egbrink, M.G.A., and Reneman, R.S., 1986, Velocity profiles of blood platelets and red blood cells flowing in arterioles of the rabbit mesentery, Circ. Res. 59:505-514.

Tangelder, G.J., Slaaf, D.W., and Reneman, R.S., 1982a, Fluorescent labelling of blood platelets in vivo, Thromb. Res. 28:803-820.

Tangelder, G.J., Slaaf, D.W., Teirlinck, H.C., Alewijnse, R., and Reneman, R.S., 1982b, Localization within a thin optical section of fluorescent blood platelets flowing in a microvessel, Microvasc. Res. 23:214-230.

Tangelder, G.J., Teirlinck, H.C., Slaaf, D.W., and Reneman, R.S., 1985, Distribution of blood platelets flowing in arterioles, Am. J. Physiol. 17:H318-H323.

Teirlinck, H.C., Tangelder, G.J., Slaaf, D.W., Muijtjens, A.M.M., Arts, T., and Reneman, R.S., 1984, Orientation and diameter distribution of rabbit blood platelets flowing in small arterioles, Biorheology 21:317-331.

Turitto, V.T., 1982, Blood viscosity, mass transport, and thrombogenesis, Progress in Hemostasis and Thrombosis 6, edited by TH Speat, New York, Grune and Stratton, pp 139-177.

Turitto, V.T., and Baumgartner, H.R., 1979, Platelet interaction with subendothelium in flowing rabbit blood: effect of blood shear rate. Microvasc. Res.17:38-54.

BLOOD FLOW IN MICROCIRCULATORY NETWORKS AND

POROUS-MEDIA-LIKE TISSUES

Heinz D. Papenfuss[1] and Holger Schmid-Schönbein[2]

[1]Ruhr-University Bochum, Institute for Thermo- and Fluid
Dynamics, D-4630 Bochum, F.R.G.

[2]RWTH Aachen, Department of Physiology, D-5100 Aachen, F.R.G.

ABSTRACT

Two theoretical models are presented which determine the hemodynamics in microvascular networks and porous-media-like tissues. The models differ completely due to the structural differences of the two systems. The network model incorporates experimental data for the hemorheology and predicts the flows, pressures, hematocrit values and viscosities for all individual vessels of arbitrary microvascular networks. As an example, the capillary bed of the cat sartorius muscle is analysed using the model. The fluid mechanics of porous media is applied to the hemochorial multivillous placenta. The mathematical approach is based on the application of potential theory. Numerical results provide a simulation of the trajectories of maternal blood and the pressure distribution in the intervillous space. In an experimental study with an enlarged mechanical model of a placental functional unit, the pressure loss and permeability could be determined. The permeability agreed remarkably well with a value predicted by the Carman-Kozeny equation.

Keywords: microcirculation, hemorheology, microvascular networks, porous media, placenta, permeability, mathematical model.

INTRODUCTION

It is well known that the microcirculation is the crucial part of the circulatory system where the exchange of oxygen, fluid and substances between blood and tissue takes place. In the circulatory system blood flows internally through conduits – the blood vessels. In the microcirculation, these vessels are the capillaries, the arterioles and the venules. There is a single exception to the internal flow of blood in vessels in the circulation. This occurs during pregnancy and is the flow of maternal blood in the hemochorial multivillous placenta. This is essentially an external flow around a large number of obstacles called "villi" which contain fetal capillaries responsible for the exchange of oxygen and nutrients with maternal blood in the surrounding intervillous space.

The exceptional structure and type of blood flow in this system make it necessary to develop a mathematical model which differs completely from that for blood flow in vascular networks. In order to describe the movement of blood in the intervillous space we shall treat this structure as a porous medium. Both systems have in common that the flow is determined by friction and pressure forces whereas inertia is neglegible. This is the regime of creeping flow which is always laminar. It is the aim of this work to elucidate the differences of the fluid dynamic models for these two cases: microvascular networks and the hemochorial multivillous placenta.

FLUID MECHANICS OF MICROVASCULAR NETWORKS

Mathematical Model

The hemorheology in microvessels is governed by the two-phase character of blood (plasma and red cells). Since the size of the red cells is comparable with the vessel diameters, non-continuum effects occur which lead to the well known Fåhraeus and Fåhraeus-Linqvist effects and to red cell plasma separation at divergent bifurcations.

Typical for the microcirculation is the distinction between discharge and tube hematocrit, H_D and H_T. H_D is the volume flow fraction of red cells in a blood vessel while H_T is the volume fraction: $0.5 \leq H_T/H_D \leq 1$.

Due to the peculiar manner in which red cells adapt their shape, two characteristic length scales come into the picture: the square root of the surface area of a red cell, $\sqrt{A_c}$, and the minimum tube diameter, D_T^*, which a red cell can traverse without enlarging its surface area (which would lead to hemolysis). In the following we refer to the work of Papenfuss and Gross (1987) concerning microhemorheology and compile the most important empirical equations which have been updated according to recent findings:

$$H_T/H_D = 1 + F_1 \cdot G_1 \quad ; \quad \eta_{rel} = 1 + H_T \cdot (F_2 + G_2) \tag{1}$$

$$d = D_T/D_T^* \quad ; \quad D = (1 - 1/d) \cdot D_T/\sqrt{A_c} \quad (D_T = \text{vessel diameter}) \tag{2}$$

$$F_1 = a_1 \cdot (d^4 - d^{-12}) + (d^{b_1} - 1) \tag{3}$$

$$G_1 = 0.5 \cdot (1 - H_D^{c_1}) \quad ; \quad F_2 = d^{a_2}/(d^{b_2} - 1) \tag{4}$$

$$G_2 = 0 \quad \text{if } D \leq 0.5 \tag{5}$$

$$G_2 = c_2 \cdot [1 + 12 \cdot \exp(d_2 \cdot H_T)] \cdot (1 - 0.5/D)^2 \quad \text{if } D > 0.5 \tag{6}$$

$$a_1 = 0.7639 \cdot 10^{-5} \quad b_1 = -1.2435 \quad c_1 = 1.5232 \tag{7}$$

$$a_2 = 1.3882 \quad\quad b_2 = 1.5618 \quad c_2 = 1.7586 \quad d_2 = 1.6409 \tag{8}$$

Eqs. (1-8) can be used for $d \leq 4$. For $d > 16$, Eqs. (1-2) should be replaced by

$$D = D_T/\sqrt{A_c} \tag{9}$$

$$H_T/H_D = 0.313 + 0.233 \cdot \ln D + (0.33 - 0.123 \cdot \ln D) \cdot H_D \tag{10}$$

$$\eta_{rel} = (1 - H_T)^{-1.969}. \tag{11}$$

The lacking regime $4 < d < 16$ should be completed by an interpolation.

The elements of vascular networks are the loops. They are formed by a number of vessels which are connected at divergent or convergent bifurcations. The fluid dynamic determinant for the vessels in a loop is the Hagen-Poiseuille relationship between the local pressure drop Δp and volumetric flow rate Q of blood:

$$\Delta p = 128 \cdot \frac{\eta_{pl} \cdot \eta_{rel} \cdot L}{\pi \cdot D_T^4} \cdot Q \tag{12}$$

The fluid dynamic determinant for bifurcations is the mass conservation of blood and its constituents. At divergent bifurcations in the microcirculation, however, a reorganisation of the red cell distribution to the daughter branches takes place. This occurs in the sense that the branch with the higher flow rate receives an unproportionally higher percentage of red cells. This effect is expressed by the following empirical equations:

$$H_{D12}/H_{D0} = 1.25 \cdot (1 - 0.1 \cdot Q_0/Q_{12}) \quad ; \quad 0.1 < Q_{12}/Q_0 < 0.9 \tag{13}$$

$$H_{D12}/H_{D0} = Q_0/Q_{12} \quad ; \quad Q_{12}/Q_0 > 0.9 \tag{14}$$

In Eqs. (13-14), the subscript 0 refers to the feeding vessel and the subscript 12 to either of the daughter branches.

The system of equations and fluid dynamic principles described above are sufficient to develop a calculation scheme which makes it possible to predict the flows, pressures, hematocrit values and viscosities for the individual vessels of an arbitrary microvascular network. The prerequisite is that the entire network architecture is given and the flows in the incoming and outgoing vessels are prescribed. In addition, the hematocrit values of the incoming vessels must be known a priori. The numerical algorithm is necessarily iterative due to the nonlinearity of the equations. A computer package which has all these features has been developed by Papenfuss and Gross (1986, 1987) and is known under the name "McNET".

Capillary Network of Skeletal Muscle: Numerical Results

We consider the microcirculatory network of the cat sartorius muscle as an example. The geometry of the network is shown in Fig. 1 and has been obtained by a video technique (Koller and Johnson, 1986). The capillaries are arranged predominantly parallel to the muscle fibers. An average diameter is 7 μm. The feeding arteriole and collecting venule run crosswise to the capillaries and constitute a physiological unit of about 700 μm in length. More units with similar networks occur periodically at both sides of this capillary bed. The periodicity makes it possible to formulate proper conditions for the vessels at the left and right boundaries of the vascular system with outgoing or incoming flows.

The vascular dimensions obtained by Koller and Johnson (1986) for this network have been used for the simulation. The results for the hemodynamic parameters which are obtained for each vascular segment are grouped together in terms of a normalized pressure drop parameter

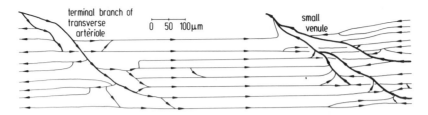

Fig. 1. Geometrical arrangement of microvascular bed of cat sartorius muscle

$$\Delta p^* = \frac{p_A - p}{p_A - p_V} . \tag{15}$$

The subscripts A and V refer to the arteriole and venule, respectively. As result of the calculation, the center of each vascular segment is labeled with a particular value for Δp^* which varies between 0 and 1. In other words, Δp^* represents the physiological distance from the feeding arteriole within the vascular network.

Fig. 2 shows the blood and red cell velocities and the blood viscosity ($\eta = \eta_{pl} \cdot \eta_{rel}$) as functions of the pressure drop parameter Δp^*. The solid curves are actually interpolations between a number of discrete numerical data points according to the number of vascular elements in the network. It is interesting to note that the venous part of this network has a larger number of vessels than the arterial part (about 60:40 %). The larger cross-sectional area of the venous part of the network leads to lower velocities. It is, further, interesting to note that the difference in blood and red cell velocity which is the manifestation of the

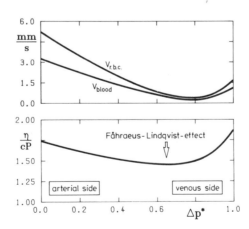

Fig. 2. Distribution of red cell velocity, blood velocity and blood viscosity in microvascular bed of cat sartorius muscle

Fåhraeus effect is a factor of 0.6. This factor which is only a mean value is used by several investigators to determine the blood velocity from
the observed red cell velocity in the microcirculation. The present model
makes it possible to determine a corresponding correction factor for each
vessel.

The blood viscosity is lowest where the vessel diameters are smallest in the network. This is the manifestation of the Fåhraeus-Lindqvist
effect. The viscosity reaches its maximum value in the venule. Although
the tube hematocrit is lower on the venular than on the arteriolar side,
the viscosity is higher there due to the fact that the vessel diameters
are larger than on the arteriolar side.

FLUID MECHANICS IN POROUS-MEDIA-LIKE TISSUES

Theory of Flow in Porous Media

The fundamental equation for the fluid dynamics in porous media is
Darcy's law. For a one-dimensional filter column of length L this law
relates the pressure drop across the column to the superficial velocity:

$$w = \frac{k}{\eta} \cdot \frac{\Delta p}{L} \tag{16}$$

In this equation, k is the permeability of the porous medium (with
the unit m^2) and η is the viscosity of the perfusate. For a three-dimensional porous system, Eq. (16) can be generalized to

$$\underline{w} = - \frac{k}{\eta} \cdot \text{grad } p \tag{17}$$

where $\underline{w}$ is now the local superficial velocity vector. Eq. (17) is linear
and together with the continuity equation replaces the fundamental equation of fluid dynamics, the Navier-Stokes equation. The fluid density is
not a variable in the flow in porous media in absence of inertia.

It is important to note that Eq. (17) simplifies the mathematical
description of the complicated flow in porous media drastically by replacing the solid-fluid system by a continuum. Application of the
principle of mass conservation, i.e., div $\underline{w}$ = 0, results in

$$\Delta^2 p = 0. \tag{18}$$

This means that the pressure obeys Laplace's equation, i.e., the
pressure distribution can be obtained from potential theory.

It is now easy to show that the velocity field is irrotational:

$$\text{curl } \underline{w} = 0. \tag{19}$$

While Darcy's law provides the mathematical framework for the experimental determination of the permeability of a porous medium, a number of
investigators have attempted to derive the permeability from geometric
quantities of the porous medium; see e.g. Dullien (1979). The following
geometric quantities are usually incorporated in the equations:

ϵ = porosity = void volume/total volume
S = specific surface area = surface area of wetted particles/
                              volume of wetted particles

d_s = diameter of equivalent sphere with the same specific surface area as the wetted particle.

With these geometric quantities the permeability can be calculated according to the Carman-Kozeny equation:

$$k = C \cdot \frac{\epsilon^3 \cdot d_s^2}{(1-\epsilon)^2} \qquad (20)$$

C = 1/180 for spherical particles. For non-spherical particles, experiments have shown that C should be approximated by 1/150.

It is important to note that in porous media the pressure and velocity fields are not coupled by Bernoulli's equation. Instead, the coupling is given by Darcy's law.

Flow in the Intervillous Space of the Human Placenta: Numerical Results

The idea of the present model for the flow through the intervillous space of the human placenta is based on the theory for porous media and has been proposed by Schmid-Schönbein (1988). Its origin is the so-called anatomical concept postulated by Moll (see the review article of 1981) in which the frictional forces in the intervillous space have been recognized as the predominant mechanism of momentum transport. A schematic representation of the anatomy of the human placenta is given in Fig. 3. See also Kaufmann (1983). It is essentially a spherical segment bounded by the base plate and chorionic plate. The space in between is filled with a high portion of strongly branched out coral-like villi which are washed by maternal blood. Taking the fifth month of pregnancy as an example, the volume of the villi is 63 cm^3 and their surface area is 14800 cm^2 (Kaufmann, 1981). The placenta is divided into 60 % villi and 40 % intervillous space.

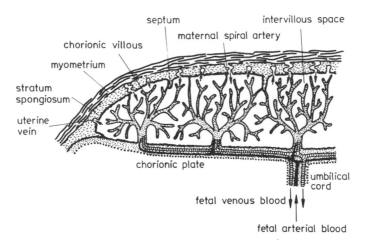

Fig. 3. Anatomy of hemochorial multivillous placenta

The terminal villi contain fetal capillaries which provide the exchange of oxygen and nutrients with the maternal blood in the intervillous space. The maternal blood is injected into the intervillous space by a number of spiral arteries which penetrate the base plate. Each of these arteries is surrounded by about eight uterine veins.

It is known that the human placenta is composed of a number of physiological units, the cotyledons. Each cotyledon has a single maternal spiral artery in its center and a number of (about eight) surrounding uterine veins in the base plate. The cotyledon can be approximated by a cylinder so that ignoring the curvature of the base plate is ignored. A typical diameter for such a cylinder is 1.6 cm if the fifth month of pregnancy is considered. A typical cylinder height is 0.88 cm.

For the mathematical model of the cotyledon we assume that the flow in all planes normal to the cylinder axis have the same flow pattern. The feeding artery and the draining veins are replaced by a line source and concentrically arranged line sinks. The outer cylinder wall is only thought but is impermeable due to symmetry conditions. It is, therefore, a stream surface.

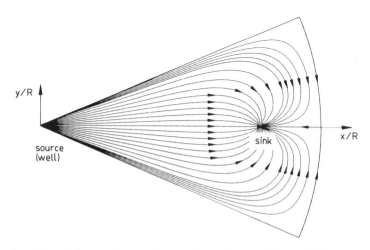

Fig. 4. Streamline pattern in sector of cotyledon of hemochorial multivillous placenta

For the geometry described above, the fluid dynamic problem becomes two-dimensional. Numerical solutions have been obtained for the potential equation with the relevant boundary conditions. The resulting streamline pattern is shown in Fig. 4. The corresponding pressure distribution is shown in Fig. 5. A dimensionless pressure drop parameter Δp^* has been used which follows exactly the definition given in Eq. (15). A relatively low pressure drop $p_A - p_V$ of about 8 mmHg within the cotyledon has been reported by Moll (1981). Streamlines and isobars are perpendicular to each other. It is important to note that the results shown in these figures are independent of the actual value of the permeability. However, in order to assign values with physiological units to the isobars, the permeability and the pressure drop between artery and veins (or arterial pressure and flow rate) must be known. In principle, the Carman-Kozeny equation could be applied to determine the permeability. Unfortunately, reliable information on the actual shape of the villi is still lacking so that a simple evaluation of that equation is presently impossible. In fact, the present theory and its future developments together with measurements of the flow in the spiral arteries could make it possible to determine the permeability and, therefore, the flow conductance of the placenta. The manner in which this can be realized will be demonstrated by means of a mechanical model of the cotyledon.

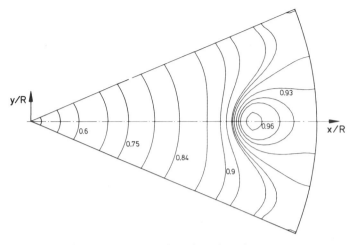

Fig. 5. Normalized pressure distribution in sector of cotyledon.
Numbers refer to Δp*

Flow in the Intervillous Space of the Human Placenta: Experimental Results in vitro

In addition to the numerical calculations, experiments with an enlarged mechanical model of the cotyledon (scale 6.5:1) made of lucite were done; see Fig. 6. The cylindrical model was tightly filled with bottle brushes with the stems parallel to the cylinder axis. The brushes with their bristles simulated the villi. By this way, the porosity could be reduced to $\epsilon = 0.73$ which, however, was larger than the porosity of the human placenta ($\epsilon = 0.4$).

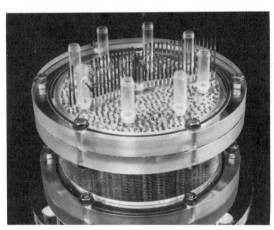

Fig. 6. Mechanical model of cotyledon filled with bottle brushes

The upper lid of the model contained a central feeding pipe ("artery") of 8 mm diameter and eight draining pipes ("veins") of 11 mm diameter. The model was perfused with a mixture of glycerol and water (volume ratio 3:1) to adjust the viscosity so that the Reynolds number of the flow in the human placenta could be simulated. The volume flow was measured with a calibrated rotameter and the pressure in the "artery" and "veins" with simple tube manometers. The measurements were done at room temperature (22 °C).

The measurements of the pressures and corresponding flow rates were used to calculate the head loss coefficient ζ and the permeability of the system. The results are shown in Fig. 7 as functions of the Reynolds number which was determined using the velocity, V_A, and diameter, D_A, of the feeding pipe. The physiological range would be $Re_A < 50$ which corresponds to an extremely low Reynolds number in the intervillous space. In the relevant Reynolds number regime, the head loss coefficient is inversely related to the Reynolds number as is typical for laminar flows. It should be noted that Δp_t in the definition of the head loss coefficient

$$\zeta = \frac{\Delta p_t}{\rho/2 \cdot V_A{}^2} \tag{21}$$

is the loss in total pressure, i.e., the sum of hydrostatic and dynamic pressure. This is important since the velocities in the artery and the veins are different.

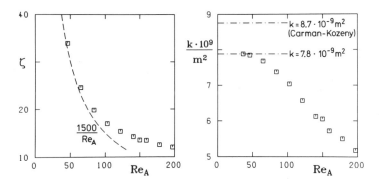

Fig. 7. Head loss coefficient and permeability of mechanical model of cotyledon as function of the Reynolds number

The permeability has been calculated by comparing the measured pressure drop with the numerical results for the dimensionless pressure distribution between artery and veins described above. The dependence on the Reynolds number shown in Fig. 7 has two reasons. First, the approach used to determine the permeability is valid only in the limiting case of very low Reynolds numbers in porous beds. Consequently, the results are meaningful only in this limiting case. Second, increasing the Reynolds number leads to a tighter packing of the soft brushes with concomitant lower permeability for the fluid.

The asymptotic value for the permeability for $Re_A \rightarrow 0$ has been determined to be $k = 7.8 \cdot 10^{-9}$ m^2. By comparison, the Carman-Kozeny equation, which only requires knowledge of geometric quantities of the packed bed, yields a value of $k = 8.7 \cdot 10^{-9}$ m^2. The two values are close. The difference is caused by the difficulty to determine the geometric quantity d_s for a bottle brush. Moreover, the constant 1/150 in the Carman-Kozeny equation is uncertain and could be 1/180 in an extreme case. In principle, two approaches now exist which can determine the permeability of the hemochorial multivillous placenta and give similar results.

CONCLUSIONS

The present work shows the different approaches to the modelling of blood flow in two different systems: microvascular networks and the hemochorial multivillous placenta. Whereas the flow in networks is channeled in conduits and is, therefore, an internal flow, the flow in the intervillous space of the placenta is essentially an external flow around a large number of obstacles, the villi. It is understandable that the mathematical descriptions of the two flow systems are completely different. The two-phase character of blood (plasma and cellular particles) leads to highly nonlinear equations in which recent experimental data on the microhemorheology are incorporated. For the given angioarchitecture of a network, the model predicts the flows, pressures, hematocrit values and viscosities for the individual blood vessels. A network analysis of the capillary bed of the cat sartorius muscle has been carried out to demonstrate the potential applicability of the model.

The second model is based on Darcy's law for the flow in porous media. In this case the flow and the pressure field result from potential theory. The flow conductance, i.e., the inverse of the flow resistance is equivalent to the permeability. A theoretical method to determine the permeability is given by the Carman-Kozeny equation which requires information of geometrical quantities of the porous medium. The mathematical model has been applied to the intervillous space of the hemochorial multivillous placenta. The numerical results can be used to determine the flow and pressure distribution when the permeability is known. By means of a mechanical model of a functional unit of the human placenta it could be shown that the Carman-Kozeny equation provides reasonable values for the permeability of the intervillous space.

The present mathematical model for the human placenta is still based on the assumption of a Newtonian fluid for blood. In further studies this assumption should be relaxed. This is important since it is known that in the case of preeclampsia or other forms of maternal dysfunction the non-Newtonian character of blood is enhanced. More specifically, there is evidence that reduced fluidity of maternal blood might critically limit the perfusion of the intervillous space and, therefore, the supply of oxygen and nutrients to the fetus.

Since the placenta in situ is inaccesible to flow visualization or flow measurements, a theoretical model can complete this gap and provide a quantitative description of the hemodynamics. It might, therefore, in the future be possible to determine with the aid of the model if the placenta is sufficiently perfused which will have considerable therapeutic implications.

LIST OF SYMBOLS

A_c	red cell surface area
a_1 b_1 c_1	constants
a_2 b_2 c_2 d_2	constants
d, D	abbreviations, Eqs. (2 and 9)
D_T, D_T^*	vessel diameter, minimum (critical) vessel diameter
F_1, G_1	abbreviations, Eqs. (3-4)
F_2, G_2	abbreviations, Eqs. (4-6)
H_T, H_D	tube hematocrit, discharge hematocrit
k	permeability of porous medium
L	vessel length or length of filter column
p, p_t	hydrostatic pressure, total pressure

Q	blood volumetric flow rate
Re_A	Reynolds number of flow in artery $= V_A \cdot \rho \cdot D_{TA}/\eta$
s	specific surface area
V	fluid velocity
w	superficial velocity
ϵ	porosity
ζ	head loss coefficient
ρ	fluid density
η	fluid viscosity
η_{rel}, η_{pl}	relative viscosity, plasma viscosity

REFERENCES

Dullien, F.A.L., 1979, Porous Media. Fluid Transport and Pore Structure, Academic Press.

Kaufmann, P., 1981, Entwicklung der Plazenta, in: Die Plazenta des Menschen, V. Becker, T.H. Schiebler and F. Kubli, eds., G. Thieme Verlag, Stuttgart-New York.

Kaufmann, P., 1983, Vergleichend-anatomische und funktionelle Aspekte des Placenta-Baues, Funkt. Biol. Med. 2: 71-79.

Koller, A. and P.C. Johnson, 1986, Methods for in-vivo mapping and classifying vessel networks in skeletal muscle, in: Microvascular Networks: Experimental and Theoretical Studies, A.S. Popel and P.C. Johnson, eds., Karger, Basel.

Moll, W., 1981, Physiologie der maternen plazentaren Durchblutung, in: Die Plazenta des Menschen, V. Becker, T.H. Schiebler, and F. Kubli, eds., G. Thieme Verlag, Stuttgart-New York.

Papenfuss, H.D. and J.F. Gross, 1986, Mathematical simulation of blood flow in microcirculatory networks, in: Microvascular Networks: Experimental and Theoretical Studies, A.S. Popel and P.C. Johnson, eds., Karger, Basel.

Papenfuss, H.D. and J.F. Gross, 1987, Mathematical simulation of blood flow and fluid exchange in the microcirculation, Proceedings of the 1987 ASME/JSME Thermal Engineering Joint Conference - Vol. 3: 543-548, P.J. Marto and I Tanasawa, eds., The American Society of Mechanical Engineers, New York.

Schmid-Schönbein, H., 1988, Conceptional proposition for a specific microcirculatory problem: maternal blood flow in hemochorial multivillous placentae as percolation of a "porous medium", in: Trophoblast Research, Vol. 3: 17-38, P. Kaufmann and R.K. Miller, eds., Plenum Publishing Corp., New York.

ERYTHROCYTE IN THE CAPILLARY - THE MATHEMATICAL MODEL

Yu. Ya. Kislyakov and A.V. Kopyltsov

The USSR Academy of Sciences, Institute for
Informatics and Automation, Center for Ecological
Safety, 39, 14th line, Leningrad 199178 USSR

ABSTRACT

A mathematical model consisting in a system of partial
differential equations solved by the method of finite
differences has been developed to describe the motion of red
blood cells through microvessels less than 8 micrometers in a
diameter. The model implies simulation of a three-dimensional
asymmetrical elastic red cell with a tanktreading flexible but
inextensible membrane and lubrication theory is used to
describe the flow of plasma between them and vessel walls. The
computations allow the estimation of the shape of the red blood
cells, their positions in the capillary tube, the frequency of
the cell membrane rotation and the pressure gradient over an
individual red cell. It is shown that the final result of the
interplay of the assumed basic conditions is a reduction in
hydrodynamic resistance.

KEYWORDS

MATHEMATICAL MODEL - RBC - FLOW IN MICROCAPILLARIES - MEMBRANE
TANKTREADING - RBC-DEFORMABILITY

INTRODUCTION

A mathematical model describing the motion of red blood
cells through capillaries with a diameter less than 8 µm is developed in order to analyze blood flow in the microcirculatory
system. In this model, the size of capillaries, the volume
visco-elastic properties the surface area and of red blood
cells and blood plasma viscosity are considered. The volume and
the surface area of red blood cells are assumed to be constant,
while the cell membrane is assumed to be perfectly flexible,
but inextensible. Plasma flow in the gaps between the cell
surface and the capillary wall is considered to be governed by

partial differential equations which can be solved by a computer using the method of finite differences.

The distribution of red blood cells and the velocity with
which they move through capillaries, both determining the conditions of respiratory gases transfer to tissue micro-domains,
depend upon the forces influencing blood flow, the shape of
circulatory channel, the size and mechanical properties of
blood cells, and plasma viscosity. In order to investigate the
influence of a complex of all these factors upon red blood
cells in microvessels, mathematical models are generally used
(Lighthill 1968; Secomb et al. 1986; Vann and Fitz-Gerald,
1982). Most of them imply an axisymmetric, parachute-shaped red
blood cell. However, there exist also more complicated two-
dimensional models considering cell asymmetry (Secomb and skalak, 1982a). For further approximation to the realistic conditions, the development of a model simulting the motion of a
three-dimensional, asymmetric red blood cell through capillaries with a diameter less than 8 μm seems to be important. In
such narrow capillaries, the influence of the size of blood
cells and their mechanical properties as well as of plasma
viscosity upon the resistance to blood flow is most pronounced.
Here we develop such a model and use it to analyze the
relations between main parameters which determine the
microhemodynamic conditions, i.e. the red blood cell and plasma
velocities, the width of a gap between the red blood cell and
the capillary wall, and the frequency of a tank-treading motion
of the cell membrane.

FORMULATION

Blood flow in a capillary may be regarded as the motion
of an elastic body (red blood cell) together with surrounding
viscous fluid (plasma) through a cylindrical tube, resulting
from a hydrostatic pressure gradient. For the modelling of this
motion a system of equations can be used which describes both
the hydrodynamics phenomena in plasma and the deformation of a
red blood cell, taken as an elastic body. A red blood cell is
assumed to have a bullet shape and to move through a capillary
with defined velocity. Its surface represents a truncated
cylinder which is limited at one side by a rotation
hemiellipsoid with hemiaxes a and b.

Plasma pressure in a gap between the red blood cell and
the capillary wall is assumed to depend on two parameters X_I
and Ψ. Then, in the cylindrical system of coordinates (X_I, r,
Ψ), pressure P (X_I, Ψ), axial u, radial v, and azimuthal w components of plasma velocity in the gap between the red blood
cell and the capillary wall, with the width h (X_I, Ψ), satisfy
the equations:

$$\frac{\partial P}{\partial X_I} = \frac{\mu}{r} \cdot \frac{\partial}{\partial r}\left(r \cdot \frac{\partial u}{\partial r}\right), \tag{1}$$

$$\frac{\partial P}{\partial \varphi} = \mu \cdot \frac{\partial}{\partial r}\left(r \cdot \frac{\partial w}{\partial r}\right), \tag{2}$$

218

$$\frac{\partial P}{\partial r} = 0, \tag{3}$$

$$\frac{\partial w}{\partial \varphi} + \frac{\partial}{\partial X_I}(ru) + \frac{\partial}{\partial r}(rv) = 0, \tag{4}$$

where μ is plasma viscosity.

While solving the system of equations (I) – (4), it is convenient to assume that the cell is motionless and that the capillary wall moves with velocity U. In this case, the boundary conditions are given by

$$u\Big|_{r=R} = V_I, \quad u\Big|_{r=R+h} = U_I,$$
$$w\Big|_{r=R} = 0, \quad w\Big|_{r=R+h} = 0. \tag{5}$$

The radial component of plasma velocity v at both boundaries satisfies kinematic conditions

$$v_w = u_w \cdot \frac{\partial R_w}{\partial X_I} + \frac{w_w}{R_w} \cdot \frac{\partial R_w}{\partial \varphi},$$

where

$$v_w = v\Big|_{r=R}, \quad u_w = V_I, \quad w_w = 0, \quad R_w = R, \text{ given } r=R,$$

and

$$v_w = v\Big|_{r=R+h}, \quad u_w = U_I, \quad w_w = 0, \quad R_w = R+h, \text{ given } r=R+h.$$

$P(X_I, \varphi)$, $u(X_I, r, \varphi)$, $v(X_I, r, \varphi)$, $w(X_I, r, \varphi)$ are the 2π-periodical functions of φ. $R(X_I, \varphi)$ is the red blood cell radius, $h(X_I, \varphi)$ is the gap width, U_I and V_I are projections of U and V (velocities of points on the cell membrane) on the axis X_I. The value of V, considering that the points on the cell membrane move in the plane which is parallel to that of cell symmetry with equal frequency f, is estimated by Secomb and Skalak, 1982b.

$$V = n \cdot \Delta (f \cdot F(X_3)),$$

where n is a single normal to the cell surface, and $F(X_3)$ is a function which satisfies equation

$$F'(X_3) = T(X_3),$$

where $T(X_3)$ are the lengths of close guidelines which indicate the motion of the cell membrane points.

Using equations (I) and (2) and considering the boundary conditions (5), plasma velocity components, u and w, can be estimated. Integrating equation (4) according to the gap width and filling in u and w into resulting equation give differential equation of elliptical type

$$A_I \cdot \frac{\partial^2 P}{\partial X_I{}^2} + A_2 \cdot \frac{\partial P}{\partial X_I} + A_3 \cdot \frac{\partial^2 P}{\partial \varphi^2} + A_4 \cdot \frac{\partial P}{\partial \varphi} + A_5 = 0 \qquad (6)$$

Quotients of equation (6), with given values for U, D (capillary diameter), μ, S (red blood cell surface area), and W (red blood cell volume), are determined by the cell's shape (a,b), its position in the capillary (β is the angle between the X_I and the capillary axis, 1 is the distance between the point at which the axis X_I intersects the cell surface and the capillary axis), and the frequency f. Equation (6) can be solved using the network procedure and considering condition of 2π -periodicity of P according to the angle φ and a boundary condition that plasma pressure before the front of the red blood cell is zero.

While estimating the red blood cells shape and its position in the capillary, it is assumed that its motion is rectilinear and uniform, and that its membrane rotates uniformly. That means that the integral sums of forces and their moments which act upon the red blood cell are zero. It is also considered that the resultant of tangent forces acting on the membrane from outside and from inside the cell is equal to zero. In addition to these conditions, a relationship which characterizes the cell elastic properties can be obtained.
To an approximation,

$$P_I - P_2 = E \cdot \Delta d/d,$$

where Δ d is the increment of d (the cell length as measured along the axis X_I), P_I and P_2 are the frontal and the lateral tensions in the cell, which both result from the external forces, and E is Young's modulus of the cell. Thus, there is a system of 5 equations including 5 unknown parameters, $a,b,\beta,1$, and f, which can be estimated with the aid of a computer.

RESULTS AND DISCUSSION

Computations using the present model allow to estimate the shape of the red blood cell and its position in the capillary, the gap width, the frequency of the cell membrane rotation, the distribution of plasma pressure and velocity in the vicinity of the cell, and the pressure gradient over the endpoints of the cell, which causes the cells motion through the capillary. Thus, the cell's velocity being 1000 μm/s and the diameter of the capillary being 4 μm, the red blood cells length is 11.94 μm and its diameter is 3.94 μm. The frequency of the cell membrane rotation f is then 7 r.p.s., which allows the points on the cell membrane surface to move with velocities

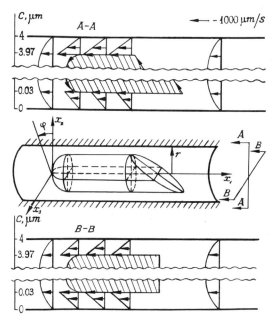

Fig.1. Plasma velocities distribution along A-A
and B-B sections. Diameter of a capillary
is 4 μm. Red blood cells velocity is 1000 μm/s.

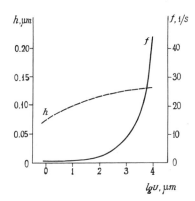

Fig.2. Alteration of the gap width (h) and the
frequency (f) of tank-treading motion of the cell
membrane as function of the cells velocity (U).
Diameter of a capillary is 5 μm.

of up to 200 μm/s. Therefore, the distribution of plasma velocities along A-A section (Fig. 1) is asymmetric. In the upper part of A-A section the cell's velocity is added to the velocity of its membrane, which leads to an approximate 20 percent increase in the velocity with which the points on the cell membrane surface move relative to the capillary surface. This velocity may thus reach 1200 μm/s. The conditions in the lower part of A-A section are just the opposite. The cell membrane velocity is subtracted from the velocity of the cell, which causes an approximately 20 percent decrease in the membrane points velocity relative to the capillary surface to an estimate of 800 μm/s. Along B-B section, since the red blood cell is planisymmetric, the distribution of plasma velocities is symmetric. An asymmetric distribution of plasma velocities due to the cell membrane rotation with the frequency f leads to a 10 to 20 percent decrease in resistance which opposes the cells motion through the capillary.

A change in the red blood cell's velocity U in the capillary causes alteration of the cells shape and position, as well as of the membrane rotation frequency f. While, at low velocities ($\sim$100 μm/s) the frequency f is only a few r.p.s., it increases rapidly with the increase in U up to 1000 μm/s or more and reaches the value of some tens of r.p.s. (Fig. 2). In a similar way, yet however to a lesser extent, there is an increase in the width of the gap between the cell and capillary wall (Fig. 2). This can be explained by the fact that the increase in U amplifies the forces which act upon the lateral surface of the cell. Thus, the cells diameter is reduced and h grows. There is an alteration of the cell surface areas above and below B-B section, which causes a disbalance between the tangent forces acting on the membrane from outside and from inside the cell. This results in an alteration of the frequency of a tank-treading motion of the cell membrane.

REFERENCES

Lighthill, M.J., 1968, Pressure-forcing of tightly fitting pellets along fluid-filled elastic tubes, J. Fluid Mech., 34:113.

Secomb, T.W., and Skalak, R., 1982a, A two-dimensional model for capillary flow of an asymmetric cell. Microvasc. Res. 24:194.

Secomb, T.W., and Skalak, r., 1982b, Surface flow of viscoelastic membranes in viscous fluids, Q. J. Mech. Math., 35:233.

Secomb, T.W., Skalak, R., Ozkaya, N., and Gross, J.F., 1986, Flow of axisymmetric red blood cells in narrow capillaries, J. Fluid Mech., 163:405.

Vann, P.G., and Fitz-Gerald, J.M., 1982, Flow mechanics of red cell trains in very narrow capillaries, Microvasc. Res. 24:296.

CLOSED FLUID QUADRILAMINA MODEL OF THE ERYTHROCYTE MEMBRANE

Reinhard Grebe and Holger Schmid-Schönbein

Parametric Image Processing and Simulations
Institute of Physiology
RWTH Aachen

ABSTRACT

Flow properties of whole blood highly depend on erythrocyte mechanical characteristics. To describe these static and dynamic properties, erythrocyte configurations are simulated using a numerical optimization method based on importance sampling and the principle of adiabatic cooling. The erythrocyte membrane is regarded as formed by a closed fluid quadrilamina. The energy function of the fluid laminae includes (i) curvature-elastic energy terms, (ii) the compression elasticity of the quadrilaminate, and (iii) the volume elasticity related to the osmotic pressure across the membrane.

KEYWORDS

Erythrocyte shape, numerical optimization, free energy function, mean curvature.

INTRODUCTION

The most striking attribute of the RBC is its smooth highly symmetric and stable shape at rest, combined with its unique "deformability", i.e. compliance resilience. To provide a basis for a mathematical treatment of its shape, the geometrical magnitude mean curvature has been used. It quantifies the local curvature of a surface as the mean of the two principal curvatures at a point. As a global measure, Canham (1970) used the 'total curvature' calculated by integrating the mean curvatures over the whole membrane. He stated that if the membrane is considered uniform total curvature becomes least for minimum bending energy.

The idea of a spontaneous curvature had been introduced in 1976 by Deuling and Helfrich in their work dealing with the optimization of RBC shapes exclusively using curvature elasticity. They considered a 'fluid' membrane with constant membrane properties. Their constraints have been constant membrane area and included volume. In order to achieve the normal RBCs shapes, they had to introduce an additional membrane property (the socalled *spontaneous* curvature or precurvature): the biconcave shape of the discocyte being found under the assumption that precurvature had a negative sign.

In a different approach Beck (1978) used simple models of stomato-, disco- and spheroechinocytes to show that this sequence of shapes is characterised by an increase of difference in area between the two laminae from $0.41 \mu m^2$ to $0.93 \mu m^2$, respectively.

Using a similar geometrical approach Svetina et al. (1982a) optimized models of RBC shapes based on four geometrical parameters with inner and outer lamina area and volume as constraints. Using variational calculus he found minima of bending energy for discocyte to stomatocyte shape transformation depending on the difference in lamina area. Therefore he took the membrane to be uniform, especially with respect to constant bending modulus over the whole membrane. In addition Svetina et al. (1982b) gave evidence that the difference in lamina area may be related to 'modifier binding' and membrane charge distribution. For these calculations he used the relationship between area difference and precurvature.

In all these models the membrane is considered uniform. Evans and Skalak (1979) used the well established theory of thin shells to treat the mechanical properties of the closed bilayer membrane. Based on this theoretical framework Evans and La Celle (1975) analysed the results of their micropipette-aspiration experiments. Using the presumptions of uniformity and constancy they derived that the shear modulus dominates the deformations in micropipetts.

At first sight, these theoretical and experimental results seem to be consistent. But there remain doubts about their validity in sight of concerning the high mobility of all the membrane components expressed in their diffusion constants. The available experimental data describing membrane properties on the molecular level make it more plausible to treat the membrane as a fluid consisting of components freely mobile in the plane. In addition, since membrane material can be transfered between the two laminae by the socalled flip-flop of cholesterol (Lange et al., 1981) and since even the high molecular weight spectrin molecules forming the membrane skeleton, probably the most stable part of the whole cell, have a diffusion constant of $10^{-12} cm^2/s$ our approach is more realistic.

We thus present a new model of the "fluid mosaic membrane" to describe the RBC membrane mechanical and shape behavior.

THE MEMBRANE: A CLOSED FLUID QUADRILAMINATE

The purpose of this paper is to present a new continuum model which is closely related to the established molecular properties of the constituents of the RBC-membrane of:

1. an outer lamina (assumed to be 4 to 20 nm thick) formed by the outer extramembraneous parts of intrinsic proteins and glycolipids, forming a velvet-like "membrane cortex".

2. two laminae of phospholipids and cholesterol forming the 4 nm thick core laminae, an incompressible fluid

3. an inner lamina, the socalled membrane-skeleton, consisting of the long (70 to 200 nm) extrinsic spectrin molecules, fibrillar but highly flexible. We assume them to have the configuration of an ionic gel (Stokke et al. 1986).

The role each of these structures plays in contributing to membrane behavior and erythrocyte shape will be focussed on now. This will provide the background for the development of the membrane model as a FLUID QUADRILAMINATE.

The Glycocalix

The outer lamina of the erythrocyte membrane is formed by the glycocalix consisting mainly of the extramembraneous parts of the intrinsic glycoproteins (glycophorins, band 3, band 4.5 etc.) and glycolipids (Viitala and Järnefelt, 1985). It exhibits a negative net surface charge mainly carried by sialic acid residues (most of which are parts of the glycophorins). Evenso fixed to this plane they are free to move in it.

Our membrane model in Fig. 1. gives a scheme of the charge carrying glycocalix. The small circles representing the negative charges the black sticks with branches illustrate their connection to the intrinsic proteins which are given by big ellipses. Surfaces containing mobile charges are characterised by a constant electrostatic surface potential. In addition constant

Figure 1. Submodel of the outer fluid lamina. The charge carrying glycocalix, represented by sticks with branches carrying circles indicating the negative charges. The bases are coupled to intrinsic protein molecules represented by big ellipses in the fluid lipid moving in an aqueous phase.

(a) Undisturbed configuration with equidistributed charges.

(b) Directly after application of an external shape changing force. Changes in local charge densities due to the geometry are induced. Increase of the charge density at the bottom of the protrusion decrease at the tip.

(c) After equilibration processes, charge carrying intrinsic molecules have moved until again an equipotential surface is reached.

mean curvature of this surface leads to constant charge density. Thus in the undisturbed and equilibrated scheme of the glycocalix (s. Fig. 1.a) the symbols of both the charges and the charge carrying molecules are distributed in a homogenous fashion.

Assume a localized external force to act on the membrane e.g. a pressure difference by a micropipette. This will produce local change of membrane shape as shown in Fig. 1.b. Local change of shape means local change of membrane curvature. As illustrated in Fig. 1.b the formerly uniform surface charge density is altered due to the changes in local curvature, the glycocalix representing no longer a surface of constant electrostatic potential. Since net coulombic forces are acting on the surface charges, a bending moment is induced due to the differences in surface tension of inner and outer membrane lamina. Thus the free electrostatic energy of the system is increased. This is the stage investigators using the invariant properties model have dealt with.

However, our knowledge about the properties of the molecular membrane components requires to go a step further: due to the high mobility of the charge carrying intrinsic proteins equilibration processes will occur. The net coulombic forces act on the charges and thereby move the intrinsic proteins until again a constant surface potential exists. The static state after completion of the equilibration processes is shown in Fig. 1.c. Again a constant surface charge density can be assumed. Thereby the density of charge carrying proteins is altered, they are no longer equidistributed in the membrane, free chemical energy of the membrane is therefore increased, directly related to inhomogeneities of local mean curvatures.

In fact, these processes will be much more complicated because a new equilibrium has to be established minimizing free electrostatic and chemical energy (Grebe et al., 1988). In addition, in order to solve three dimensional electrostatic problems, long range coulombic forces have to be taken into account also (Grebe et al., 1987), neglected in this first approach.

The Phospholipid-Cholesterol Core Laminae

The core of the membrane is build up by two asymmetric laminae of phospholipids facing each other by their hydrophobic molecular parts (hydrocarbon chains). Each of the core laminae is build up by about $2 * 10^8$ molecules, whose shapes vary between rod likes, and cone sections with a great variety of negative and positive inclines (Kleinfeld, 1987). The shapes not only vary between different types of molecules but also are subject to temperature influences.

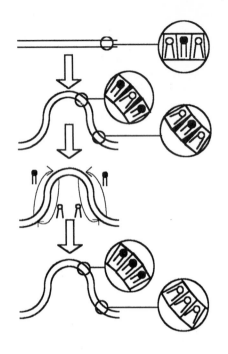

Figure 2. Simplified model of the outer core lamina consisting of two different cone shaped phospholipid molecules one with positive and the other with negative incline.

(a) The undisturbed configuration with the alternating sequence of different shaped phospholipids.

(b) Immediately after application of an external shape changing force. The induced surface tension is indicated by the black wedges between the lipid cones.

(c) During equilibration phospholipid molecules change places until the sum of free areas is minimized which means minimization of surface tension.

In the model we treat only one lamina, the *outer* of the core laminae. For the inner core lamina the same line of arguments is applicable. As a first approach this lamina shall be composed of two kinds of cone shaped phospholipids only, one facing with its thick base, the other with its narrow tip to the core. Fig. 2.a shows an undisturbed equilibrated planar part of the cell model which is stress free.

The action of an external force again leads primarily to local changes of membrane shape and change in mean curvature: on the molecular level this means the molecules don't fit any longer due to their geometry. In Fig. 2.b the induced free volumes are indicated by black wedges. In previous models of the membrane, all molecules were assumed to remain in place (topological conservation) and thus to store energy caused by local bending and shear. Localized changes in internal lateral membrane pressure were assumed while the changes in remainder of the membrane were neglected (mechanical conservation).

This again is in clear contradiction to the fluid character of the membrane, rendering the fluid lamina to an equipotential surface with relaxation of tension everywhere. Inducing local change in lateral pressure and differences in tension between neighbouring membrane parts, in bound induce a corresponding movement with a finite time constant according to the membrane viscosity. In our simple fluid model of the membrane therefore, molecules are forced to move between the basis of the protrusion and the tip as indicated by the arrows in Fig. 2.b (topological and mechanical equilibration).

This process spreads out over the *whole* membrane. The final stage of this process is given in Fig. 2.c. The model lipids have rearranged until no more free volumes occur. Again lipid membrane components are no longer equidistributed. However, free chemical energy increases, again directly related to inhomogeneities of local mean curvatures.

The Spectrin-Lamina

The fourth lamina is a protein lamina mainly formed by the densely crowded, loosely connected peripheral proteins spectrin and actin, connected via ankyrin (band 2.1) and band 4.1 protein to the intrinsic proteins band 3 and glycophorin or directly to other integral proteins

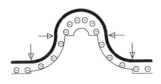

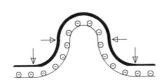

Figure 3. The inner lamina representing the spectrin molecules covering the endoplasmatic surface of the membrane as an ionic gel.

(a) Undisturbed configuration with constant thickness of lamina and of charge density.

(b) Shape changed configuration. Thickness of lamina and charge density are altered due to changes in curvature.

(c) After termination of equilibration process. Spectrin molecules have moved until constant thickness of lamina and of charge density have been reestablished.

(Bennett, 1989). The long, flexible spectrin molecules carry a negative net charge. Attached to the intrinsic domain by high-affinity bonds, they undergo dynamic processes of dissociation and re-association. Although the exact spatial arrangement of the spectrin molecules remains unclear, it is adequate to introduce a fourth lamina of finite thickness and of homogeneous negative charge distribution as a first approximation for a continuum model (Fig. 3.a). If an external force changes this equilibrated state, inhomogeneities in the thickness of the laminae as well as in charge density are induced. As indicated by the arrows in Fig. 3.b, such inhomogeneities occur in regions of changing curvature. Again these processes induce an increase of free electrostatic and chemical energy of the membrane, bound to lead to equilibration processes. The highly compliant spectrin lamina will rearrange until density of charge and thickness of lamina will be uniform again (Fig. 3.c).

The Fluid Quadrilaminate as a Dissipating and Conserving System

When shape-altered by an external force, the membrane undergoes equilibration processes of its components. During this process the former equidistribution of charge carrying intrinsic proteins, as well as of various types of phospholipids and of spectrin molecules and their anchoring proteins will be redisturbed, a process associated with the dissipation of mechanical energy. At the same time free chemical energy in the membrane will increase. Thus overall equilibrium between all the engaged membrane components has to be established which means new local compositions of membrane components due to the locally induced mean curvatures and the molecular response to external forces.

Due to the fluid character of the membrane, these equilibration processes are bound to spread out over the entire membrane until again an equipotential state is reached over the whole surface (probably on a different level). So in principle one has to exclude measurable and computable *local* membrane properties. The logic of our approach leads to the conclusion that the only "properties" which can be measured experimentally and which can be theoretically evaluated are *global* ones. All magnitudes have to be integrated over the entire membrane: local areas, volumes, mean curvatures and local inhomogeneities in mean curvature.

In this first approach we restrict ourselves to a simplified model of the membrane comprising only one closed singly-connected viscoelastic lamina. In addition we treat static states here i.e. after termination of all equilibration processes.

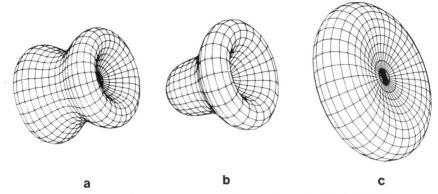

a **b** **c**

Figure 4. Shapes exhibiting free local (a and b) and the probable global (c) energy minima obtained by optimization with constrainted values of volume $90 \mu m^3$, surface area $134 \mu m^2$, and MMC $0.317 \mu m^{-1}$. E_r gives the free energy related to the one of the discocytic shape (c).
(a) Local free energy minimum $E_r = 5.5$ (established after ≈ 6000 runs) .
(b) Local free energy minimum $E_r = 3.27$ (established after ≈ 8000 runs) .
(c) Probable global free energy minimum $E_r = 1$ (≈ 14000 runs) exhibiting a discocytic shape.

THE MATHEMATICAL SIMULATION

The aim of the simulations is to determine the most probable shapes under different sets of constraints. Therefore a set of equations has to be introduced describing the free energy of the membrane as a whole. According to the principle of minimum energy the shape representing the membrane with least free energy has to be found.

Geometrical models representing typical erythrocyte shapes, exhibiting axial symmetry, can be built up from conic sections. They provide the geometrical data needed to calculate the total free energy of the membrane (chemical plus residual mechanical).

Contributions to the energy function for the fluid quadrilaminate are the compression elasticity of the laminate, a term depending on the volume enclosed by the fluid which is related to the osmotic pressure across the membrane, the curvature-elastic energy which depends on mean mean curvature and a novel feature due free chemical energy. This is a quadratic term treating the homogeneity of the local cuvatures. It limits the fluctuations of the local mean curvature and therefore controls curvature homogeneity.

These free energy terms are summed up giving the so-called 'cost function' which has to be minimized by an optimization procedure. We used both of the two following optimization methods:

- The standard Metropolis Monte Carlo method, a stochastic method which generates new configurations of a given system by means of an energy criterion.

- The gradient relaxation method, a deterministic method based on Newton's equations.

For a detailed description of the simulations see Grebe and Zuckermann (1990).

RESULTS

The aim of the optimization procedure is to find the one configuration exhibiting the least free energy in the entire membrane with respect to the given constraints. While there are numerous configurations which exhibit local minima (some of which occur during every run of the optimization procedure) the principle aim of the method is to verify the one configuration which represents the global minimum.

During a typical optimization run volume and area are first optimized followed by optimization of the mean mean curvature. This follows from the cost function used to describe the free energy of the membrane. Thence, after a reasonable number of steps, every configuration exhibiting a local minimum will show small deviations from the constraints (volume, area and mean mean curvature) at the expense of relatively high values in the curvature homogeneity term.

In order to bias the search as little as possible, the starting configurations in these optimizations have been chosen to be nearly spherical, (volume $135\mu m^3$, surface area $134\mu m^2$ and a MMC of $0.317\mu m^{-1}$) since the sphere is the most homogeneous body in terms of mean curvature and the initial configuration with the global minimum of the free energy for the given constraints. If one or more constraints are changed (e.g. if the volume is decreased to $90\mu m^3$), there is a corresponding jump in total free energy due to the volume elasticity; obviously the sphere no longer represents the global minimum.

Fig. 4.a and b show configurations obtained during such optimization processes representing each a *local* free energy minimum after (a) ≈ 6000 runs and (b) ≈ 8000 runs, respectively. The values for area, volume and MMC reached by these simulations show errors less than 1%. Fig. 4.c shows the most probable *global* free energy minimum first reached after ≈ 14000 runs exhibiting a discocytic shape.

The graphs in Fig. 5 display the homogeneity of the local mean curvatures, as depicted functions of the perimeter (L) of the cross-section of the solid of revolution. The homogeneity term depends on the difference in neighbouring local mean curvatures as discussed above. It is weighted by the length of the circumference where this difference occurs. So it is given by $2\pi r * \frac{dc_m}{dl}$. Fig. 5.a to c gives an impression of how curvature homogeneity of the cell surface changes with proceeding optimization. As is to be seen, any decrease in free energy is related to decrease of frequency and maxima of deviation in curvature homogeneity.

However, the related free energy of the body in Fig. 5.a is 4.8 times and in Fig. 5.b 2.8 times higher than that of the discocytic shape shown in Fig. 5.c which represents the probable *global* minimum of this optimization procedure. Until now we have not been able to discover any configuration exhibiting a lower free energy although various types of optimization procedures have been tested. Thus, we feel justified to state that the typical discocyte presents the most probable configuration if the constrainted values of the bodies are the volume of $90\mu m^3$, surface area of $134\mu m^2$ and of the MMC of $0.317\mu m^{-1}$.

DISCUSSION

Any attempt to model the mechanical behavior of the erythrocyte membrane must take into account the well known facts about chemical composition, physicochemical properties of the constituents, the compound nature of the membrane lipids, (cholesterol, phopholipids, glycolipids) and the extrinsic and intrinsic membrane proteins. Most importantly, however, the established *mobility* of all these constituents forming a two-dimensional fluid suspension must be taken into account, but also the constraints of a finite number of molecules in a finite volume. The latter fact leads to the unusual rheological fact that the unusually high modulus of isotropic elasticity of the membrane also resides in the fluid phase. This is but one of the membrane's peculiarities in conflict with one's intuitive notion about elastic ("conservative") and viscous ("dissipative") reactions to external forces.

The main problem in modeling the erythrocyte membrane lies in the fact that in the macroscopic world no adequate examples can be found for this kind of laminated fluid, thus there are no established physical laws describing it. We have earlier critizised the application of "shell theory" developed to explain the behavior of clearly elastic thin, laminated materials due to its inconsistency with established facts about constituent mobility. In our attempt to develop an alternative model, we present the "closed fluid quadrilaminate" model, in which

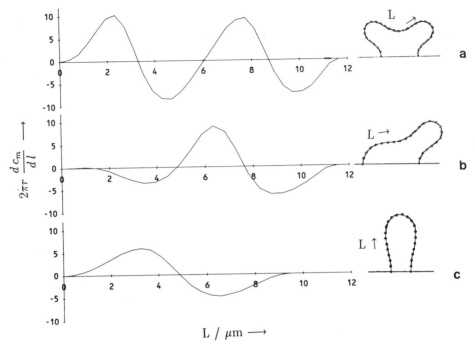

Figure 5. Free energy minima obtained by optimization of the shape of a body with volume $90\mu m^3$, surface area $134\mu m^2$, MMC $0.317\mu m^{-1}$. Given are the crossections of the solids of revolution and the related mean curvature homogeneity criteria as function of the perimeter L.

(a) Local free energy minimum ($\approx$ 6000 runs) showing high deviation and values in the homogeneity term.

(b) Local free energy minimum ($\approx$ 8000 runs) showing decreased deviation and values in the homogeneity term.

(c) Probable global free energy minimum exhibiting a discocytic shape.

compressible, elastic, chemical and electrostatic forces, and variable relaxation rates in different subsystems are incorporated, taking up in part approaches of previous authors.

Electrostatic properties have already been examined by Lopez et al. (1968) using a numerical optimization method for a continuum model of the membrane based on electrostatic field theory in the vacuum. Grebe et al. (1987) reconstructed erythrocyte shapes using charged conducting laminae in charge-carrying media. Here erythrocyte shapes were built up from toroidal and spherical components. Electrostatic properties of the surface charges and the aqueous interior and exterior media of the cell had been included in these models. The influence of molecular geometry has been investigated by Markin (1980). He used a two-dimensional model in which lateral redistribution of membrane molecules is allowed minimization of which leads to erythrocyte-like twodimensional outlines. Starting from observation in genetically spectrin deficient mouse RBC, Schmid-Schönbein et al. (1983) proposed a membrane model which exhibits overall fluid characteristics and models the influence of the spectrin lamina in terms of a dense periodic colloidal system.

Each of these models is based on the well-established fluid characteristics of the membrane. These approaches are unified in the proposed membrane model of a closed fluid quadrilamina in this paper in which the steric, electrostatic and chemical properties of glycocalix, core lamina, and the spectrin network is incorporated.

We start here with the analysis of the resting shape: Our optimization procedures used the closed fluid quadrilamina model, numerical calculations were performed for axially symmetric erythrocyte configurations. We show that the only constraints required to explain the naturally ocurring shapes are the constraints for volume, area, mean mean curvature, and homogeneity of curvature. Treating the membrane as a finite fluid quadrilaminate is the novel aspect of this model over previous ones and leads to the conclusion that a closed fluid lamina is able to exhibit global elasticity.

Compared to former continuum models, volume of cell and membrane, and surface area remain unchanged as global constraints. We redefine, however, previous concepts about "precurvature" or "spontaneous curvature" (Deuling and Helfrich, 1976) in that this variable is no longer taken as a global, but rather a local one. If in an experiment of thought, one were to take out a piece of the RBC membrane, the theory of "precurvature" would predict its bending to mechanical equilibrium. Our theory, would predict the components to be distributed such that each piece of the membrane is already in equilibrium as predicted from the set of equations used.

As a new feature of our model we introduce the demand for greatest possible curvature homogeneity (or, more precisely, to a principle of minimum inhomogeneity). Obviously in the spherical shape (which could be taken as the one in the absence of ionic pumps controlling cell volume) curvature and chemical composition is uniform. Any factor inducing curvature inhomogeneity and change in mean mean curvature is bound to induce "architectural inhomogeneity", "chemical inhomogeneity" and thence gradients of chemical energy. Conversely, any change in local or global chemical energy density is bound to induce changes in mean mean curvature.

Because every dissipative system aims to minimize a free energy the closed membrane strives to keep the sum of all local deviations of mean curvatures over the entire cell surface as small as possible. This has been demonstrated in Fig. 5. The graphs show for bodies with decreasing free energy (Fig. 5.a to c) a decrease in number and size of deviation from MMC.

The importance of the numerical value of the mean curvature is emphasised by the present results which are in full agreement with those of Grebe et al. (1987) for the case of axial symmetry. Thus the MMC offers not only an easy and direct way for calculating, for example, the difference in core laminae during normal erythrocyte shape change but also makes it possible to quantify the change in erythrocyte shapes obtained from experiments with "cup-formers" and "echinocyte formers" (Schönfeld and Grebe, 1989).

Our numerical optimization procedure is not restricted to any particular cell geometry in whole or in part but can extended from rotational symmetry to full tree dimensions. Thus the proposed closed fluid quadrilamina as the model of the membrane provides a basis for further development of a complex theory taking into account the secondary structure of the core laminae molecules, the electrostatic properties originating from the distribution of charges and dipoles, the concentration and mobility of various ions, the pH value, the viscosities of the different laminae which depend on molecular dimension and binding constants etc. All these magnitudes will be introduced into the free energy function as variables and may lead to different membrane configurations for the energy minima or different dynamic behavior of biological membranes.

CONCLUSION

A new three-dimensional continuum model of the erythroyte membrane is introduced which takes into account the wellknown fluid characteristics of the erythrocyte membrane. Basically it is a closed laminated fluid exhibiting global viscoelastic characteristics. The simplified one lamina model is described by a finite set of conic sections. So in this work we restrict ourselves to axial symmetric shapes which are optimized due to the free energy of the membrane. The energy function of the membrane laminate includes curvature-elastic energies, compression elasticity and volume elasticity which is related to the osmotic pressure across the membrane.

The free energy of an arbitrary starting configuration is optimized using numerical methods based on importance sampling in conjunction with the principle of adiabatic cooling or a multidimensional gradient evaluation.

The probable global minimum determined by this optimizations exhibit the wellknown shape of an discocyte for standard values of surface area $134\mu m^2$, volume $90\mu m^3$ and MMC $0.317\mu m^{-1}$. So erythrocyte shape might very well be determined by the physical properties of an ideal fluid forming a single connected closed lamina as described by the model presented in this paper.

ACKNOWLEDGMENTS

This work has in parts been supported by the DFG Gr 902/3.

REFERENCES

Bennett,V., 1989, The Spectrin-actin Junction of Erythrocyte Membrane Skeletons, Biochim Biophys Acta, 988(1):107-121.

Beck,J.S., 1978, Relations between Membrane Monolayers in some Red Cell Shape Transformations, Theor Biol, 75:487-501.

Canham,P.B., 1970, The Minimum Energy of bending as a possible Explanation of the Biconcave Shape of the Human Red Blood Cell, J Theor Biol, 36:61,

Deuling,H.J., and Helfrich,W., 1976, Red blood cell shapes as explained on the basis of curvature elasticity, Biophys J, 16:861-868.

Evans,E.A. and La Celle,P.L., 1975, Intrinsic Material Properties of the Erythrocyte Membrane Indicated by Mechanical Analysis of Deformation, Blood, 45(1):29-43.

Evans,E.A. and Skalak,R., 1979, Mechanics and Thermodynamics of Biomembranes: Part 1 and 2. CRC Crit.Rev.Bioeng, 3(4):181-418.

Fung,Y.C., 1981, Biomechanics: Mechanical Properties of Living Tissue. Springer-Verlag, New York.

Grebe,R., Peterhänsel,G. and Schmid-Schönbein,H., 1987, Change of local charge density by change of local mean curvature in biological bilayer membranes. Mol.Cryst.Liq.Cryst. 152:205-212.

Grebe,R., Zuckermann,M., and Schmid-Schönbein,H., 1988, Erythrocyte Shape is influenced by free Electric and Chemical Energy, in: Electromagnetic Fields and Biomembranes, M.Markov and M.Blank, eds., Plenum Press.

Grebe,R. and Zuckermann,M., 1990, Erythrocyte Shape Simulation by Numerical Optimization, Biorheology, submitted.

Kirkpatrick,S., Gelatt,Jr. and Vecchi,M.P., 1983, Optimization by Simulated Annealing, Science 220(4598):671-680.

Kleinfeld, A.M., 1987, Current Views of Membrane Structure, Membrane Structure and Function 29:1-27.

Lange,Y., Dolde,J., and Steck,T.L., 1981, The Rate of Transmembrane Movement of Cholesterol in the Human Erythrocyte, Biological Chemistry, 256(11):5321-5323.

Lopez,L., Duck,I.M. and Hunt,W.A., 1968, On the shape of the erythrocyte. Biophys.J., 8:1228-1235.

Markin,V. and Glaser,R., 1980, Forces and Membrane Particle Displacement in the Elastic Fluid Mosaic Model of Cell Membranes. Studia Biophys 3:201-211.

Schmid-Schönbein,H., Grebe,R. and Heidtmann,H., 1983, A New Membrane Concept for Viscous RBC Deformation in Shear: Spectrin Oligomer Complexes as a Bingham-Fluid in Shear and a Dense Periodic Colloidal System in Bending. Ann.N.Y.Acad.Sci. 416:225-254.

Schönfeld,M. and Grebe,R., 1989, Automatic shape quantification of freely suspended red blood cells by isodensity contour tracing and tangent counting. Computer Methods and Programs in Biomed., 28:217-224.

Sheetz,M.P. and Singer,S.J., 1974, Biological Membranes as Bilayer Couples. A Molecular Mechanism. Proc.Nat.Acad.Sci USA, 11:4457-4461.

Stokke,B.T., Mikkelsen,A., and Elgsaeter,A., 1986, The human erythrocyte membrane skeleton may be an ionic gel. I. Membrane mechanochemical properties, Eur Biophys J, 13(4):203-218.

Svetina,S., Ottova-Leitmannova,A., and Glaser,R., 1982a, Membrane Bending Energy in Relation to Bilayer Couples Concept of Red Blood Cell Shape Transformations, Theor Biol, 94:13.

Svetina,S., Pastushenk,V., and Zeks,B., 1982b, Difference in areas of the membrane monolayers due to modifiers of red blood cell shape, Periodicum Biologorum, 84:204-206.

Viitala,J., and Järnefelt,J., 1985, The Red Cell Surface revisited, Biochem Science, 10(10):392-395.

ESR AND ERYTHROCYTE STUDIES

Ana Margarida Damas[*], Maria Strecht Almeida[*], and Alexandre
Quintanilha[*,**]

* Sector de Biofísica, I.C.B.A.S., Universidade do Porto, Portugal
** Lawrence Berkeley Laboratory, University of California at
Berkeley, U.S.A.

ABSTRACT

Electron spin resonance is an important spectroscopic method for the study of
biological systems. This technique and its applications on the determination
of the intracellular microviscosity and volume of human erythrocytes are
described. Recent progress in the following areas is discussed: (a) dependence
of intracellular microviscosity and volume on the extracellular osmotic
pressure, (b) influence of aging on the anomalous osmotic behaviour of red
cells.

Key words: Spin-label/ESR/Erythrocyte/Microviscosity/Volume

INTRODUCTION

The use of electron spin resonance in conjunction with spin labels is
one of the interesting and important spectroscopic methods for the study of
biological systems.

The rotation of small spin label probes as a mesure of media viscosity
was introduced by Keith and Snipes (1974) who measured the rotational
correlation time of the spin label Tempone in different cells.

Spin label techniques have also been developed for measuring the
aqueous volume of red blood cells (Matos, 1984). The method consists of
quantitating the probe signal in a cell suspension by measuring the ESR line
heights of the labels inside the cells. The erythrocyte membrane is permeable
to these label compounds and the signal due to the nitroxide outside the cells
is quenched with a membrane impermeable paramagnetic line broadening
agent.

The small spin label Tempone was used. This is a water soluble and diffusible molecule and it was chosen because it is rapidly membrane permeable (Erikson et al., 1986) and its narrow linewidths are efficiently quenched by minimal concentrations of paramagnetic agents.

A good knowledge of erythrocyte internal microviscosity seems to be of considerable importance for the understanding of some of the complex rheological properties of the red blood cell itself and of whole blood (Dintenfass, 1968). It is probable that this microviscosity is related to cell volume. Since increasing the concentrations of osmoticum in the external medium reduces the volume of erythrocyte and, conversely, decreasing the external osmoticum increases the volume of vesicles, we looked at the behaviours of microviscosity and volume as a function of the extracellular osmolality.

The influence of the cell "age" on the osmotic response of both, the microviscosity and volume, were also investigated.

EXPERIMENTAL SECTION

Erythrocyte preparation. Human erythrocytes were isolated from blood samples obtained by venipuncture from healthy donors (EDTA was used as anticoagulant). After centrifugation at 2500 x g the plasma and buffy-coat were removed. The erythrocytes were then washed three times with an isotonic Na Cl solution. The pH was adjusted to 7.4 with 10 mM sodium phosphate buffer.

Separation of erythrocyte fractions. Red blood cells were separated by density centrifugation using the method described by Hall and Ellory (1986). Briefly, washed eythrocytes adjusted to a haematocrit of approx. 80% were centrifuged for 60 min. at 2500 x g in a bench centrifuge. The top and bottom 25% of the cells were collected and resuspended in buffer. Both fractions were treated as before and the final top and bottom fractions were saved for the experimental work. We will refer to the lighter fraction of cells as "young" and the denser fraction as "mature" cells respectively.

Variation of the osmolality of the medium. The effect of osmolality on the internal microviscosity and aqueous volume of labelled red blood cells was studied in Na Cl medium.

The osmolality was varied between 200 and 1000 mosmol/kg by using the molar concentrations of the solutes given by Waest and Astle (1978).

Spin labelling of the red blood cells suspensions. Washed red blood cells were ressuspended in isotonic phosphate buffered saline (pH 7.4) at 60% haematocrit. Then, the spin labelled cell suspensions, with different osmolalities, were prepared. Briefly, to 100 μl of the cells suspension 275 μl of phosphate buffered saline solution (variable NaCl concentration; pH 7.4) and 1.5 μl of a 100 mM aqueous solution of the spin label Tempone (2,2,6,6-tetramethylpiperidine-N-oxyl-4-one) were added. In the case of the cell suspensions in the presence of a quenching agent, the phosphate buffered solution contained $K_3[Fe(CN)_6]$ in a proper amount to a final extracelular concentration of 35 mM. Each sample was taken up into a glass cappilary and submitted to spectroscopic analysis.

Recording of the electron spin resonance spectra. ESR spectra were recorded with a Varian E-109 spectrometer at the following settings: modulation amplitude at 0.25 G; power 10mW; scan time 4 min.; and time constant 0.128 s.

Determination of the spectral parameters. The spin label ESR spectrum is extremely sensitive to the nature and rate of motions that the label undergoes. This motion was quantified by the rotational correlation time which was determined using the formula given by Keith et al. (1974)

$$t_c = 6.5 \times 10^{-10} \, \Delta H \, [\sqrt{(h_0/h_{-1})} - 1]$$

with ΔH the width of the central line (in gauss) and h_0 and h_{-1} respectively the amplitudes of the central and high field lines.

The procedure for absolute aqueous volume measurements consists of obtaining a Tempone spectrum in the presence of blood cells and a second spectrum with added quenching agent. Both signals are directly related with the number of unquenched spins in the sample.

Then the second aqueous signal is expressed as a fraction of the initial and converted to internal aqueous volume:

$$v_{aq} = (1/n) \, (h_2/h_1)$$

with h_1 and h_2 respectively the amplitudes of the central line of the spectra without and with the quenching agent, and n the number of cells per unit volume. The number of cells per unit volume suspension was determined by microscopic count in Neubauer chambers.

RESULTS

Spectra obtained after labelling of erythrocytes with Tempone. Fig. 1 shows the spectra of a suspension of red blood cells labeled with Tempone. As expected, narrow lines are observed indicating that the labels are tumbling quickly. When added to a cell suspension Tempone rapidly diffuses across the plasma membrane and the ESR spectrum is the sum of the intra and

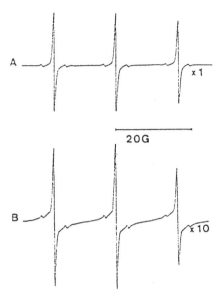

Figure 1   ESR spectra of Tempone (0.4 mM) in a suspension of cells (A) in the absence of the quenching agent (relative gain = 1); in the presence of 35 mM $K_3[Fe(CN)_6]$ (gain = 10).

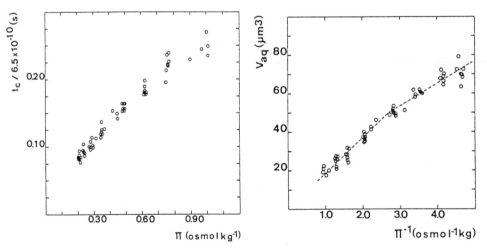

Figure 2   The variation of erythrocyte internal microviscosity as a function of the osmolality of the external medium.

Figure 3   The aqueous volume of red blood cells as a function of the (osmolality)$^{-1}$ of the external medium

extracellular components (fig. 1A). In a mixture of cells, spin label and broadening agent, the resultant electron spin resonance signal arises only from the spin labels which are not in close contact with the broadening agent; this is the intracelular aqueous space (fig. 1B). It is clear that the extracelular component was effectively removed by addition of 35 mM potassium ferricyanide.

Determination of the cytoplasmic microviscosity of normal human red blood cells. Under certain conditions, which are satisfied in our experiments, the rotational correlation time of a spin label provides a reasonably accurate estimate of the microviscosity of the system. Using the line-width and height of the spectra recorded on labeled erythrocytes, the rotation time of the label was calculated and the effect of osmolality on the internal viscosity of labelled red blood cells was studied in Na Cl medium.

The variations in the red cell volume are followed as a function of the external osmolality and the corresponding variations in the internal microviscosity were calculated (fig. 2). As expected, the internal viscosity of the erythrocytes increases with increasing osmotic pressure.

Determination of aqueous volume of normal human red blood cells. In contrast to light scattering techniques, the spin probe method will provide accurate measurements of cell volume that are independent of cell shape. The effect of osmolality on erythrocyte aqueous volume was studied according to the procedure described in the experimental section and is presented in fig. 3.

The total volume was calculated assuming that the volume occupied by hemoglobin and other cytoplasmatic solutes is 30% of the erythrocyte volume (mean value for the whole population) in an isotonic medium. This non aqueous volume was assumed to be constant when the cells pass trough variations of the osmolarity of the external medium. In an isotonic medium, the value for the aqueous volume is (mean±SD): $V_{aq}=60.6\pm1.2$ μm^3 and the total volume is $V=87.1\pm1.7$ μm^3. The error associated with each individual volume determination was 5%.

Microviscosity and aqueous volume of density separated human red cells. It has been claimed that during in vivo aging, human red blood cells lose membrane fragments and water, leading to a decrease in volume and an increase in density. A blood sample therefore contains a distribution of erythrocytes with different biochemical, functional and structural properties (Linderkamp and Meiselman, 1982; Rifkind et al., 1983; Sutera et al., 1985; Zanner and Galey, 1985; Hall and Ellory, 1986; Endre et al., 1986; Dhermy et al., 1987; Jain, 1988).

Cells were density separated and the behaviours of the microviscosity and volume were calculated for each of these populations.

Fig. 4 shows the variation of the microviscosity in "young" and "mature" red cells suspended in media of different osmolalities. Comparison of the behaviour of t_c in both populations, clearly shows that the "young" cells have a lower microviscosity. The variation of volume in those populations is presented in fig. 5. The behaviour of "mature" cells is closer to the behaviour of the whole population, while "young" cells behave in a very different way.

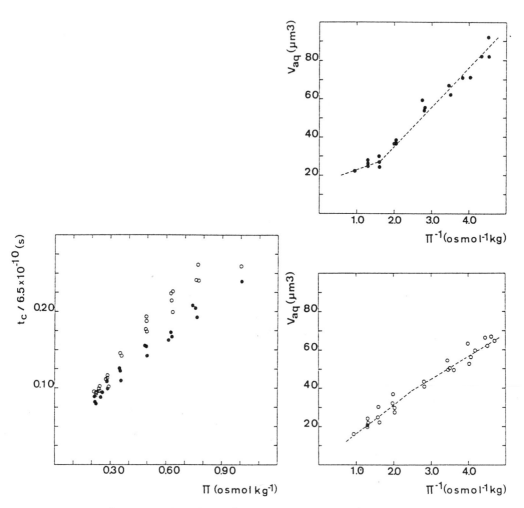

Figure 4   The microviscosity of "young" (●) and "mature" (o) red blood cells as a function of the osmolality of the external medium.

Figure 5   The aqueous volume of "young"(●) and "mature"(o) red cells as a function of the (osmolality)$^{-1}$ of the external medium.

The cell water volume of "younger" cells in an isotonic medium is 64.7 ± 3.2 μm^3.

DISCUSSION

In the normal circulation, erythrocytes traverse osmolalities ranging from 287mosmol/kg in the general circulation to 1200mosmol/kg in the vasa recta at the tip of renal medulla (Endre and Kuchel, 1986). Consequently, human erythrocytes, while circulating, undergo a vast range of variations of osmolalities. We have studied the cytoplasmic viscosity and aqueous volume of normal human red blood cells in media with different osmolalities.

The intracelular microviscosity apart from depending on the extracelular osmotic pressure is influenced by the cell "age". This dependence is explained on the basis of alterations of the red blood cell volume. In fact, comparison of "mature" and "young" cells shows that the latter have a lower microviscosity. This should be expected because of their larger volume. The volume of "younger" cells is approximately 7% greater then the value of the whole population. The value calculated for the red blood cell volume, $V=87.1\pm1.7$ μm^3, is in agreement with data reported in the literature, which has a spread from 85 to 100 μm^3 for unfractionated red cells (Sutera et al., 1985).

The results demonstrate that the osmotic response of the human erythrocytes is not the same as that of a perfect osmometer. This is in agreement with the results of Heubush et al. (1985) who measured the volumes of human erythrocytes using a Coulter Channelyzer.

The basic volume osmotic pressure relationship for a red cell which contains a portion of its volume, b, osmotically inactive is according to Gary-Bobo et al. (1968):

$$V = b + \frac{\pi_{iso}}{V_{iso} - b} \cdot \frac{1}{\pi}$$

From the experimental values of the erythrocyte aqueous volume, we see that there is a break in the linear relationship between V_{aq} and $1/\pi$. The slope is different when $1/\pi$ is greater then or less then 2.7 $osmol^{-1}Kg$. This means that the parameter b is not constant and the osmoticaly inactive volume of the cell varies when π passes by 0.37 osmol Kg^{-1} ($1/\pi= 2.7$ $osmol^{-1}Kg$) . The amount of osmotically inactive water may be associated with the cell's hemoglobin content and also the area or the shape of the membrane.

The fact that different values of the constant b will fit different regions of the curves shown in Figure 5 of the aqueous volume of erythrocytes as a function of the extracellular osmolality both in "young" and "mature" cells, suggests that the osmotic behaviour of these cells is quite different and that therefore their volume regulation mechanisms are also different. The possibility that there are substantial fractions of water in these cells that may be insensitive to the extracellular osmolality and that could include "bound water" (near the membrane surface and hemoglobin) should be further investigated.

ACKNOWLEDGEMENTS

We are grateful to Prof. M. de Sousa and collaborators for their help. This work was partially supported by a grant from INIC.

REFERENCES

Dhermy, D., Simeon, J., Wantier, M. P., Boivin, P., and Wautier, J. L., 1987, Role of membrane sialic acid content in the adhesiveness of aged erythrocytes to human cultered endothelial cells, Biochim. Biophys. Acta, 904:201.

Dintenfass, L., 1968, Fluidity (internal viscosity) of the erythrocytes and its role in physiology and pathology of circulation, Haematologia, 2:19.

Endre, Z.H., Kuchel, P.W., 1986, Viscosity of concentrated solutions and of human erythrocyte cytoplasm determined from NMR measurement of molecular correlation times. The dependence of viscosity on cell volume, Biophys. Chem., 24, 337.

Eriksson, U. G., Tozer, T. N., Sosnovsky, G., Lukszo, J., and Brasch, R. C., 1986, Human erythrocyte membrane permeability and nitroxyl spin-label reduction, Journal of Pharmaceutical Sciences, 75:334.

Gary-Bobo, C.M. and Solomon, A.K., 1968, Properties of hemoglobin solutions in red cells, J. Gen. Physiol. 52, 825.

Hall, A. C., and Ellory, J. C., 1986, Evidence for the presence of volume sensitive KCl transport in "young" human red cells, Biochim. et Biophys. Acta, 858:317.

Heubusch, P., Jung, C. Y., and Green, F. A., 1985, The osmotic response of human erythrocytes and the membrane cytoskeleton, J. Cell. Physiol., 122:266.

Jain, S. K., 1988, Evidence for membrane lipid peroxidation during in vivo aging of human erythrocytes, Biochim. Biophys. Acta, 937:205.

Keith, A. D., and Snipes, W., 1974, Viscosity of cellular protoplasm, Science (Wash. DC), 183: 666.

Linderkamp, O., and Meiselman, H. J., 1982, Geometric, osmotic and membrane mechanical properties of density-separated human red cells, Blood, 59:1121.;

Matos, M. A., 1984, "Aplicação da Ressonância Paramagnética Electrónica ao Estudo do Eritrócito, Dissertação de Mestrado em Química Teórica", Universidade do Porto, Porto.

Rifkind, J. M., Araki, K., and Hadley, E. C., 1983, The relationship between the osmotic fragility of human erythrocytes and cell age, Archives of Biochem. Biophys., 222:582.

Sutera S. P., Gardner, R. A., Boylan, C. W., Carroll, G. L., Chang, K. C., Marvel, J. S., Kilo, C., Gonen, B., and Williamson, J. R., 1985, Age-related changes in deformability of human erythrocytes, Blood, 65:275.

Waest, R.C., Astle, M.J. (eds) , 1978, "Handbook of Chemistry and Physics", 59th Edn., pp. D299, D300 and D308, CRC Press, Cleveland, OH.

Zanner, M. A., and Galey, W. R., 1985, Aged human erythrocytes exhibit increased anion exchange, Biochim. Biophys. Acta, 818:310.

MODELLING HEMODYNAMICS IN SMALL TUBES (HOLLOW FIBERS) CONSI-

DERING NON-NEWTONIAN BLOOD PROPERTIES AND RADIAL HEMATOCRIT

DISTRIBUTION

Dietmar Lerche

Humboldt University, Charité
Inst. Med. Physics & Biophysics
Berlin, DDR-1040
German Dem. Rep.

KEYWORDS / ABSTRACT

Blood flow, non-Newtonian blood viscosity, streaming profile,
radial hematocrit profile, pressure drop

Velocity profiles and capillary volumetric flow were
deduced from Navier-Stokes-equation accounting the non-Newtonian blood viscosity and its hematocrit dependence. Special
attention was paied to the influence of the radial hematocrit
distribution on radial flow pattern and the axial pressure
drop.

INTRODUCTION

Much attention has been centered on modelling the dynamics of microcirculation in biological capillaries and very
small tubes (Flaud and Quemada, 1980; Perkkiö and Keskinen,
1983) or on computing velocity fields in large geometrics
(cp. van Steenhoven et al., 1990). In the present paper we
shall characterize microhemodynamics, i.e., flow-pressure-re-
lationship and radial velocity profile, in small tubes or
small vessels yet on the basis on continuum mechanics. Two
experimentally proved facts will be incorporated into the
model. Firstly, the radial concentration distribution of
normal, deformable human erythrocytes (radial hematocrit
profile) in small tubes (P. Aarts, 1985; Nobis et al., 1985).
Secondly, we take into account the non-Newtonian behaviour of
blood at medium and higher local RBC volume fraction (equated
to hematocrit in this paper). Hematocrit dependence and the
influence of microrheological red blood cell (RBC) properties
on the blood viscosity should be explicitly described by the
chosen viscosity equation.

On the basis of the HUNT-equation (Weizel, 1949) and the
continuity law a modified Navier-Stokes-Equation was derived,

allowing to treat the viscosity as a function of radial position in dependence on radial shear rate and chosen hematocrit distribution. Radial velocity profiles and volumetric flows were calculated and the influence of hematocrit profile as well as blood flow properties (aggregation, deformation) was assessed computationally. Finally, we compared our theoretical velocity profiles and experimentally determined in vivo profiles and deduced in vivo radial hematocrit profiles.

BASIC ASSUMPTIONS AND EQUATIONS OF THE MODEL

We applied the hydrodynamic basic equation – the so-called HUNT-equation – on a stationary blood flow through rigid, cylindrical tubes at low Reynold's numbers. For a symmetric cylindric tube flow we obtain:

$$\frac{1}{r} \frac{\partial}{\partial r} \left(\eta(r) \cdot r \cdot \frac{\partial v(r)}{\partial r} \right) = \frac{\partial p}{\partial l} \tag{1}$$

where r determines the radial position, p and l are the pressure and tube length. The apparent blood viscosity η is approached by the Quemada model (Quemada 1981) and reads

$$\eta = \eta_{pl} (1 - 0.5 \cdot k(\dot{\gamma}, H) \cdot H)^{-2} \tag{2}$$

$$\text{with } k(\dot{\gamma}, H) = k_\infty + \frac{k_0 - k_\infty}{1 + (\dot{\gamma}/\dot{\gamma}_c)^{0.5}}$$

where η_{pl} is the viscosity of plasma and H equals the volume fraction of RBC, approximated in this paper by the hematocrit. The intrinsic viscosity $k(\dot{\gamma}, H)$ depends on k_∞, closely related to RBC orientation and deformation at very high shear rate as well as on k_0, related to RBC – RBC interaction (rouleaux formation and particle crowding). $\dot{\gamma}$ and $\dot{\gamma}_c$ denote the incident shear rate depending on radial position and a critical shear rate. Due to the minor influence of $\dot{\gamma}_c$ on the computation the structural parameter $\dot{\gamma}_c$ was set equal to a mean value of 1 s^{-1} (Lerche et al., 1980; Cokelet, 1987). The hematocrit dependence was approximated on the basis of literature viscosity data in dependence of hematocrit (Dufaux et al., 1980; Quemada, 1981; Cokelet, 1987):

$$k_0 = 55 \cdot H^{0.7} e^{-6} \cdot H + 1.9$$
$$k_\infty = 1.65 (H + 0.05)^{-0.3} \tag{3}$$

There are no theoretically derived models to describe the radial hematocrit profile in small vessels. To avoid a discontinuous two layer model (Perkkiö and Keskinen, 1983; Fenton et al., 1985) we used an empirical equation (Lerche and Lüders, 1987) to describe arbitrary smooth radial hematocrit profiles $H(r)$:

$$H(r) = -\frac{H_0 \cdot (n+1)(n-1)}{2} \left[\frac{r^n}{n} - \frac{2r^{n-1}}{n-1} + \frac{r^{n-2}}{n-2} - \frac{2}{n(n-1)(n-2)} \right] \tag{4}$$

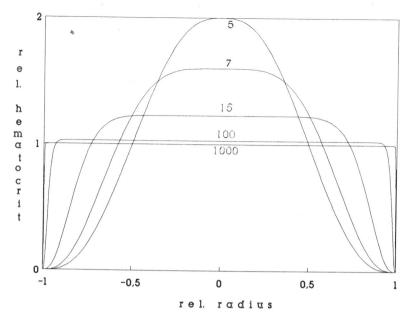

Fig. 1. Hematocrit profile according to equ. 4 in dependence
on the indicated parameter n. Profile is displayed
as H(r) divided by the mean capillary tube hematocrit H_0.

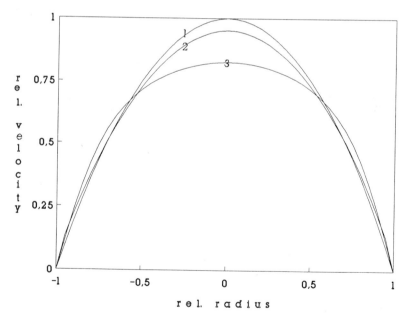

Fig. 2. Relative velocity profiles computed for different
radial hematocrit profiles (1: n = 9000; 2: n = 100;
3: n = 7) according to equ. 4 (H_O = 0.35; mean blood
and plasma viscosity equals 2.5 mPa.s and 1.2 mPa.s,
respectively).

where H_O is the mean tube hematocrit. This equation is valid for n = 5 up to n → ∞ and gives profiles as shown in Fig. 1.

RESULTS

Computed Velocity Profiles: Fig. 2 shows the influence of the chosen radial hematocrit profile on radial velocity profile. It becomes evident that the higher the RBC depletion near the wall and the concentration in the central tube part, the more pronounced is the deviation from the parabolic velocity profile. It should be stressed that an inversion of this phenomon can be obtained by increasing the value of relative hematocrit in the centre to H_{centre}/H_O-values near 2. That means the central velocity increases again despite a further increase of the shear rate near the wall. In other words, there occurs a qualitative change of the shape of velocity profile (Fig.3). Detailed studies further revealed that with the same characteristic profile (same n-parameter) blunting is more strongly marked at higher mean hematocrit H_O and that even with very high n-parameters (n = 1000) a small deviation from the parabolic profile in the centre and near the wall was obtained due to the applied non-Newtonian blood approximating viscosity law. Based on the viscosity law applied, the influence of altered rheological properties upon the velocity profile can be investigated. As an example, the intrisic viscosity k_{∞} - characterizing the ability of RBC to orient and deform under shear stress - was investigated (Fig. 4). The lower this parameter, the more the velocity profile is blunted.

Flow and Pressure Gradients: The above derived model was applied to calculate the flow-pressure relationship in capillaries (hollow fibers). Without going into detail (cp. Lerche and Oelke, 1990), it was shown that the effective hydrodynamic resistance of a capillary (r = 100μm) with the same mean hematocrit strongly depends on the radial hematocrit profile. The higher the hematocrit in the centre the higher is the flow at the same pressure gradient. The flow increased by about 50% at H_{centre}/H_O from 1 to 2. The hematocrit profile also influences radius dependent pressure gradient of the capillary. With the same flow a 10 % decrease of the capillary radius results in a higher pressure drop the higher the H_{centre}/H_O-value is (40 % difference beween H_{centre}/H_O = 1.003 and 1.6). Assuming a hydraulic conductivity of the capillary wall, it was shown that small hydraulic coefficients (e.g. L = 0.69 . 10^{-8} m/kPa . s) do not significantly change these gradients, whereas high coefficients (L = 4.6 . 10^{-8} m/kPa . s) decrease the pressure drop by about 10 % (Lerche and Oelke, 1990).

Comparison with Experimental in vivo Velocity Profile: Recently Tangelder et al., 1985 determined in vivo velocity profiles in rabbit mesenteric arterioles (diameter 17 - 32 μm). The shape of these profiles was also flattened as compared to a parabola, both in the systole and the diastole. The above authors ruled out that the flattening of the profiles can be explained by the depth of the optical section, by incomplete development of laminar flow profile, by curved vessels or by the pulsatility of the flow. As an explanation

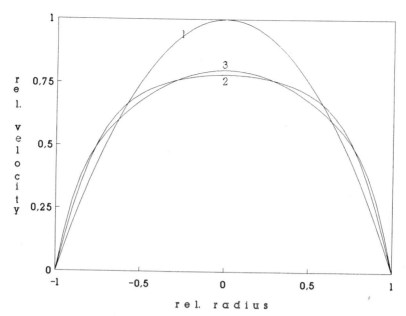

Fig. 3. Relative velocity profiles computed for radial hematocrit profiles (1: n = 9000; 2: n = 15; 3: n = 7).
For other data compare Fig. 2.

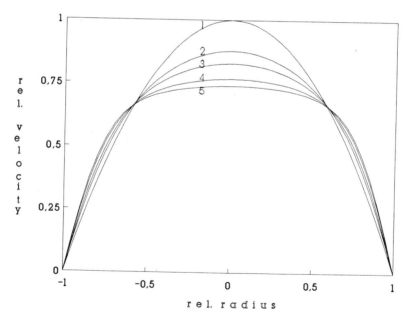

Fig. 4. Influence of rheological flow behaviour (modelled by
relative intrinsic viscosity $k_{\infty}/k_{\infty,normal}$ = 0.7
(plot 2), = 1.0 (plot 3), 1.4 (plot 4) and 1.6 (plot
5)) as compared to the parabolic profile (plot 1).
Other parameters: n = 7, H_O = 0.25, mean and plasma
viscosity 2.0 mPa.s and 1.2 mPa.s, respectively.

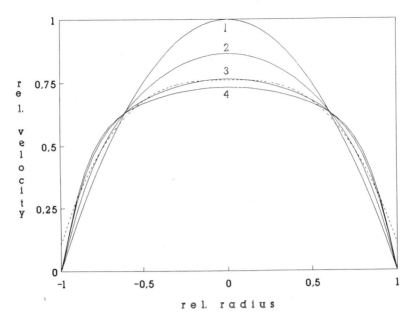

Fig. 5. Comparison of in vivo determined radial velocity
profile (- - -) and computed profiles for a radial
hematocrit profile characterized by n = 10
(H_{centre}/H_0 = 1.42) in dependence on relative intrinsic viscosity ($k_{\infty}/k_{\infty,normal}$ = 0.8 (plot 2), = 1.4
(plot 3) and 1.6 (plot 4).

for these blunted velocity profiles we investigated the influence of radial hematocrit profiles. Fig. 5 proves that
with an appropriate radial hematocrit profile the theoretical
velocity profile approaches very closely the experimentally
determined in vivo profile (mean velocity differences less
than at least 5 %). In general, the higher the mean tube
hematocrit the better the agreement.

From this fitting procedure an in vivo radial hematocrit
profile similar to profiles with an n-parameter 8 < n < 15
(1.25 < H_{centre}/H_0 < 1.5) can be deduced. Pries et al. (1989)
obtained similar profiles on the basis of a microcirculation
network analysis. It seems important to note that the wall
shear rate is about 50 % - 70 % higher than for corresponding
parabolic profiles causing the same flow.

CONCLUSION

1. Computed radial velocity profiles show a marked blunting and are similar to experimentally determined profiles, provided that a radial hematocrit profile
   with H_{centre}/H_0 > 1 is assumed. For $H_{centre}/H_0 \approx 1$
   profiles are nearly parabolic.

2. An increase of mean vessel tube hematocrit and of intrinsic viscosities (impaired rheological behaviour)
   results in a more pronounced plaque flow.

3. For uniform radial hematocrit distribution ($n > 9000$, $H_{centre}/H_O \approx 1$) non-Newtonian blood flow properties themselves will only result in minor deviations of the radial velocity profile from the parabolic one.

4. The higher the centrical RBC concentration the higher is the flow at the same pressure gradient.

5. If non-uniform hematocrit profiles characterized by $1.25 < H_{centre}/H_O < 1.5$ are assumed, velocity profiles may be computed, which are quite similar in shape as experimental ones. The apparent wall shear rate is about 50 % to 70 % higher than that of the corresponding parabolic profile resulting in the same flow.

ACKNOWLEDGEMENTS

The efforts of Mr. Flieger in computer work as well as of Mrs. Neumann and Mrs. Raddatz in typing this manuscript are greatly appreciated. This study was financially supported by the VEB MLW Keradentawerk Radeberg.

LITERATURE

Aarts, P.A., 1985, "The role of hemodynamic factors in the interaction of blood platelets with the vessel wall", Proefschrift, Rijksuniversiteit , Utrecht, The Netherlands.

Cokelet, R.G, 1987, The rheology and tube flow of blood, in: "Handbook of Bioengineering", R. Skalak and S. Chien (eds.), McGraw-Hill Book Company. New York.

Dufaux,J., Quemada, D., and Mills, P., 1980, Determination of rheological properties of red blood cells by Couvette viscometry, Rev. Phys. Appl., 15:1367.

Fenton, B.M., Wilson, D.W., and Cokelet, G.R., 1985, Analysis of the effects of measured white blood cell entrance times on hemodynamics in a computer model of a microvascular bed, Pflügers Arch. 403:396.

Flaud, P., and Quemada, D., 1980, Role des effets non newtoniens dans l'ecoulement pulse d'un fluide dans un tuyau viscoelastique, Rev. Phys. Appl., 15:749.

Lerche, D., Dufaux, J., Quemada, D., and Glaser,R., 1980, Rheological characterization of washed red blood cell suspensions in dependence on the medium human albumin concentration, in: "Proceedings of the 1. European Symposium on Hemorheology and Diseases", J. Stoltz and P. Drouin (eds.), Doin Editeurs, Paris.

Lerche, D., and Lüders, H., 1987, Modulation of tube velocity profiles and hematocrit depending non-Newtonian blood properties, in: "Microcirculation - an update.", M. Tsuchiya, M. Asano, Y. Mishima and M. Oda (eds.), Elsevier Science Publishers B. V., Amsterdam.

Lerche, D., and Oelke, R., 1990, Theoretical model of blood flow through hollow fibers considering hematocrit-dependent non-Newtonian blood properties, Int. J. Artificial Organs, in press.

Nobis, U., Pries, A.R., Cokelet, G.R., and Gaehtgens, P., 1985, Radial distribution of white cells during blood flow in small tubes, Microvasc. Res., 29:295.

Perkkiö, J., and Keskinen, R., 1983, On the effect of the concentration profile of red cells on blood flow in the artery with stenosis, <u>Bull. Math. Biol.</u> 45:259.

Pries, A.R., Ley, K., Claassen, M., and Gaehtgens, P., 1989, Red cell distribution at microvascular bifurcations, <u>Microvasc. Res.</u>, 38:81.

Quemada, D., 1981, A rheological model for studying the hematocrit dependence of red cell - red cell interaction in blood, <u>Biorheol.</u>, 18:501.

Tangelder, G.J., Slaaf, D.W., Muijtjens, A.M.M., Arts, T., Egbrink, M.G.A., and Reneman, R.S., 1986, Velocity profiles of blood platelets and red blood cells flowing in arterioles of the rabbit mesentery, <u>Circ. Res.</u>, 59:505.

Van Steenhoven, A.A., Rindt, C.C.M., Janssen, J.D., and Reneman, R.S., 1990, Numerical and experimental analysis of carotid artery blood flow, cp. this volume.

Weizel, W., 1949, "Lehrbuch der Theoretischen Physik", Springer Verlag, Berlin, Göttingen, Heidelberg.

BLOOD-WALL INTERACTIONS

3-D VISUALIZATION OF ARTERIAL STRUCTURES AND FLOW PHENOMENA

R.I. Kitney, C.J. Burrell and D.P. Giddens*

Biomedical Systems Group, Imperial College
Exhibition Road, London SW7, U.K.
*School of Aerospace Engineering, Georgia Institute
Atlanta, Georgia 30332, U.S.A.

SUMMARY

In this chapter we describe the reconstruction of arterial structures using three-dimensional solid modelling. The alternative approaches to three dimensional modelling are discussed and the voxel space system we use for intra-arterial imaging, based on ultrasonic data, is described. The data, acquired with a purpose-built, catheter-mounted ultrasound probe, is used to recreate 3-D computer models of arterial sections in vitro and in vivo. Examples to illustrate the power and flexibility of voxel space modelling in terms of post-processing and software manipulation are given. Preliminary work on tissue differentiation, using arterial models and colour-coding of the image is included. In addition to 3-D arterial visualization, we present early work on theoretical three dimensional flow field representation.

INTRODUCTION

The use of mechanical devices introduced percutaneously for the treatment of atheromatous arterial disease continues to increase. In the first half of 1989, there were in excess of one thousand PTCA procedures per million population performed in the U.S.A. Prototype devices for removal (atherectomy), splinting (stents) and thermal moulding (laser angioplasty) of atheromatous stenoses are now under evaluation. This evaluation and the appropriate use of these techniques depends on imaging the vessel wall and assessing the extent and type of atheromatous deposit present, both before and after treatment. Traditional contrast angiography is usually employed but provides only intermittent, negative images of the lumen of the vessel.

It is now possible to visualize the inner surface of the vessel and the surface of the atheromatous plaque directly by fibre-optic angioscopy, but only whilst the blood is replaced with a translucent, non-respiratory fluid. On the other hand, visualization using ultrasound has the great advantages in this context of 'seeing' through both the blood and the vessel wall, so that images of the full thickness of the artery can be obtained.

We have developed a complete ultrasound system based on catheter-mounted ultrasonic crystals for data acquisition with the aim of reconstructing detailed 3-D images of long arterial segments.

3-D ANATOMICAL VISUALIZATION

Principles of 3-D Modelling

Computer graphics are used to represent and manipulate 3-D objects. The main strategies employed are to represent the object in the form of either a wire model or a solid model. From a technical standpoint the first option is far easier to implement, but for many applications it is relatively unrealistic. Hence, an increasing amount of work has been undertaken on the development of computer-based solid models of 3-D objects. Such modelling strategies are based on three approaches:

        (i)   Constructive Solid Geometry (CGS) Modelling
        (ii)   Boundary Representation or B-Rep
        (iii) Voxel Space Modelling

    (i) Constructive Solid Geometry (CGS) Modelling. CGS modelling is based on the assumption that man-made objects can be built using simple geometric shapes. In CGS these basic shapes are called 'primitives' and objects are represented by sets of operators (e.g. union and intersection) applied to primitives positioned by means of translation and rotation. Figure 1 illustrates a chess pawn generated using this approach.

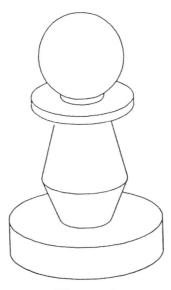

Figure 1

A Constructive Solid Geometry (CGS) chess pawn

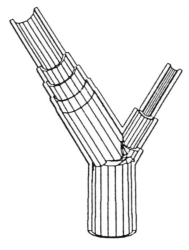

Figure 2

CGS carotid artery simulation with nearly
160 blocks in the complete segment

The number of parameters required to define each
primitive is dependent upon its geometry. Simple primitives
require three co-ordinates. One for the centre plus three to
describe angles of rotation. A sphere requires one additional
value, its radius, while a cylinder requires two, one for
radius and one for length.

Since CGS objects are defined as conglomerates of
primitives, any operation performed on an object must be
performed on each of its constituent parts and the relative
positions of each of the components maintained. This operation
can be assisted by determining the intersection between a
plane and the 3-D model. This intersection is calculated by
finding the intersection between every primitive and the given
plane.

CGS models must be computed each time they are used.
Computing intersections and unions between primitives is very
time consuming. Hence, the processing time grows not only as a
function of the number of primitives, but more importantly, as
a function of the number of interactions between primitives.
Figure 2 shows a CGS model of a carotid artery simulation
using approximately 160 primitives organised in the form of
rings.

CGS modelling forms the basis a large number of CAD/CAM
packages when man-made objects are to be modelled, but it is
not a good approach to the 3-D modelling of biological
structures because of the large number of geometric
primitives required.

(ii) <u>Boundary Representation (B-Rep)</u>. The B-Rep approach to 3-D modelling is a very natural one as it is closely related to the methods used in two dimensional drawing. In the 2-D case a curve is represented by a sequence of x,y points connected in the simplest form by a straight line or lines. Boundary Representation is a direct extension of this concept to three dimensions. In the 3-D case, a volume is defined by the surfaces which delimit it. The surfaces are described by sets of points (x,y,z) which are connected, again in the simplest form, by straight lines. However, whereas in the 2-D case the sequence of points gives the order of connection, in 3-D such a simple connection cannot be implemented since any point on the boundary surface is connected to more than one other point. Hence the connection order must be explicitly given by using either explicit polygons, explicit edges or by keeping a list of all the points and their connections. Usually the region within the polygons on the model surface is assumed to be planar i.e. a facet, hence the reconstruction is sometimes referred to as 'faceted'.

Data input to B-Rep systems is usually performed by either defining a series of contours in different planes and then allowing the system to connect them, or by defining primitives in a similar manner to that used in CGS modelling. In the latter case models are not stored as CGS models, but as facets and vertices computed from the primitives as they are defined. The method is very suitable when the structures to be modelled are smooth i.e. when not too many facets are required to represent the volume adequately.

Performing any operation on a B-Rep model means processing all the facets individually. For Boundary Representation models, finding the intersection between any given plane and the 3-D model implies finding the intersection between the plane of each facet and the intersecting plane. If the intersection occurs within the facet, then the intersection line must be computed, otherwise the facet is not intersected by the plane. For models with many facets this operation is very time consuming.

The strength of the boundary representation method lies in the fact that it uses computer memory as required, so that simple models are accurately represented by small amounts of memory. This subsequently leads to fast processing, one of the reasons why boundary representation is so popular with 3-D modelling packages designed for microcomputers.

Since B-Rep models are represented by their boundary surfaces, which in turn are made of facets, it is necessary that each facet be oriented in order to convey the information both "inside" and "outside" the volume. This is usually done by keeping track of the direction of the vector normal to each facet, assumed to point outwards from the model (Foley and van Dam, 1984). The models are defined by lists of vertices, their connections and normal vector directions. These lists must be kept in very strict order. If for any reason a single value in any of these lists is suppressed, or even modified, the resulting model will bear little or no resemblance to the original model.

Using B-Rep, it is possible to define models which cannot be

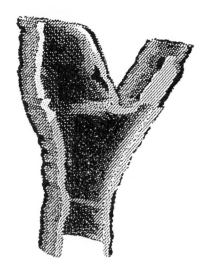

Figure 3

A Boundary Representation
carotid bifurcation simulation

defined in the real world, such as the famous Escher objects
(Pipes, 1985). Most CAD/CAM systems make use of this
technique to define the solid model.

Historically, B-Rep models were the first to be
implemented on computers, and consequently most of the theory
and the computer techniques developed for handling and viewing
3-D models were initially developed using B-Rep objects. An
example of a B-Rep rendering from our work on carotid
bifurcations is illustrated in Figure 3.

(iii) <u>Voxel Space Modelling</u>. The voxel, or "volume
element", can be seen as an extension to three dimensional
space of the familiar two dimensional digital pixel, or
"picture element". In the same way that 2-D space can be
divided into a large number of square pixels, so in voxel
modelling is the 3-D space divided into a large number of
cubes, called voxels, arranged side by side (Figure 4).

A voxel, v, can be defined by the x,y,z co-ordinates of
its centre, where x, y and z are assumed to be integers in the
interval [1, N_v], where N_v is the number of voxels in the
space.

Hence, voxel space, VS, is defined by:

$$VS = \{ V / 1 < x < N_v \; {}^\wedge \; 1 < y < N_v \; {}^\wedge \; 1 < z < N_v\} \quad (1)$$

where the symbol { denotes "The Set Of", / denotes "So
That", and ^ denotes the Boolean operator "And".

A number V(v) associated with each voxel is called the
"voxel value" and can be used to determine whether a voxel is
part of an object. In its simplest form a voxel, v, may be
said to pertain to an object O if V(v) is non-zero i.e.

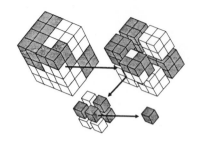

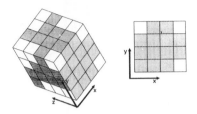

Figure 4

Figure 5

Voxel Space - made up from
cubic volume elements

Relationship between Voxel
Space and digital images

Any plane of the Voxel Space reconstruction can be
seen as a digital image. In particular, voxel planes
parallel to the x,y plane can represent series of
square (pixel) images taken at different depths.

O = {v/V(v) ≠ 0ˆv*VS}, where the symbol * stands for "In".

The voxel space can be represented by a 3-D array in the
same manner as the digital image is represented by a 2-D
array. Indeed, any plane of the voxel space can be seen as a
digital image. In particular, a plane defined by z = k, called
a voxel plane, can be associated with a digital image D as
follows:

$$D(x,y) = V(x,y,z), \quad 1<x<N\underline{v}ˆ1<y<N\underline{v}ˆz = k \quad (2),$$

where D(x,y) is the pixel value at the pixel (x,y) on the
image D and V(x,y,z) is the voxel value at the voxel space
co-ordinates (x,y,z). This concept is illustrated in Figure 5.

Voxel Space representations tend to be extremely computer
intensive because the number of voxels required to define
complex, lifelike objects is large. In fact the number of
voxels in Voxel Space - the voxel space resolution -
determines the model complexity. Even though objects can be
encoded in order to reduce the amount of memory required to
store them, the processing time is ultimately dependent on the
number of voxels.

Voxel Space representation is relatively new when
compared with boundary representation. The concept of Voxel
Space was first applied in the mid seventies (Robb et al,
1974) and its use for medical purposes began seriously only 10
years ago (Herman and Liu, 1979). However, as volumes rather
than surfaces are represented, voxel space models are always
physically consistent and this approach therefore has major
advantages for modelling complex biological structures.

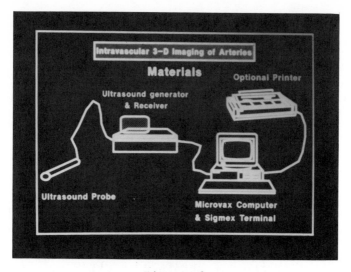

Figure 6

Intravascular ultrasound imaging system comprising an
ultrasound probe, ultrasound transceiver and computer system.

Ultrasonic Imaging System Configuration

We have developed a complete experimental system for
reconstructing solid, voxel space arterial models from
ultrasonic data collected from within the vessel lumen (Kitney
et al, 1988, 1989a, Burrell et al, 1989). The system
comprises three elements: an ultrasonic probe, an ultrasonic
transceiver, and a Microvax II computer (with a Sigmex 6264
colour graphics terminal) (Figure 6).

The Ultrasonic Probe. The ultrasonic probe consists of 12
piezo-electric crystals mounted on a specially designed tip
fixed to the distal end of a standard 8F coronary catheter.
The probe is capable of working in the frequency range 5-20
MHz and comprises a phased array, the firing sequence of which
is controlled by the ultrasonic transceiver.

The Ultrasonic Transceiver. The crystals are connected to
the ultrasonic transceiver via integral wiring in the
catheter. The transceiver has been specially designed to
capture data at high speed from the probe. Analog data is
typically acquired in 700 microseconds, converted to digital
form by a customized A-to-D converter and subsequently
downloaded to the Microvax.

The Computer and Display Unit. The ultrasonic
transceiver is interfaced with a standard DEC Microvax II
computer (a 32 bit machine with a 40 MHz clock). The
associated graphics terminal is a Sigmex 6264 with a
resolution of 1448 x 1024 (vertical) pixels with a palette of
256 colours and 24 image planes. The computer system runs
under the DEC-VMS operating system. Three dimensional
visualization is undertaken by software which has been written
for this purpose.

Our current software, which produces full voxel space images, allows display and manipulation in both 2-D and 3-D, with the ability to easily transfer from one to the other (described in Kitney et al, 1987). The design of the system also allows a wide range of operations to be performed on the 3-D solid models. Image manipulation routines include rotation in all three dimensions, cross-sectional display at any depth, shading of images using depth code, planar cuts so that any section of the model can be removed and X-ray projection, where a 3-D structure can be made semi-transparent.

Data Acquisition

Parallel 2-D slices are acquired from the artery under examination using the catheter-mounted ultrasonic probe (Figure 7a-c). The time taken to acquire the data is small in comparison to the cardiac cycle, hence the slices can be aligned without significant movement error. The 2-D morphological information obtained in this way is converted into a 3-D solid model of the section of artery under investigation using full voxel space (Figure 7d).

Figure 7a

Figure 7b

The ultrasound probe is placed within the lumen of a vessel – reflected signals are received from each interface within the vessel wall

Acquisition of a single slice of the artery under investigation using a phased-array

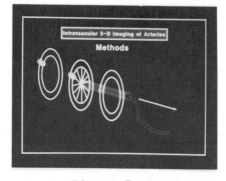

Figure 7c

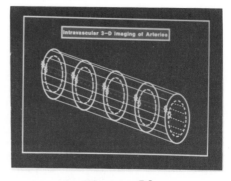

Figure 7d

Withdrawal of the probe to acquire multiple 2-D slices

3-D voxel space reconstruction from the 2-D slices

In order to understand how the 3-D model is constructed, let us assume that the 2-D image of an arterial cross-section is represented by M x M picture elements or pixels. This concept is extended to three dimensional visualisation. The depth, or z-dimension, of the object can be constructed from a series of N 2-D pixel planes, by laying them one behind the other. The 2-D picture elements (pixels) now become volume-elements (voxels) in forming the 3-D image. Hence a 3-D object can be modelled in M x M x N voxel space. The strategies used in this system for contour definition and 3-D interpolation are described in Kitney et al (1987,1989b).

Simple Intra-arterial 3-D Visualisation

In this section we will describe the reconstruction of arteries by means of 3-D voxel space models. Arterial phantoms and _in vitro_ arterial segments have been modelled from data acquired intra-luminally with the specimen suspended in a water bath. Clinical data acquisition to create 3-D models from _in vivo_ sections of femoral artery is also demonstrated. In the examples, various features of the software including rotation, slicing and semi-transparent representation of the voxel models will be illustrated.

Phantoms. A series of 2-D slices of a perspex phantom, into which a symmetrical, concentric stenosis has been machined (shown in Figure 8a), were acquired using the catheter-mounted ultrasound probe placed within the lumen of the model.

By the method described above, the series of 2-D slices was used to reconstruct a 3-D model of the phantom in full voxel space (Figure 8b). Referring to the figure, the model is rotated in 30 degree increments from side view to end view in the left hand panel. In the right hand panel the model is shown in the same orientations after bisection of the volume. In the bisected volume, the stylised stenosis and detail of the inner wall are easily appreciated. Geometric measurements of the original phantom and its visualisation confirm the apparent close correlation.

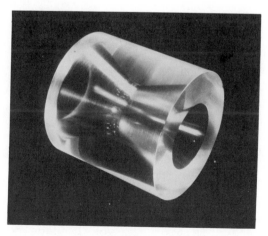

Figure 8a A perspex phantom with a symmetrical,
concentric stenosis

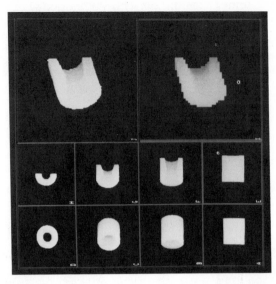

Figure 8b Voxel space reconstructions from
intra-luminal ultrasound data

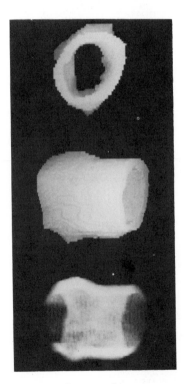

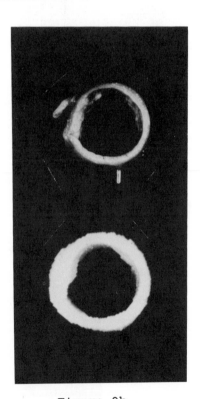

Figure 9a

Section of human atheromatous
aorta reconstructed from
intraluminal data acquisition.
End-on, 75 degree rotated and
semi-transparent versions.

Figure 9b

Photograph of the original
aortic specimen and the 3-D
voxel space model seen from
the same end (opposite to
that in Figure 9a).

In Vitro Specimens. In a similar fashion, 3-D models of
aortic, iliac and carotid arteries have been reconstructed
from human in vitro specimens. A reconstruction of a section
of aorta is shown in Figure 9a. Starting at the top of the
figure, the model is shown first end-on and then rotated
through 75 degrees. A semi-transparent version is shown below
in this same orientation. Figure 9b shows a photograph of the
original aortic specimen, and the 3-D model viewed from the
same end (opposite to that illustrated in Figure 9a).

As in the case of the phantom, geometric measurements
on the arterial specimen showed good correlation with the
model. A typical example of diameter measurements from a 2-D
slice of the model and the corresponding histological
cross-section of the original specimen are shown in Figure 10.

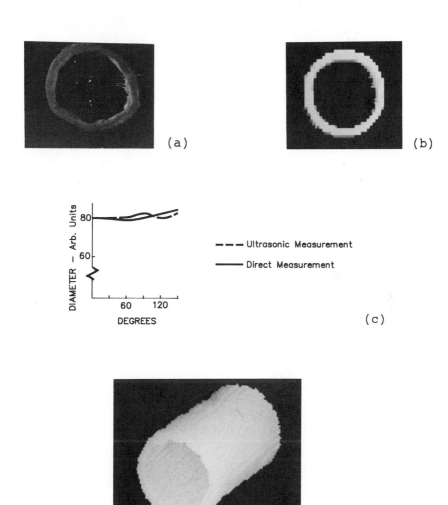

Figure 10   (a) Histological section and (b) 2-D (B-mode) slice
             of aorta at the same level, with (c) diameter
             measurements comparisons; and (d) complete 3-D
             voxel space reconstruction of aortic segment.

263

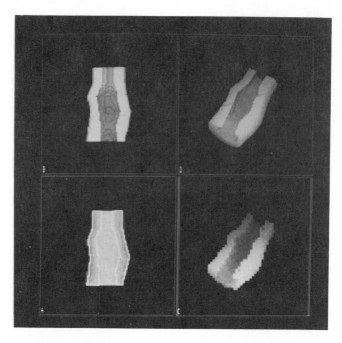

Figure 11

Carotid artery reconstruction demonstrating the
full resolution version (panel A), planar slicing
(B), low resolution version for rapid rotation (C)
and full resolution, sliced and rotated version (D)

Another feature of the software is planar slicing. This
technique, when applied to arterial models, enables close
inspection of the intimal surface, the initial site of
arterial disease. The combination of planar slicing and
rotation of a 3-D arterial model is demonstrated on a carotid
artery reconstruction (Figure 11). In panel A, the full
resolution model is shown together with the outline contours
of the proposed section. After removal of the section, the
intimal lining of the vessel is seen (panel B). In order to
facilitate rapid rotation of the model, the resolution is
reduced to one eighth of the original (panel C). Finally, the
full resolution image, after planar slicing and rotation to
the desired orientation, is shown (panel D).

In Vivo Images. The same techniques can be applied to
the acquisition of data in clinical situations. For example,
data acquired from a normal femoral artery in vivo, by
advancing the catheter mounted probe a few centimetres from
the end of a standard femoral sheath used during cardiac
catheterisation, were used to reconstruct the model shown in
Figure 12a. A semi-transparent (X-ray simulation) version of
this reconstruction (Figure 12b) has an oblique section
selected for removal. The full resolution image with this
section removed is reconstructed in Figure 12c.

a

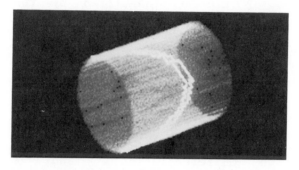

b

c

Figure 12 (a) 3-D model of human femoral artery from in vivo ultrasound data acquisition; (b) Semi-transparent (X-ray simulation) version: an oblique section is selected for removal; (c) same model after removal of oblique section.

Advantages of an Ultrasonic 3-D Digital System

It is important to note that the data in the in vivo examples given above was acquired in the femoral artery without obstructing blood flow. One of the major advantages of ultrasonic visualisation is that it is highly effective in a blood filled medium. This is in marked . contrast to other direct methods for visualising the arterial wall, such as fibre optic angioscopy, which requires replacement of the blood with a translucent medium.

The Voxel Space 3-D software illustrated above, which has been used for reconstruction and manipulation of the images, also represents a major advance in itself, in terms of medical imaging. Using Voxel Space modelling, it can be seen that a very detailed and flexible image can be achieved. One specific, inherent advantage in this context is the ease with which length, area and volume measurements can be made. Once the system has been calibrated, the exact position in 3-D space of all the voxels is known for any orientation of the model.

Furthermore, because each voxel can be addressed individually, progressively more sophisticated images can be developed. For example, colour coding can be used to represent different types of tissue, identified by their ultrasonic characteristics. This type of more complex application to intra-arterial imaging will now be discussed.

Complex Intraluminal 3-D Visualisation

The reconstruction of simple 3-D arterial models represents an important new diagnostic technique for arterial disease. However, diagnostic accuracy is likely to be improved by further differentiation of normal and diseased tissue. For ultrasonic data, this is a two stage process. Firstly, the identification of the boundaries between different types of tissue (tissue differentiation), and secondly the characterisation of those tissue types (tissue characterisation).

The analysis of ultrasonic data rapidly identifies interfaces between tissues of different acoustic impedance. Identification of these interfaces and colour coding of the resultant layers in the 3-D model enables differentiation of tissue types. To demonstrate this principle an arterial model was fashioned from a small (20mm diameter) plastic beaker

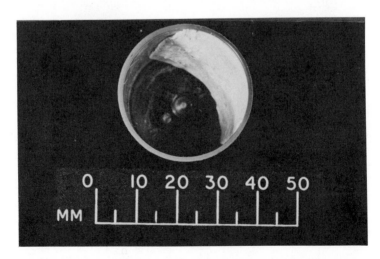

Figure 13a Arterial model with simulated atheromatous plaque encroaching into the lumen of the vessel

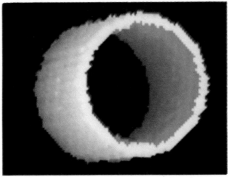

Figure 13b Computer representation of the arterial model in 3-D using colour coding to identify the "plaque"

Figure 13c "Stretch volume" representation of arterial model

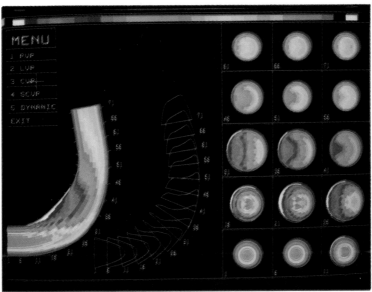

Figure 14 Longitudinal cross-section (in the left column) and horizontal cross-sections (in the right column) of flow in a curved tube, calculated from solutions of the Navier-Stokes equation. White and purple represent regions of low velocity; yellow and orange, regions of high velocity. In the centre column, calculated velocity profiles for a plane through the centre of the artery are shown.

within which adhesive putty was used to simulate atheromatous plaque encroaching into the lumen of the vessel (Figure 13a). Using the catheter-mounted ultrasound probe positioned within the lumen of this model, a series of 2-D slices of the model were acquired.

By the method described above, these data were then used to reconstruct a 3-D image of the model in full voxel space (Figure 13b). Colour-coding is used to identify the "plaque" separately from the wall of the vessel. The software can again be applied in a number of other ways in order to demonstrate regions of interest within the reconstructed 3-D volume. For example, the model can be split longitudinally and opened out ("stretch volume" representation) to show the location and amount of the "plaque" more clearly (Figure 13c).

3-D BLOOD FLOW REPRESENTATION

So far we have discussed the visualization of anatomical structures using three dimensional solid modelling. However, the detrimental effects of atheromatous plaque depend not on its appearance but on the disturbance of blood flow it causes. An understanding of the fluid dynamic changes which occur during the development and following the treatment of atheromatous stenoses is necessary in order to understand and predict these effects.

Classical ultrasonic imaging systems, like duplex scanners, are based on the concept that as an atheromatous plaque develops, flow becomes progressively more disturbed. Inherent in this concept is the idea that there is a relatively simple relationship between flow changes at a single point in the flow field and the development of the disease. Hence, theoretically, by estimating the velocity waveform at a single point it should be possible to determine the severity of the stenosis e.g. 50 or 75%. This is typically done by estimating two parameters: the peak velocity during the cardiac cycle, and the degree of spectral broadening during the deceleration phase of the waveform. It has been found in clinical practice that such measurement gives a reasonably robust, but rather non-specific estimate of the degree of stenosis. As a result, there has been a desire to develop ultrasonic scanners which give a much clearer image of the flow field.

Colour flow mapping scanners have been commercially available for several years, but there is still considerable debate regarding their accuracy. The colour flow map represents the three dimensional flow field in only two dimensions. Hence, because of the cosine in the Doppler equation, any change in the spatial orientation of the artery can, and often does, result in a change in the colour of the velocity estimate. It is therefore often difficult to determine whether or not a change in colour within the colour flow image results from a change in the spatial orientation of the artery or is truly a change in velocity. The only effective way to overcome this fundamental disadvantage of 2-D colour flow mapping systems is to present the full 3-D flow field.

We have begun to develop techniques for the estimation and representation of the entire 3-D flow field for the section of artery under investigation. To date the flow fields have been calculated from numerical solutions of the Navier-Stokes equations using a finite difference approach. In order to solve the Navier-Stokes equations by a flow solver under defined conditions, two principal pieces of information are necessary: the profile of the inlet velocity waveform and the shape of the external grid structure of the flow field.

The next stage is to represent the data. Although representation of 3-D flow is an extremely complex problem, clinically useful 3-D information can be presented effectively by using a combination of 2-D cross-sectional displays. The example which follows demonstrates the use of our 3-D visualisation software to create such 2-D displays.

<u>Presentation of 3-D Flow</u> In this example, longitudinal and horizontal cross-sections of the flow in a curved tube representing the aortic arch are illustrated (Figure 14). On the right hand side of the figure, the horizontal cross-sections correspond to the numbered levels on longitudinal cross-section, shown on the left hand side. On the computer screen velocity is colour coded, such that white and purple represent regions of low velocity, and yellow and orange regions of high velocity. Flow is laminar at the inlet (bottom left of the longitudinal cross-section), but as the bend is approached the flow becomes non-laminar, with high velocity at the outer wall and very low velocity at the inner wall.

Having calculated the flow field, it is now a simple matter to calculate and plot the velocity profiles for a series of points along the centre line of the artery. These velocity profiles are shown in the centre panel of figure 14. It can be seen that 2-D colour flow maps and the velocity profiles may be combined to give a considerable amount of information relating to the 3-D flow field.

CONCLUSION

Ultrasound has a number of advantages over other methods for arterial imaging. In particular, the data can be acquired within the blood-filled lumen without obstructing blood flow. We have shown how digital, three dimensional images can be reconstructed in full voxel space from ultrasonic data, and the power of the digital approach, in terms of post-processing, image manipulation and the development of progressively more complex images has been demonstrated. Finally, it has been seen that an approach to the three dimensional presentation of flow information, based on theoretical calculations from the 3-D visualisations, may be possible.

The combination of 3-D visualization of an artery using tissue differentiation techniques, coupled to the presentation of the associated 3-D flow field data, is likely to result in significant clinical benefit in the assessment of arterial disease. This type of approach is likely to be of most benefit in combination with intraluminal devices for treating athero-sclerosis, where detailed information about relatively short arterial segments is demanded for accurate assessment of the effects of treatment and the diagnosis of complications. It is however clear that these techniques generate a large amount of data. Hence, a key factor in their clinical usefulness will be how the data can be best presented to allow the clinician to extract the maximum amount of useful information.

Acknowledgements   We warmly acknowledge the contributions of L. Moura, A. Kardan, X. Zhenrong and A. Mazher to the work presented in this chapter.

References

Burrell, C.J., Kitney, R.I., Straughan, K., McDonald, A.H., Rothman, M.T., 1989, Intravascular 3-D Imaging of Arteries: A Novel Ultrasound System, Br Heart J., 61:476.

Farrell, E.J., 1987, Visual Interpretation of Complex Data, IBM Systems Journal, 26:174.

Foley, J.D. and van Dam, A., 1984, Fundamentals of Computer Graphics, Pub. Addison Wesley.

Herman, G.T. and Jayaram, K.U., 1983, Display of 3-D Digital Images : Computational Foundations and Medical Applications. IEEE Trans C.G. and A., 39.

Herman, G.T., 1986, Computerised Reconstruction and 3D Imaging in Medicine. Ann Rev of Comput Science:153.

Kitney R.I., Moura, L., and Straughan, K., 1987, Three Dimensional Solid Modelling of Arterial Structures using Ultrasound, Proc IEEE (IXth Conf. on Engineering in Medicine and Biology), 1:401.

Kitney, R.I., Moura, L., Straughan, K., Burrell, C.J., McDonald, A.H. and Rothman, M.T., 1988, Ultrasonic Imaging of Arterial Structures using 3-D Solid Modelling. Proc.IEEE:Computers in Cardiology, 15:3.

Kitney, R.I., Burrell, C.J., Moura, L., Straughan, K., McDonald,A.H., Rothman, M.T., 1989a, A Catheter Mounted Ultrasound Probe for 3-D Arterial Reconstruction. SPIE Proc. (Medical Applications of Lasers and Optics), 1068:185.

Kitney, R.I., Moura, L., Straughan, K., 1989b, 3-D Visualization of Arterial Structures using Ultrasound and Voxel Modelling, Int J Cardiac Imaging, 4:135.

Pipes, A, 1985, Solid Modellers: This Year's Style. CAD/CAM International, 9:20.

Robb, R.A., et al, 1974, 3-D Visualisation of the Intact
    Thorax and Contents: A Technique for Cross-sectional
    Reconstruction from X-ray Views, Computers and
    Biomedical Research, 7:395.

METHODOLOGICAL STRATEGIES TO INVESTIGATE BLOOD-WALL INTERACTIONS

Ch. Baquey, L. Bordenave, J. Caix, F. Lespinasse, N. More, and B. Basse-Cathalinat

INSERM - U. 306
Universite de Bordeaux II
146, rue Leo-Saignat, 33076 Bordeaux Cedex (France)

ABSTRACT

The haemocompatibility of a biomaterial can be considered as the best expectable result of its interactions with blood. These interactions are led by at least two sets of parameters : a first one related to morphological physical and chemical characteristics of the surface exposed to blood ; a second one, related to blood flow conditions, these latter acting as modulators of the role played by the first set of parameters. Authors describe methodologies, based on the use of radioactive tracers, which allow direct studies of these interactions, and give a few examples which illustrate their interest in both experimental and clinical contexts.

Key words : Biomaterials - Blood-wall interactions - Red blood cells - Platelets - Granulocytes - Plasma proteins - Extracorporeal circulation - Vascular prostheses - Haemodynamics.

INTRODUCTION

Blood-wall interactions are led by at least two sets of parameters : a first one, related to morphological, physical and chemical characteristics of the surface exposed to blood ; a second one, related to blood flow conditions, these latter acting as modulators of the role played by the first set of parameters. Indeed, these ones are of particularly critical importance, when the arterial wall is in a pathological state in case of vascular diseases, or when segments of the natural artery have been replaced by artificial substitutes. So, it is quite important to design experimental protocols, aimed at forecasting the behaviour of flowing blood in contact with biomaterials on the one hand, and clinical investigations able to reveal any consequence of blood-material interactions on blood elements, on the other. Methodologies based on the use of radioactive tracers, which allow direct studies of these interactions as illustrated by a few examples from experimental and clinical contexts, are described here below.

IN VITRO STUDIES

These studies involve experimental models which are designed to allow a simplified approach of the interaction of blood with artificial or bio-artificial surfaces, in so far as simpler fluids than blood are used. According to a common principle, these fluids are circulated by various means through open or closed circuits, portions of which consist in tubular samples either made of a reference material or of the material under test.

Two examples are presented here dealing respectively with the role of the chemistry and of the morphology of the material surface which comes in contact with blood.

Role of surface chemistry

Medical grade tubing (2.9 x 4 mm) obtained from Clay Adams (a division of Becton & Dickinson, Parsipanny NJ, USA) was supplied and treated according to an original procedure (Migonney et al., 1986) by the Laboratoire de Chimie des Macromolécules at the Université de Paris-Nord. The inner surface of this tubing was bearing functionnal residues which made it able to adsorb antithrombin III (AT III) and catalyse the inhibition of thrombin by the latter. The experimental procedure described here can be used as a quality control since it allows a determination of the average amount of AT III which can be adsorbed on such surfaces ; the homogeneity of the adsorbed protein can be assessed as well as its ability to be still recognized by a related monoclonal antibody.

Untreated tubing (C) was used as a control to which two kinds (A and B) of treated tubing, as shown in Table 1, were compared. 30 cm lengths of each of these three kinds of tubing were mounted in series in a closed circuit filled with Michaelis'buffer, which is allowed to circulate for 3 hours, flow (10 ml.mn^{-1}) arising from a peristaltic pump. The circuit is fitted with two three-way valves (T-valves), which are used in order to avoid any air-liquid interface, to exchange the circulating buffer by a fresh solution twice again and finally by a buffered solution of AT III (one unit per ml). This solution is added with an aliquot of ^{123}Iodine labelled antithrombin III (AT III*).

Antithrombin III of human origin was isolated and purified by the blood bank in Lille (CTS - Lille, France). Each vial (special batch n° 78302) contained 600 U to be made up in 20 ml of distilled water. Iodination was carried out according to a method (Caix et al., 1987) some what related to the Iodo-Gen technique (Fraker and Speck, 1978).

During the period of circulation of the AT III/AT III* solution, the radioactivity of the circuit was sequentially measured by means of a gamma-camera and scintigrams of each portion of interest (A, B and C) of the circuit could be recorded. After the first three minutes of circulation, the radioactivity became homogenously distributed into the bulk of the flowing fluid and a value Rp could be measured for each of these portions of interest. This value Rp corresponds to the whole amount Q of AT III available for adsorption on the wall of the circuit. At the end of a 30 mn exposure, the circuit was washed with saline, and the residual radioactivity Rw of the same portions of the circuit was measured. Rw corresponds to the amount q of AT III which was adsorbed on the inner surface of tubing A, B or C. From the ratio Rw/Rp the average value of C_s, the surface concentration of adsorbed AT III, can be determined using the following equation :

$$C_s = \frac{Cr}{2} \quad \frac{Rw}{Rp-Rw}$$

where C is the bulky concentration of AT III

and r the inner radius of the tubing

Values of C_s respectively obtained for tubing A, B and C are given in Table 1. One can see that tubing A and B demonstrate the same affinity for AT III.

In order to get some information about the conformationnal state of the adsorbed protein, its affinity for a monoclonal antibody (MAB-AT III) which uses to bind to AT III in solution, has been checked. Tubing on which AT III had previously adsorbed have been incubated with a flowing buffered solution of (MAB-AT III), which contained an aliquot of [111]Indium labelled (MAB-AT III) playing the role of tracer. At the end of the incubation period, tubing were rinsed with saline, and the amount of MAB-AT III which was retained on the surface of each kind of tubing could be determined according to the scheme presented for AT III. Corresponding values are given in Table 1. It can be noticed that tubing A and B are able to retain an equivalent amount of AT III, which is significantly greater than the amount adsorbed on the control tubing (tubing C). But only AT III adsorbed on tubing A is able to be recognized by the MAB-AT III, which means that AT III adsorbed on tubing B has lost its original conformation.

Role of material surface morphology

We have already published (Baquey et al., 1989) a study dealing with the role of the structure of carbon-carbon composite on their in vitro interaction with blood platelets. The main differences between the two types of composites of interest came from a different organization of the carbon fibers. For type A these fibers were woven and rolled while non-woven fibers were reinforcing the type B material. In spite of extensive densification with pyrolytic carbon, both types featured an open porosity which occupied a larger volume for type B than for type A.

Table 1

	Tubing under study								
	A			B			C		
	PE — PS $\underset{\textstyle SO_2\text{-ASP}}{\overset{\textstyle SO_3^-}{<}}$			PE — PS — SO_3^-			PE		
Retained concentration of AT III* in pmol/cm2 of inner surface	67.3	69.6	69.6	60.3	67.3	64.9	18.6	23.2	20.8
Retained concentration of anti AT III* antibody in pmol/cm2 of inner surface	69.6	69.6	71.9	6.9	9.2	6.9		0	

PE = Polyethylene
PS = Polystyrene

The experimental procedure has been fully described (Baquey et al., 1989), but we may recall here the rationale of the methodology. A blood cell suspension containing 3.10^5 platelets per mm^3 and red blood cells (RBC) in order to get an hematocrit set at 40 %, is prepared from freshly collected human blood. Platelets are labelled with an ^{111}In-oxine complex according to a method derived from THAKUR's (Thakur et al., 1976) and red blood cells are labelled with ^{99m}Tc sodium pertechnetate in the presence of stannous pyrophosphate (Ducassou et al., 1976). 80 to 100 ml of this radioactive blood cell suspension are aspired by a syringe pump through twin lines made of medical grade silicon elastomer tubing, along which tubular samples (I.D. = 3.5 mm) made of the carbon-carbon materials are inserted. The flow rate is set at 5.7 $ml.mm^{-1}$. In order to avoid any air-suspension interface both lines are initially filled with Michaelis' buffer. When all the available volume of suspension is near to being consumed, the lines are fed with saline in order to rinse them extensively. Then the carbon-carbon samples are disconnected from the lines and their radioactivity is measured and analyzed, in order to determine the respective amounts of platelets on the one hand and of RBC on the other, which are retained by the material.

RBC are supposed to be retained only for mechanical reasons and we make the assumption that this amount of RBC, corresponds to an equivalent volume of suspension, the platelet content of which can be easily determined. The corresponding platelet subset is said to correspond to platelets entrapped into the porosity or morphology irregularities of the materials.

This subset must be substracted out of the total amount of retained platelets to determine the real amount of adherent or agregated platelets to the material of interest. Thus, it has been possible to determine that material B was able to scavenge ten times more platelets than material A, as it was expected from the comparison of the respective porosity of both materials. However, the subset of mechanically retained platelets is relatively less important for material B than for material A (Table 2).

In addition to these results the subset of retained platelets which are reversibly bound to the material can be quantified through an experiment aimed at determining the fraction of the retained platelets pool which is readily exchangeable by flowing platelets. Such an experiment consists in a second circulation stage through the samples of interest following the rinsing stage previously mentionned, and along which a blood cell suspension (same composition as for the first stage) without any radiotracer is used. After a second rinsing stage, the residual radioactivity of the samples is measured and analyzed and it appears that the recorded values are smaller than those recorded after the first rinsing stage. That means that some of the labelled platelets and their non-labelled counterparts entrapped into the material during the first dynamic incubation stage have been replaced by non-labelled platelets during the second dynamic incubation stage. Thus it can be calculated under given experimental conditions, that 46 % of entrapped platelets are exchangeable for material A compared to 12.5 % only, for material B.

To summarize, it was found that platelets entrapped into material B are more numerous (and mostly irreversibly bound) than platelets entrapped in material A. For material B the subset of exchangeable platelets among those which are entrapped, is slighly larger than the subset of platelets which are mechanically retained. On the contrary the subset of exchangeable platelets among those which are entrapped into material A, is three times larger than the subset of mechanically retained platelets. Then it clearly appears that the mechanisms according to which platelets interact with each type of carbon-carbon composites are different.

Table 2

Carbon-carbon composite type	Platelets/$10^4 \mu m^2$ (total number)	Platelets mechanically retained (%)	Exchangeable platelets (%)
A	30	16	46
B	600	7	12.5

Role of flow conditions

Experiments about which we report here above, do not take into account the pulsatile character of the blood flow, as the flow used in our models keeps constant during a given experiment. So we have tried to improve our models and the set-up shown on fig. 1 can be operated in order to generate a pulsatile flow with characteristics (pressure wave, flow rate wave) equivalent to physiological ones, femoral artery flows for instance. Basically this set-up is derived from a device which has been developed at the Institut de Mécaniques des Fluides in Marseille (France) by R. PELISSIER's team (Pelissier et al., 1983) in cooperation with E. KRAUZE's team at the Aerodynamisches Institut in Aachen (West Germany). Briefly, the fluid inside the primary circuit (fig. 1) is moved according to a preset forward and backward mode established by appropriate settings on the synthetizer, in order to periodically press and depress the artificial ventricule which is the active part of the secondary circuit. Then the fluid inside the secondary circuit which is mainly made of flexible materials and fitted with adjustable equivalents of compliancy and peripheric resistance, may flow according to a quasi physiological pattern (fig. 2). Tubular samples of materials under study may be mounted in series on the D and D' parts of the circuit which must keep identical to each other. These sections can be open to the flow alternatively. When the fluid goes through the D' part, the flow parameters are adjusted as precisely as possible, and then it is allowed to flow through the D' part and the study of the fluid-wall interactions may be undertaken, using the same methodology as described here above.

EXPERIMENTAL IN VIVO STUDIES

These studies involve an ex-vivo model which has been described elsewhere (Basse-Cathalinat et al., 1980, Baquey et al., 1981). Briefly, radiotracers either prepared from homologous plasma proteins, or from autologous blood cells, are injected intravenously to an anaesthetized dog bearing an arterio-venous shunt (AV shunt) or an arterio-arterial by pass (A-A by-pass). As soon as blood is allowed to flow through the extracorporeal circuit (ECC), the radioactivity of the latter is sequentially recorded and analyzed with a high energy resolution detector (the same as for in vitro studies) in order to assess the specific contribution of any tracer involved in the experiment to this radioactivity, the corresponding values being proportionnal to the respective amount of the blood species (which have been traced) contained in the ECC, whether they are circulating in the bulk of the flow, or retained by the ECC wall. As the bulk contribution can be determined from direct measurements carried out on collected blood samples, the specific contribution of the ECC wall can be known as well as the kinetics of adsorption and of adhesion phenomena responsible for the radioactivity of this wall.

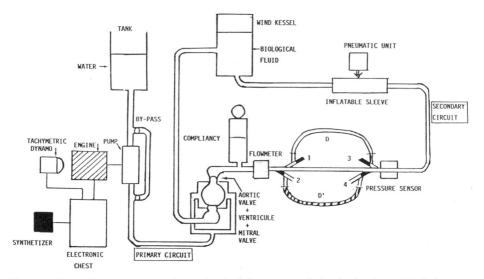

Fig. 1. Hydrodynamic generator adapted from an original device which has
been developped at the Institut de Mécanique des Fluides in
Marseille (France) ; D : standardization loop ; D' : measurement
loop ; 1-2-3-4 : special connectors avoiding turbulences to occur.

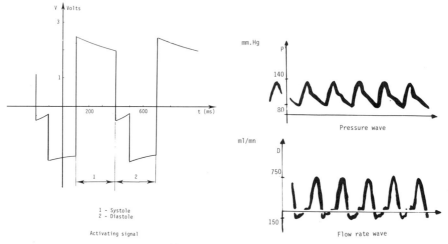

Fig. 2. This graphs show a possible shape for the activating signal of
the pumping unit (see figure 1) and the corresponding pressure
and flow rate waves recorded for the fluid in the secondary
circuit (see figure 1).

276

From the slope of curve A (fig. 3) for instance, it is possible to determine the average number of platelets τ_s which adhere to the material which flowing blood is in contact with, per time unit and per surface unit during a given period of exposure ; τ_s is expressed as a number of platelets/mm^2/s. It is also possible to determine the average weight μ_s of a given plasma protein which adsorbs onto the material per time unit and per surface unit during the same period of exposure from the slope of curve C (fig. 3) ; μ_s is expressed as ng/cm^2/s. Direct assessment of such kinetics parameters is the main advantage of the methodology presented here. But the latter offers other advantages as demonstrated on fig. 3 ; curve A shows a steep increase of the platelet density deposited onto the wall of the ECC, followed by a sudden drop which corresponds to embolization of platelet agregates. At the same time the cross section of the ECC becomes larger, which explains a slightly greater amount of RBC seen by the detector. The occurence of such phenomena would have never been observed if the method used did not involve direct measurement of events happening at the wall level.

CLINICAL STUDIES

Haemodialysis membranes (HM) may activate the complement system (Craddock et al., 1977), giving rise to complement fragments responsible for an increase of the adhesiveness of granulocytes (polymorphonuclear leucocytes or PMN), which then attach to vessel and capillary walls and marginate more easily. This phenomenon is particularly effective in the lung parenchym and can affect the respiratory function. In order to evaluate this effect for different HM types, the following protocol has been undertaken. Six long term chronic haemodialysis patients, without any cardiorespiratory disease, undergoing maintenance haemodialysis for 4 hours three times a week, have been studied. Three different artificial kidney membranes have been compared in this group of patients : two cellulosic membranes CUPROPHAN (CUP) and HEMOPHAN (HEM), and a poly-acrylanitrile membrane, (PAN).

Before a dialysis session, patients received through the I.V.route, ^{111}In-labelled autologous granulocytes (PMN) and ^{99m}Tc labelled RBC ; blood samples were collected 5 mn before and 10, 20, 40, 60, 90, 105 and 120 mn after the beginning of the dialysis session, in order to reveal an eventual leucopenia. During the dialysis session the flowing blood and the lung parenchym radioactivities were respectively monitored with a high purity germanium diode-based detector alternatively set above the right thigh region and above the right lung. Sometimes it has been also possible to get the patient lying under a gamma-camera in order to record scintigraphies of several areas of interest. One of these areas obviously corresponded to the right lung which does not suffer any shadow from the heart. From the data related to radioactivity counting and/or from digital processing of recorded scintigraphies, the variation, during the dialysis session, of the granulocyte content on the one hand, and of the RBC content on the other for tissues or organs involved by every explored area, could be assessed. Figures obtained for a given zone of interest (and more precisely the ratio ^{111}In-PMN counts/^{99m}Tc-RBC counts) and with a given dialysis membrane could be compared to those obtained with another membrane, (fig. 4, Tables 3 - 4) and to those obtained for the general circulation. It was then possible to demonstrate a transient PMN entrapping in the lung parenchym which concerned a larger amount of cells when Cuprophan and, in a lesser extent, Hemophan membranes were used than when PAN membranes were used ; and it was possible also to correlate this phenomenon with simultaneous peripheric neutropenia. Moreover it has been observed that these phenomena are more or less important according to the patient of interest.

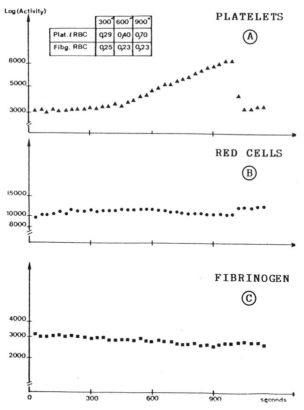

Fig. 3. Recorded radioactivities for the shunt and
related to each of the involved tracers
during extracorporeal circulation of blood
through a synthetic tubing.

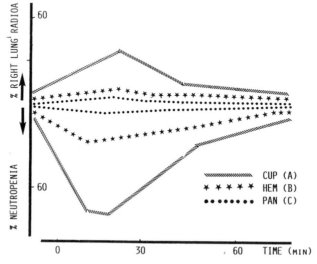

Fig. 4. Relative variations of the neutropenia
(bottom part) and of the Indium 111–WBC
related right lung radioactivity (top part)
during a session dialysis for a given patient.

Table 3. Changes of WBC count (as mm^{-3}) between the beginning of the dialysis session (t_o) and the twentieth minute. The residual value is expressed for each patient as percent of the initial value (4th, 7th and 10th columns).

1	A			B			C		
	t_o	t+20'	%	t_o	t+20'	%	t_o	t+20'	%
1	7100	1600	22	6500	4600	70	4500	3700	82
2	5700	1600	28	7800	5400	69	6000	5220	86
3	4400	1100	25	4500	2650	58	4600	4550	97
4	8690	4230	48	9130	7240	79	7630	7000	92
5	11000	2500	22	8900	3700	41	8600	6600	76
6	5300	1800	33	5200	4600	88	5200	4800	92
m %	$29,66 \pm 9,9$ $p < 0,02$			$67,5 \pm 16,4$ $p < 0,02$			$87,5 \pm 7,68$ $p < 0,02$		

Table 4. Levels of radioactivity in the right lung $\frac{^{111}In}{^{99m}Tc}$ ratio increase is expressed, for each of the patients as percent of the background value.

	A	B	C
1	24	13	8
2	36	7	5
3	8	4	1
4	21	8	5
5	17	1	13
6	82	37	2
m	31.3 ± 24.1 $p < 0.05$	11.6 ± 11.9 NS	5.6 ± 3.9 $p < 0.05$

This study shows that clinical radiotracers based investigations may bring objective parameters which should be taken into account for dialyser evaluation, and to improve haemodialysis therapy in order to reduce cardiopulmonary dysfunctions in cell chronic dialysis patients. Moreover such a methodology offers an opportunity to better understand the different side effects coming from haemodialysis membranes, which is the key for the production of materials with a better biocompatibility.

CONCLUSION

As far as they behave like their non-labelled counterparts, biological radiotracers are among the best tools one can expect to investigate blood-wall interactions. As far as protein adsorption or cell adhesion phenomena are concerned, they allow a non-invasive quantification of these phenomena and assessment of their kinetics. Non-invasiveness make them particularly useful for in vivo investigations ; furthermore the spatial distribution of the tracers of interest can often be assessed whereas blood-wall interactions are developping or after they have been interrupted ; in both cases an opportunity to know about the homogeneity, of the wall surface in terms of morphology and chemistry, is offered. Lastly, molecular probes such as labelled antibodies or experiments aimed at determining the ability of adsorbed or adhered biological species (cells or biomolecules) to exchange with their free equivalents in solution or suspension, give a unique mean to investigate the nature of their binding to the surface.

The main objective of blood-wall interaction studies is a better understanding of cardio-vascular diseases, and of the lack of haemo-compatibility featured by vascular prostheses and other biomedical devices used in relation with blood circulation. In this respect, methodologies based on the use of radioactive tracers may bring a lot of advantages as recalled here above so far as experimental models and protocols are correctly designed.

REFERENCES

Baquey Ch., Bordenave L., More N., Caix J. and Basse-Cathalinat B., 1989. Biocompatibility of carbon-carbon materials : blood tolerability. Biomaterials, 10, 435-440.

Baquey Ch., Masson B., Basse-Cathalinat B., Janvier G. et Serise J.M., 1981. Etude directe par traceurs radioactifs de l'interaction du sang circulant avec les prothèses vasculaires. Chirurgie, 191, 107, 418-423.

Basse-Cathalinat B., Baquey Ch., Llabador Y. and Fleury A., 1980. Direct in vivo study of flowing blood-artificial surface interactions - an original application of dynamic isotopic techniques. Int. J. Appl. Rad. Isot., 31, 747-751.

Caix J., Perrot-Minnot A., Beziade A., Vuillemin L., Belloc F., Baquey Ch. and Ducassou D, 1987, Conditions for radioiodination of antithrombin III retaining its biological properties. Appl. Radiat. isot., 38 (n° 12), 1003-1006.

Craddock P.R., Fehr J., Brigmam K.L. et al., 1977. Complement and leukocyte-mediated pulmonary dysfunction in hemodialysis. N Engl. J. Med., 296, 769-774.

Ducassou D., Arnaud D., Bardy A., Beydon J., Hegesippe M. and Baquey Ch., 1976. A new stannous agent kit for labelling red blood cells with ^{99m}Tc, its clinical applications. Br. J. Radiol., 49, 344-347.

Fraker P.J. and Speck J.C., 1978. Protein and cell membrane iodination with a sparinglay soluble chloroamide, 1,3,4,6-tetrachloro-3a,6a-diphenylglycolmil. Biochem. Res. Comm., 20, 849-857.

Migonney V., Fougnot C. and Jozefowicz M., 1986. Heparin-like treated tubes : preparation and characterization. Biological an Biomechanical Performance of biomaterials edited by P. Christel, A. Meunier and A.J.C. Lee (Elsevier Science Publishers, Amsterdam), 201-206.

Pelissier R., Cassot F., Farahifar D., Bialonski W., Issartier P., Bobart H. et Friggi A., 1983. Simulateur cardiovasculaire. Innov. Techn. Biol. Med., 4 (n°1), 33-45.

Thakur M.L., Welch M.J., Joist J.H. and Coleman R.E., 1976. Indium 111 labeled platelets : studies on preparations and evaluation of in vitro and in vivo functions. Thromb. Res., 9, 345-357.

INFLUENCE OF FLUID MECHANICAL STRESSES ON VASCULAR CELL ADHESION

Timothy M. Wick, *Sherry D. Doty, and *Robert M. Nerem

Schools of Chemical and *Mechanical Engineering, Georgia
Institute of Technology, Atlanta, GA 30332-0100 USA

ABSTRACT

Elucidation of the response of endothelial cells to flowing blood will indicate the role of hemodynamics in the pathogenesis of atherosclerosis. *In vitro*, physiological levels of laminar shear stress influence endothelial cell shape, cytoskeletal organization, protein synthesis, and receptor-mediated endocytosis. In this paper, steady laminar shear stress is reported to alter the endothelial cell extracellular matrix fibronectin distribution. The fibronectin distribution in unsheared endothelial cells consists of short, randomly oriented fibers. When subjected to shear stress, the fibronectin fibers elongate, thicken, and become oriented with the flow direction in a time and shear stress-dependant manner. Thus, fluid mechanical forces exert a transmembrane signal extending into the extracellular matrix. Abnormal blood cell/endothelial cell interactions appear to be important determinants of the clinical severity of vaso-occlusive diseases such as malaria or sickle cell anemia. We have also investigated the ability of several endothelial cell-derived adhesive proteins to promote sickle red blood cell adhesion to endothelial cells under venous flow conditions. Unusually large von Willebrand factor (vWF) multimers (synthesized by endothelial cells and not normally present in human plasma) are potent promoters of sickle red cell adhesion to endothelial cells. It is hypothesized that unusually large vWF multimers promote sickle red cell adhesion to endothelial cells *in vivo*, leading to microvascular obstruction contributing to pain crises.

KEY WORDS

Endothelial cells, shear stress, atherosclerosis, adhesion, sickle cell anemia, red blood cells.

INTRODUCTION

The endothelium is a monolayer of cells forming a continuous contact with flowing blood. The fluid mechanical forces exerted by blood on endothelial cells have an opportunity to affect endothelial cell structure and function. In microvessels, blood cells are in close proximity with the vessel wall and interact intimately with the endothelium. Therefore, fluid mechanical forces, adhesive proteins, blood cells, and endothelial cells interact in complex ways to maintain normal hemostasis and blood fluidity. It is also likely that qualitative or quantitative alterations in any of these parameters contribute to vascular pathophysiology. We have chosen to investigate the possible relationships between these parameters and their roles in atherosclerosis and sickle cell disease.

The primary clinical manifestations of sickle cell anemia are periodic localized vaso-occlusive crises. These crises are the result of chronic sludging of rigid deoxygenated sickle erythrocytes in the microcirculation (Ferrone, 1989). When deoxygenated, sickle hemoglobin polymerizes and forms rigid red cells incapable of traversing the microcirculation (Breihl et al., 1987). Morphologic sickling is delayed sufficiently after deoxygenation such that red cells do not usually sickle until after they pass through the microcirculation (Mozzarelli et al., 1987). Thus, it has been hypothesized that sickled red cell obstruction of the microcirculation does not normally occur unless other factors prolong red cell passage through the microvascular network (Ferrone, 1989; Mozzarelli et al., 1987).

Sickle red cells are abnormally adhesive to cultured human endothelial cells under static and physiological flow conditions (Hebbel et al., 1980a; 1980b; Mohandas and Evans, 1985; Barabino et al., 1987; Wick et al., 1987). Red cell adhesion to endothelial cells correlates positively with the clinical severity of a patient's disease and is hypothesized to be one factor which periodically delays sickle red cell microcirculatory transit (Hebbel et al., 1980b). Certain plasma factors, including collagen binding proteins (Mohandas and Evans, 1985), promote sickle red cell adhesion to endothelial cells. When sickle red cell adhesion to endothelial cells is studied under flow conditions, the quantity of red cells adhering is maximum at shear stresses approximating those in post-capillary venules *in vivo* (Barabino et al., 1987). This has also been observed in the living microvascular network of a denervated rat (Kaul et al., 1989), further indicating a role for sickle red cell adhesion to endothelial cells in the pathophysiology of sickle cell disease.

Atherosclerotic plaque formation has been noted to be localized to specific regions of the circulation characterized by low shear stress (Schettler et al., 1983; Stehbens, 1983; Nerem et al., 1984). *In vivo*, endothelial cell morphology is reported to be a function of the local wall shear stress (Levesque et al., 1986; Kim et al., 1989). *In vitro*, steady laminar shear stresses ranging from 30 dynes/cm^2 to 85 dynes/cm^2 induce cell elongation and alignment in the direction of flow (Dewey et al., 1981; Eskin et al., 1984; Levesque et al., 1985). The rate and extent of elongation as well as the degree of alignment in the flow direction are a function of both the magnitude and the duration of the applied shear stress (Levesque and Nerem, 1980; Levesque et al., 1989). Endothelial cell F-actin fiber morphology is also altered by the application of steady laminar shear stress (Levesque et al., 1989). Under shear stress, the randomly-oriented, dense peripheral F-actin bands disappear and stress fibers appear (Levesque et al., 1989). After prolonged exposure to shear, stress fibers align in the direction of flow (Levesque et al., 1989). Laminar shear stresses also increase endothelial cell prostacyclin biosynthesis (Frangos et al., 1985) and enhance the rate of receptor-mediated LDL endocytosis (Sprague et al., 1987).

Fibronectin is a 440 kDa disulfide-bond linked dimer which has many important adhesion related functions *in vivo*. Fibronectin is known to promote cell adhesion to substrate and cell spreading, as well as influence cell shape (Yamada et al., 1987). Fibronectin, plasma vWF, and plasma fibrinogen also selectively increase sickle red cell adhesivity (Hebbel et al., 1981; Patel et al., 1985). In this report we examine the relationship between hemodynamics and adhesive proteins in a study of endothelial cell extracellular matrix organization and sickle red blood cell adhesion to endothelial cells.

METHODS

Fibronectin Experiments

Bovine aortic endothelial cells (BAEC) were harvested from adult bovine aortas, cultured on tissue culture polystyrene, and used between passage eleven and fifteen as previously described (Levesque et al., 1989). Confluent BAEC monolayers made up the base of a parallel-plate flow chamber as described in detail by Levesque and Nerem (1985). For the experiments reported here, flow was laminar and fully developed (Levesque and Nerem, 1985).

When assembled into the chamber, BAEC monolayers were exposed to steady-laminar shear stresses of 10.0 to 80.0 dynes/cm^2. After exposure to shear stress, cells were removed from the flow chamber but remained attached to the petri dish substrate. Fibronectin was visualized as follows. First, cells were washed with phosphate buffered saline (PBS), and fixed in 4% gluteraldehyde. Cells were permeabilized with 1% Triton-X 100 in PBS and incubated with rabbit anti-human plasma fibronectin. Cells were then washed three times with PBS and incubated with fluorescein isothiocyanate-conjugated (FITC) mouse anti-rabbit IgG. In negative control cultures, sheared endothelial cell monolayers were handled identically but incubated with 1% rabbit serum. Fibronectin alignment with the direction of flow was visualized as follows. A silicon intensified target (SIT) camera furnished images of cells or fluorescently-labeled fibronectin fibers to a microprocessor-controlled digital image processing system (Perceptics Inc., Knoxville, TN). Images were converted to binary data, and the fibronectin fibers were identified as white fibers on a black background by thresholding the image (Doty et al, 1989). In applying the thresholding routine, regions of the image that were darker than a certain grey level (the threshold) were transformed into black areas. Regions of the image brighter than the threshold (fibers) were transformed into white regions. Prior to thresholding, the local brightness of the original images were rendered uniform to eliminate possible artifacts in the data due to uneven field illumination. Adjusting the average brightness of the original image ensured that the proper threshold value was used to identify every fibronectin fiber.

To analyze the orientation of the cells or the fibronectin fibers, cell outlines or fibronectin fibers were reduced to lines of one pixel width by an image processing software routine supplied with the image processor (Doty, 1989). To quantify the orientation of the cells or fibers, the image was analyzed on a pixel-by-pixel basis. For each pixel determined to be part of the fibronectin extracellular matrix or the cell periphery (e.g. white pixels), neighboring pixels were examined to determine if they were also part of the fibronectin fiber. Two neighboring pixels parallel to the direction of flow were considered as one 'count' in the direction of flow (P). Two neighboring pixels orientated normal to the flow direction were considered as one count in the direction normal to the flow (N). Two neighboring pixels oriented 45° degrees off the flow direction were considered as one-half of a count for both N and P (Doty, 1989). P and N were summed over an entire image (on the order of 100 endothelial cells) and the orientation factor was defined as (Doty, 1989):

$$\text{Orientation Factor} = \frac{P - N}{P + N}$$

The value of the orientation factor is 1.0 if all skeletal lines are parallel to the flow direction and -1.0 if all of the skeletal lines are normal to the flow direction.

Sickle Cell Adhesion Experiments

Human umbilical vein endothelial cells (HUVEC) were harvested and cultured onto 75x38mm glass slides as previously described (Wick et al., 1987).

Unusually large von Willebrand Factor (vWF) multimers synthesized by endothelial cells were collected as an endothelial cell supernatant by incubating endothelial cells with serum-free medium (SFM) for 48 hours (Wick et al., 1987). The supernatants contained 5-20% of the normal plasma antigenic activity of vWF (determined by immunoradiometric assay [Counts, 1975]). The presence of unusually large vWF multimers was verified by agarose gel electrophoresis and autoradiography (Moake et al., 1984). Blood was drawn from normal donors or patients with sickle cell anemia. All patients studied were in steady state. Red cells were separated from the plasma and the buffy coat, washed three times in SFM, and resuspended to 1% hematocrit in either SFM or endothelial cell supernatant as previously described (Wick et al., 1987).

Confluent monolayers of endothelial cells made up the base of a parallel-plate flow chamber. The chamber was mounted on an inverted phase-contrast microscope to visualize endothelial cells and flowing red blood cells. Adhesion assays were performed as follows (Wick et al., 1987). After chamber assembly, the endothelial cell monolayer was rinsed for 5 minutes. Then, red cells suspended to 1% hematocrit in SFM or endothelial cell supernatant were perfused over the monolayer for 10 minutes. This was followed by a 20 minute rinse with SFM to remove nonadherent red cells. Finally, the number of red cells adhering to the endothelial cell monolayer was enumerated for 24 random microscopic fields ranging over the entire slide and the average number of red cells adhering to the endothelium was calculated. At all times during the adhesion assay, the fluid flow was steady and was set such that the shear stress exerted on the endothelial cells was 1.0 dynes/cm^2. This shear stress is similar to that reported for endothelial cells in post-capillary venules (Turitto, 1982).

RESULTS

Fibronectin Alignment

Figure 1a demonstrates the randomly oriented, cobblestone morphology typical of unsheared BAEC monolayers. When BAEC monolayers are exposed to 10 dynes/cm^2 of shear stress for 24 hours, the shape and orientation of the endothelial cells remains qualitatively unchanged (Fig 1b). As the shear stress is increased, BAECs elongate and align in the direction of flow. Fig 1c shows a BAEC monolayer after 24 hours of exposure to 50 dynes/cm^2.

When sheared endothelial cells are fixed, permeabilized, and stained for fibronectin, changes in the organization of the cytoskeletal fibronectin are also observed (Figs 1d-1f). Unsheared, confluent BAEC monolayers exhibit a diffuse and randomly oriented distribution of short fibronectin fibers (Fig 1d). When a shear stress of 10 dynes/cm^2 is applied for 24 hours, the distribution of the fibronectin fibers appears more heterogeneous and an increased amount of fibronectin is organized into thick fibers (Fig 1e). Some of the thicker fibers align in the flow direction; however, aligned fibers are not evident in all images. When the endothelial cells are exposed to 50 dynes/cm^2 for 24-hours, the distribution of the extracellular fibronectin fibers is observed to change markedly (Fig 1f). Most of the fibronectin is organized into thick fibers that tend to be strongly oriented in the flow direction.

Cell and fibronectin orientation factors vary with both the magnitude and duration of applied shear stress. The orientation factors for cells or fibronectin fibers are greater than zero for all sheared cultures, indicating that the cells and fibronectin align with the flow direction (as qualitatively indicated in Fig 1). Cell orientation factor values increase from -0.002 for cells in static culture to 0.538 for cells exposed to 60-80 dynes/cm^2 (high shear) for 24 hours. Intermediate values of shear stress applied for 24 hours lead to intermediate values of the extent of fibronectin or cell alignment. Cell and fibronectin orientation factor also increase with time of exposure to steady shear. For example, when confluent monolayers are exposed to 50 dynes/cm^2 the cell

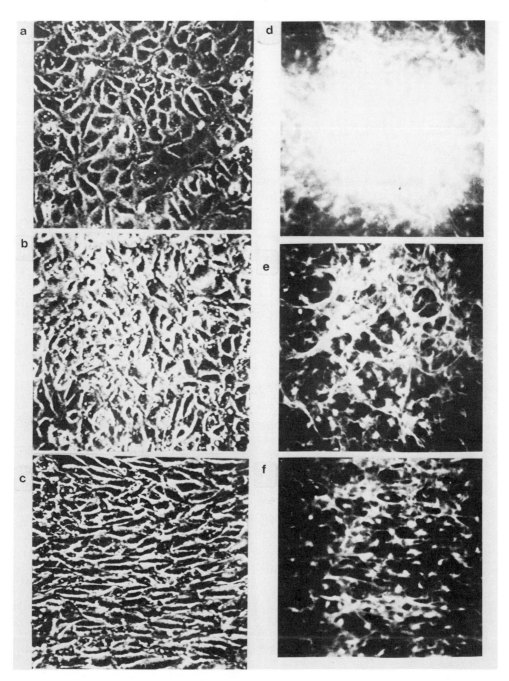

Figure 1. Cell (a-c) and fibronectin (d-f) images after exposure to 0 (a,d), 10 (b,e), or 50 dynes/cm^2 (c,f) steady shear stress for 24 hours.

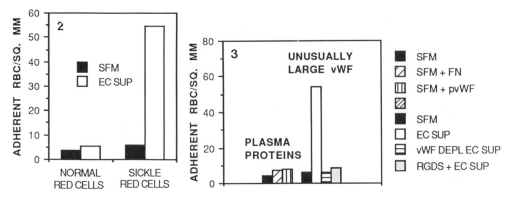

Figure 2. Effect of unusually large vWF (in endothelial cell supernatant; EC SUP) on sickle and normal red blood cell adhesion to endothelial cells.

Figure 3. Effect of plasma fibronectin (SFM + FN), plasma vWF (SFM + pvWF), and unusually large vWF (EC SUP) on sickle red cell adhesion to endothelial cells.

orientation factor increases from zero for unsheared cultures to 0.420 for cultures sheared for 48-hours. Similarly, the fibronectin orientation factor increased from zero to 0.202 after 48-hours exposure to laminar shear stress.

Sickle Cell Adhesion Studies

Figure 2 compares the effects of unusually large vWF multimers on the adhesion of sickle and normal red blood cells to endothelial cells. Normal red blood cells suspended in SFM (not containing adhesive proteins) only adhere to human endothelial cells in low quantities (less that 5.0 RBC/mm^2). When normal red cells are suspended in endothelial cell supernatant (containing vWF multimers, including unusually large forms) normal red cell adhesion to endothelial cells increases only slightly (and not significantly). Sickle red cells suspended in SFM are also only minimally adhesive to endothelial cells under venous flow conditions (Fig 2). When sickle red cells are suspended in endothelial cell supernatant however, the adhesion to endothelial cells increases significantly. For the 14 donors studied, the presence of the unusually large vWF multimers led to an average 10-fold increase in sickle red cell adhesion to endothelial cells.

Figure 3 compares the ability of plasma fibronectin (FN), plasma vWF (which does not contain unusually large vWF multimers), and endothelial cell supernatant (containing small amounts of unusually large vWF multimeric forms) to promote sickle red cell adhesion to endothelial cells. As demonstrated in the lanes labelled SFM + FN or SFM + pvWF, twice the physiological concentration of plasma fibronectin or plasma vWF increase sickle red cell adhesion to endothelial cells less than 2-fold. When endothelial cell supernatants are immunologically depleted of all vWF multimers, red cell adhesion to endothelial cells is decreased nearly 99% (lane vWF DEPL EC SUP). The cell binding region of fibronectin and vWF is the tripeptide amino acid sequence arginine-glycine-aspartic acid (RGD) (Gardner and Hynes, 1985; Plow et al., 1985b). When sickle red blood cells are preincubated with an excess of a synthetic RGD-containing tetrapeptide and then washed (to remove unbound peptide) prior to being suspended in endothelial cell supernatant, unusually large vWF-mediated sickle red cell adhesion to endothelial cells is inhibited 95% (RGDS + EC SUP in Fig 3).

DISCUSSION

Steady laminar shear stress induces endothelial cell elongation and alignment in a shear-stress and time-dependant manner (Levesque and Nerem, 1980; Dewey et al., 1981; White et al., 1982; Levesque et al., 1986; 1989). As shown here, shear stress also induces qualitatively similar changes in the extracellular matrix fibronectin distribution. The degree of alignment and orientation of fibronectin varies in a time and shear-stress dependent manner as previously demonstrated for cell elongation (Levesque and Nerem, 1980) and cytoskeletal organization (Levesque et al., 1989). These results of fibronectin organization and alignment are in agreement with another report (Wechezak et al., 1989) where shear stresses of 6 dynes/cm^2 or 26 dynes/cm^2 were applied to BAEC monolayers for 24 hours.

Since the fibronectin is not directly exposed to fluid flow, the reorganization must occur as a result of a signal transmitted through the cells. Vinculin-fibronectin focal contacts have been hypothesized as membrane sites responsible for cell-substrate attachment (Singer, 1982). During cell attachment and spreading, the F-actin distribution and the fibronectin distribution have been observed to qualitatively correlate (Hynes and Destree, 1978). Thus, the relationship between cell mobility changes and cell structure might be associated with alterations in the fibronectin-cell adhesion points (Singer, 1989). When cells are sheared, however, this coincidence is not as evident (Sato et al., 1987). Thus, it appears that the fluid mechanical signal perceived by the endothelial cell is transmitted to the abluminal surface where fibronectin reorganization and alignment occur.

In vivo, endothelial cells are exposed to a pulsatile flow environment. Several other reports have investigated the effect of flow pulsatility on endothelial cell structure and function and have obtained qualitatively and quantitatively similar results. *In vitro*, cell elongation occurs more rapidly and to a greater extent when the applied shear stress is pulsatile (Levesque et al., 1989). Similarly, endothelial cell monolayers synthesize approximately 250% more prostacyclin under pulsatile shear conditions (Frangos et al., 1985). *In vivo*, cell elongation and orientation occurs in much the same manner as demonstrated *in vitro* (Levesque et al., 1986; Kim et al., 1989). Thus, steady shear experiments appear to qualitatively mimic the changes observed under more physiologically relevant pulsatile flow conditions.

Unusually large vWF multimers are potent promoters of sickle red cell adhesion to endothelial cells under physiological flow conditions (Fig 2). Large vWF multimers bind to platelet membranes via glycoprotein (GP) Ib (Coller et al., 1983) and GPIIb/IIIa complexes (Plow et al., 1985a). Arginine-glycine-aspartic acid sequences are involved in vWF binding to GPIIb/IIIa receptors (Plow et al., 1985b). Endothelial cells contain both GPIb (Sprandio et al., 1988) and GPIIb/IIIa-like (Charo et al., 1986) receptors. These types of receptors have not been identified on sickle red blood cells, but the results of Fig 3 (lane RGDS + EC SUP) as well as other data (Wick et al., 1988) suggest that sickle red blood cells also contain vWF receptors.

Sickle red cell adhesion to endothelial cells has recently been observed in living, denervated rat microvessels (Kaul et al., 1989). Under steady perfusion conditions, the red cells adhere almost exclusively in the venules, vessels in which the endothelial cells are exposed to approximately 1.0 dynes/cm^2 shear-stress as modeled in our experiments. Furthermore, when the microvasculature was perfused with vasopressin (which stimulates the release of unusually large vWF from endothelial cells *in vivo* [Ruggeri et al., 1987]), increased quantities of red blood cells adhered to venular endothelial cells (Tsai et al., 1990). Thus, our *in vitro* results indicating that sickle red cell adhesion occurs under venous shear flow conditions and that unusually large vWF multimers are potent promoters of sickle red cell/endothelial cell interactions are confirmed in a living microvascular network.

CONCLUSIONS

It is not clear from our data whether unusually large vWF multimers are more adhesive than plasma multimeric forms as a result of an increased number of RGD sequences exposed on the surface of vWF, or whether unusually large vWF is structurally different from plasma multimeric forms. However, given the fact that these two populations of vWF promote vastly different quantities of sickle red cell adhesion to endothelial cells under flow conditions, it is interesting to speculate on a role for vWF in sickle cell vaso-occlusive crises. Crises often occur in conjunction with vigorous exercise, pregnancy, increased stress, or infection. These stresses could lead to chemical or mechanical stimulation of endothelial cells and accelerated release of unusually large vWF multimers into the vessel lumen. *In vivo,* bacterial endotoxin leads to rapid increases in plasma vWF concentrations and the appearance of unusually large multimeric forms (Gralnick et al., 1989). If stimulation occurs in the microvessels (where red cells and endothelial cells are closely apposed and the fluid shear forces opposing adhesion are low), released unusually large vWF multimers can bind simultaneously to sickle red blood cells and the endothelium, leading to adhesion. This adhesion would subsequently delay red cell passage through the microcirculation, increasing the likelihood of intracapillary sickling, leading to further occlusion.

REFERENCES

Barabino, G. A., McIntire, L. V., Eskin, S. G., Sears, D. A., and Udden, M. A., 1987, Endothelial Cell Interactions with Sickle Cell, Sickle Trait, Mechanically Injured, and Normal Erythrocytes Under Controlled Flow, Blood 70:152-157.

Breihl, R. W. and Mann, E. S., 1989, Hemoglobin S Polymerization: Fiber Lengths, Rheology, and Pathogenesis, Ann NY Acad Sci 565:295-307.

Charo, I. F., Fitzgerald, L. A., Steiner, B., Rall, S. C., Jr., Bekeart, L. S., and Phillips, D. R., 1986, Platelet Glycoproteins IIb and IIIa: Evidence for a Family of Immunologically and Structurally Related Glycoproteins in Mammalian Cells, Proc Natl Acad Sci USA 83:8351-8355.

Coller, B. S., Peerschke, E. I., Scudder, L. E., and Sullivan, C. A., 1983, Studies with a Murine Monoclonal Antibody that Abolishes Ristocetin-Induced Binding of von Willebrand Factor to Platelets: Additional Evidence in Support of GPIb as a Platelet Receptor for von Willebrand Factor, Blood 61:1456-1461.

Counts, R. B., 1975, Solid-Phase Immunoradiometric Assay of Factor VIII Protein, Br J Haematol 31:429-436.

Dewey, C. F., Jr., Bussolari, S. R., Gordon, E. J., and Gimbrone, M. A., Jr., 1981, The Dynamic Response of Vascular Endothelial Cells to Fluid Shear Stress, ASME J Biomech Engr 103:177-186.

Doty, S. D. 1989, Fluid Shear Stress Effects on Fibronectin in Endothelial Cells, MS Thesis, School of Mechanical Engineering, Georgia Institute of Technology, Atlanta, GA.

Eskin, S. G., Ives, C. L., McIntire, L. V., and Navarro, L. T., 1984, Response of Cultured Endothelial Cells to Steady Flow, Microvasc Res 28:87-94.

Ferrone, F. A., 1989, Kinetic Models and the Pathophysiology of Sickle Cell Disease, Ann NY Acad Sci 565:63-74.

Frangos, J. A., McIntire, L. V., Eskin, S. G., and Ives, C. L., 1985, Flow Effects on Prostacyclin Production by Cultured Human Endothelial Cells, Science 227:1477-1479.

Gardner, J. M. and Hynes, R. O., 1985, Interaction of Fibronectin with its Receptor on Platelets, Cell 42:439-448.

Gralnick, H. R., McKeown, L. P., Wilson, O. M., Williams, S. B., and Elin, R. J., 1989, von Willebrand Factor Release Induced by Endotoxin, J Lab Clin Med 113:118-122.

Hebbel, R. P., Yamada, O., Moldow, C. F., Jacob, H. S., and Steinberg, M. H., 1980, Abnormal Adherence of Sickle Erythrocytes to Cultured Vascular Endothelium: Possible Mechanism for Microvascular Occlusion in Sickle Cell Disease, J Clin Invest 65:154-160.

Hebbel, R. P., Boogaerts, M. A. B., Eaton, J. W., and Steinberg, M. H., 1980, Erythrocyte Adherence to Endothelium in Sickle-Cell Anemia: A Possible Determinant of Disease Severity, N Engl J Med 76:992-995.

Hebbel, R. P., Moldow, C. F., and Steinberg, M. H., 1981, Modulation of Erythrocyte-Endothelial Interactions and the Vasocclusive Severity of Sickling Disorders, Blood 58:947-952.

Hynes, R. O. and Destree, A. T., 1978, Relationship Between Fibronectin (LETS Protein) and Actin, Cell 15:369-377.

Kaul, D. K., Fabry, M. E., and Nagel, R. L., 1989, Microvascular Sites and Characteristics of Sickle Cell Adhesion to Vascular Endothelium in Shear Flow Conditions: Pathophysiological Implications, Proc Natl Acad Sci USA 86:3356-3360.

Kim, D. W., Gotlieb, A. I., and Langille, B. L, 1989, In Vivo Modulation of Endothelial F-Actin Microfilaments by Experimental Alterations in Shear Stress, Arteriosclerosis 9:439-445.

Levesque, M. J. and Nerem, R. M., 1980, The Study of Rheological Effects on Vascular Endothelial Cells In Culture, Biorheol 26:345-357.

Levesque, M. J. and Nerem, R. M., 1985, Elongation and Orientation of Cultured Endothelial Cells in Response to Shear, ASME J Biomech Engr 34:341-347.

Levesque, M. J., Leipsch, D., Moravec, S., and Nerem, R. M., 1986, Correlation of Endothelial Cell Shape and Wall Shear Stress in Stenosed Dog Aorta, Arteriosclerosis 6:220-229.

Levesque, M. J., Sprague, E. A., Schwartz, C. J., and Nerem, R. M., 1989, The Influence of Shear Stress on Cultured Vascular Endothelial Cells: The Stress Response of an Anchorage-Dependant Mammalian Cell, Biotechnology Progress 5:1-8.

Moake, J. L., Byrnes, J. J., Troll, J. H., Rudy, C. K., Weinstein, M. J., Colannino, N. M., and Hong, S. L., 1984, Abnormal VIII:von Willebrand Factor Patterns in the Plasma of Patients with the Hemolytic-Uremic Syndrome, Blood 64:592-598.

Mohandas, N. And Evans, E., 1985, Sickle Erythrocyte Adherence to Vascular Endothelium: Morphologic Correlates and the Requirement for Divalent Cations and Collagen Binding Proteins, J Clin Invest 76:1605-1612.

Mozzarelli, A., Hofrichter, J., and Eaton, W. E., 1987, Delay Time of Hemoglobin S Polymerization Prevents Most Cells from Sickling in Vivo, Science 237:500-506.

Nerem, R. M. and Levesque, M. J., 1984, Fluid Dynamics as a Factor in the Localization of Atherogenesis, Ann NY Acad Sci 416:709-719.

Patel, V. P., Ciechanover, A., Platt, O., and Lodish, H. F., 1985, Mammalian Reticulocytes Lose Adhesion to Fibronectin During Maturation to Erythrocytes, Proc Natl Acad Sci USA 82:440-444.

Plow, E. F., McEver, R. P., Coller, B. C., Woods, Jr., V. L., Marguerie, G. A., and Ginsberg, M. H., 1985a, Related Binding Mechanisms for Fibrinogen, Fibronectin, von Willebrand Factor, and Thrombospondin on Thrombin-Stimulated Human Platelets, Blood 66:724-727.

Plow, E. F., Pierschbacher, M. D., Ruoslahti, E., Marguerie, G. A., and Ginsberg, M., 1985b, The Effect of Arg-Gly-Asp-Containing Peptides on Fibrinogen and von Willebrand Factor Binding to Platelets, Proc Natl Acad Sci USA 82:8057-8061.

Ruggeri, Z. M., Mannucci, P. M., Federici, A. B., Lombardi, R., and Zimmerman, T. S., 1987, Multimeric Composition of Factor VIII/von Willebrand Factor Following Administration of DDAVP: Implications for Pathophysiology and Therapy of von Willebrand Disease Subtypes, Blood 70:173-176

Sato, M., Levesque, M. J., and Nerem, R. M., 1987, Micropipette Aspiration of Cultured Bovine Aortic Endothelial Cells Exposed to Shear Stress, Arteriosclerosis 7:276-286.

Schettler, G., Nerem, R. M., Schmid-Schonbein, H., Morl, H., and Deihm, C., eds., 1983, "Fluid Dynamics as a Localizing Factor in Atherosclerosis", Springer-Verlag, New York.

Singer, I. I., 1982, Association of Fibronectin and Vinculin with Focal Contacts and Stress Fibers in Stationary Hamster Fibroblasts, J Cell Biol 92:398-408.

Singer, I. I., 1989, Fibronectin-Skeleton Relationships, in "Biology of Extracellular Matrix: Fibronectin (A Series)", Mosher, D. F., editor, Academic Press, Inc., New York, 139-161.

Sprague, E. A., Steinbach, B. L., Nerem, R. M., and Schwartz, C. J., 1987, Influence of Laminar Steady-State Fluid-Imposed Wall Shear Stress on the Binding, Internalization, and Degradation of Low-Density Lipoproteins by Cultured Endothelium, Circulation 76:648-656.

Sprandio, J. D., Shapiro, S. S., Thiagarajan, P., and Montgomery, R. R., 1988, Cultured Human Umbilical Vein Endothelial Cells Contain a Membrane Glycoprotein Immunologically Related to Platelet GPIb, Blood 71:234-237.

Stehbens, W. E., 1983, Hemodynamics and Atherosclerosis, Biorheology 19:95-101.

Tsai, H. M., Sussman, I. I., Nagel, R. L., and Kaul, D. K., 1990, Desmopressin (DDAVP)-Induced Adhesion of Normal Human Red Cells to the Endothelium of a Perfused Microvasculature, Blood 75:261-265.

Turitto, V. T., 1982, Blood Viscosity, Mass Transport, and Thrombogenesis, Prog Hemostasis Thromb 6:139-177.

Wechezak, A. R., Viggers, R. F., and Sauvage, L. R., 1989, Fibronectin and F-Actin Redistribution in Cultured Endothelial Cells Exposed to Shear Stress, Lab Invest 53:639-647.

White, C. E., Fujiwara, K., Shefton, E. J., Dewey, Jr., C. F., and Gimbrone, Jr., M. A., 1982, Federation Proceedings 41:321.

Wick, T. M., Moake, J. L., Udden, M. M., Eskin, S. G., Sears, D. A., and McIntire, L. V., 1987, Unusually Large von Willebrand Factor Multimers Increase Adhesion of Sickle Erythrocytes to Human Endothelial Cells under Controlled Flow, J Clin Invest 80:905-910.

Wick, T. M., Moake, J. L., Udden, M. M., and McIntire, L. V., 1988, Unusually Large vWF Multimers Bind to GPIb-Like and Integrin Receptors on Sickle and Young Non-Sickle RBC and on Endothelial Cells (EC): A Mechanism for Sickle and Other Young RBC Adhesion to EC, Blood 72:76a.

Yamada, K. M., Humphries, M. J., Hasegawa, T., Hasegawa, E., Olden, K., Chen, W.-T., and Akiyama, S., 1987, Fibronectin: Molecular Approaches to Analyzing Cell Interactions with the Extracellular Matrix, In "The Cell in Contact", Adhesions and Junctions as Morphogenetic Determinants", Thiery, J. P., editor, J. Wiley and Sons, New York, 1987.

NON-INVASIVE ASSESSMENT OF VELOCITY PATTERNS IN AND WALL PROPERTIES OF THE CAROTID ARTERY BIFURCATION IN MAN BY MEANS OF A MULTI-GATE PULSED DOPPLER SYSTEM

Robert S. Reneman, Tiny Van Merode, Frans A.M. Smeets, and Arnold P.G. Hoeks

Departments of Physiology and Biophysics, Cardiovascular Research Institute Maastricht, University of Limburg Maastricht, The Netherlands

ABSTRACT

Multi-gate pulsed Doppler systems have been developed which allow the non-invasive on-line recording of instantaneous velocities, simultaneously at various sites along the ultrasound beam, and, hence, of velocity profiles at discrete time intervals during the cardiac cycle. With these systems one can also record on-line the relative arterial diameter changes during the cardiac cycle, providing insight into arterial wall properties in general and local differences in particular. Although the technique is not ideal yet, with multi-gate pulsed Doppler systems valuable information can be obtained about flow patterns in arteries and arterial wall properties under normal and pathological circumstances in man.

Keywords: multi-gate pulsed Doppler - carotid arteries - compliance - distensibility - age - hypertension - atherosclerosis

PRINCIPLES AND LIMITATIONS OF MULTI-GATE PULSED DOPPLER SYSTEMS

In Doppler systems the ultrasound can be transmitted continuously (CW Doppler devices) or intermittently (pulsed Doppler devices). In the latter devices pulses are emitted with a certain frequency, the pulse repetition frequency (PRF). The crystal receives the backscattered signals from the red blood cells and the vessel wall during the

Biomechanical Transport Processes, Edited by F. Mosora *et al.*
Plenum Press, New York, 1990

interval between pulses. In pulsed Doppler systems single-gate and multi-gate instruments can be distinguished. In single-gate pulsed Doppler systems an electronic gate allows the selection of scattering either from the vessel wall or the red blood cells at a given distance from the transducer. This makes it possible to determine the velocity as an instantaneous function of time in a small sample volume at various sites in an artery. Since during one cardiac this velocity can only be determined at one site in the vessel, velocity profiles can only be synthesized by assessing the instantaneous velocities at various sites in an artery during consecutive heart beats.

More recently multi-gate pulsed Doppler systems have been developed that have the ability to detect simultaneously and instantaneously velocities over the full range of interest (Hoeks et al, 1981; Hoeks et al, 1984; Reneman et al, 1986a). These systems allow the on-line recording of the velocity as an instantaneous function of time, simultaneously at various sites along the ultrasound beam (Fig. 1A), and, hence, of velocity profiles - that is the velocity distribution along the cross-section of the vessel- at discrete time intervals during the cardiac cycle (Fig. 1B). Processing of the velocity

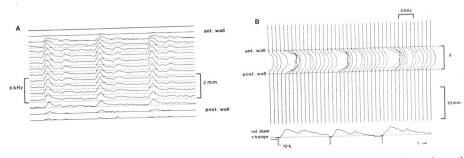

Fig 1. A. Instantaneous velocity tracings as simultaneously recorded at various sites along the ultrasound beam in the common carotid artery of a young volunteer. Three cardiac cycles are depicted. B. Axial velocity profiles at discrete time intervals during the cardiac cycle as derived from the velocity information in Fig 1A. The dotted lines indicate the width of the profiles during systole from which the internal arterial diameter (d) during systole can be estimated. The relative arterial diameter changes during the cardiac cycle are depicted as well. The negative deflection represents the trigger of the R wave of the ECG. After Van Merode et al, 1988. With permission.

information occurs in a serial way, which means that all the signals are processed after each other by the same circuitry in between the emission of pulses. Processing occurs on a digital base because of the high dynamic range of the signals. With the system in use 256 velocity profiles can be preserved simultaneously covering a few cardiac cycles. To obtain detailed velocity information along the ultrasound beam and, hence, reliable velocity profiles, the sample resolution has to be high and the sample distance along the ultrasound beam must be small. The multi-gate system, as used in our laboratory for peripheral applications, has a sample distance of 0.5-0.7 mm and a sample volume of 1.2-1.7 mm3, depending on the application and the emission frequency used.

With multi-gate pulsed Doppler systems one can also determine on-line the relative changes in arterial diameter ($\Delta d/dx100\%$) during the cardiac cycle (Fig. 1). This assessment is based upon the processing of low frequency Doppler signals, origi- nating from the sample volumes coinciding with the anterior and posterior walls (Hoeks et al, 1985). Integration of these velocity signals delivers displacement. The relative ar- terial diameter changes are independent of the angle of interrogation and can be determined with an absolute accuracy of 0.5% (Hoeks et al, 1985). This means that for a given relative change in diameter of, for instance, 7% a relative change in diameter between 6.5% and 7.5% can be measured. From the width of the velocity profile the systolic internal arterial diameter can be determined (Fig. 1; Reneman et al, 1986a and 1986b). The internal diameter, as obtained in this way, is dependent on the angle of interrogation. Although in multi-gate pulsed Doppler systems the vessel of interest can be localized easily without the use of an image of the vessel, the combination of a multi- gate system and a B-mode or 2-D echo imager is preferred for proper localization of the site of sampling with respect to artery structures (e.g. bifurcations) and to be informed of the angle of interrogation.

A major limitation of pulsed Doppler devices is that the Doppler bandwidth of the system is limited. The maximum Doppler velocity that can be detected unambiguously depends on the distance between probe and vessel, and the transmitter frequency (Reneman and Hoeks, 1977). This distance sets an upper bound to the PRF, which theoretically should exceed the maximum Doppler frequency at least twice. However, re- cently a frequency estimator for sampled Doppler signals has been developed that tracks frequencies beyond the Nyquist frequency (half the PRF), significantly increasing the maximum Doppler frequency that can be detected unambiguously (Hoeks et al, 1984).

VELOCITY PATTERN IN THE CAROTID ARTERY BIFURCATION

Under normal circumstances the velocity profile in the common carotid artery is symmetric and has the shape of a flattened parabola (Fig. 1B and Fig. 2A). In the carotid artery bulb the axial blood flow velocities, as recorded in the plane of the bifurcation with an angle of interrogation of 60%, are highest on the side of the flow divider (Fig. 2B). The skewness is most pronounced early in systole. In young healthy volunteers (20-30 years of age) regions of flow separation and recirculation are seen on the side opposite to the flow divider. Flow separation is not continuously present throughout the cardiac cycle, but starts during the deceleration phase (Fig. 2C).   Similar   patterns   of

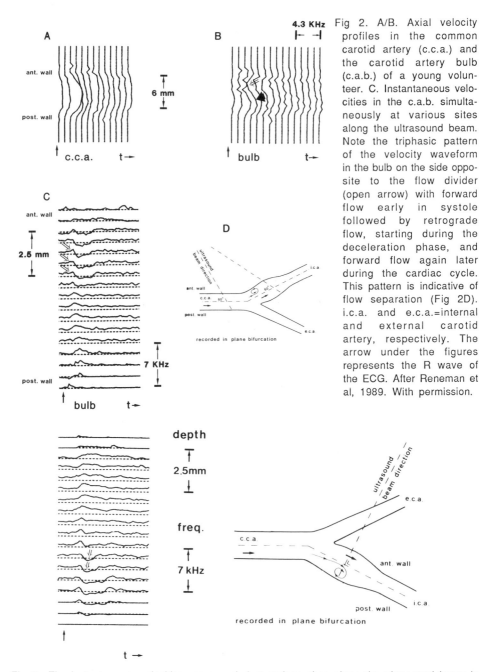

Fig 2. A/B. Axial velocity profiles in the common carotid artery (c.c.a.) and the carotid artery bulb (c.a.b.) of a young volunteer. C. Instantaneous velocities in the c.a.b. simultaneously at various sites along the ultrasound beam. Note the triphasic pattern of the velocity waveform in the bulb on the side opposite to the flow divider (open arrow) with forward flow early in systole followed by retrograde flow, starting during the deceleration phase, and forward flow again later during the cardiac cycle. This pattern is indicative of flow separation (Fig 2D). i.c.a. and e.c.a.=internal and external carotid artery, respectively. The arrow under the figures represents the R wave of the ECG. After Reneman et al, 1989. With permission.

Fig 3. The instantaneous velocities, as recorded at various sites along the ultrasound beam in the plane of the bifurcation perpendicular to the flow axis in the carotid artery bulb of a healthy young volunteer. In this direction mainly radial flow components are recorded. The arrow represents the R-wave of the ECG. Note the triphasic pattern of the velocity waveform on the side opposite to the flow divider (open arrow) with flow away from the probe early in systole followed by flow towards the probe, starting during the deceleration phase, and flow away from the probe again later during the cardiac cycle. This pattern is indicative of flow separation and recirculation as depicted in the schematical drawing.

recirculation are observed when the instantaneous velocities are recorded in the plane of the bifurcation and the artery is interrogated perpendicular to the vessel axis (Fig. 3). In this position mainly radial flow components are recorded. The finding that blood recirculates in both the axial and radial direction suggests the presence of helical flow, which is in agreement with observations in model bifurcations (Ku and Giddens, 1983)

It is an interesting observation that these regions of recirculation are less common and less pronounced in older (50-60 years) presumed healthy volunteers. This difference between younger and older volunteers cannot readily be explained on the basis of a more favorable diameter transition from common to internal carotid artery in the younger subjects or by differences in the angle between these arteries (Reneman et al, 1985). It has been hypothetized that differences in distensibility at the transition from common to internal carotid artery between young and old volunteers (see below) may explain the differences in flow separation and recirculation between both age groups (Reneman et al, 1985).

ESTIMATION OF CAROTID ARTERY WALL PROPERTIES IN VIVO

Introduction

To obtain reliable information about arterial wall properties it is essential to assess the relevant parameters in vivo, because arteries show a stiffer behavior in vitro than in vivo (Gow and Hadfield, 1979). Accurate determination of such parameteres as distensibility and compliance in vivo has been hampéred by the absence of reliable techniques. One apprach is to assess in an arterial segment the pulse wave velocity (Laogun and Gosling, 1982), which is equivalent to the square root of the reciprocal value of the distensibility times density (Mc Donald, 1974). This means that a diminished distensibility results in a higher pulse wave velocity. This method does not consider changes in arterial diameter, if any, which are taken into account in the calculation of compliance. Therefore, it is prefered to assess distensibility and compliance, preferably locally, to describe arterial wall properties. This approach also offers the possibility to determine and appreciate local differences in arterial wall properties within a vessel segment.

Distensibility can be defined as the relative increase in arterial volume ($\Delta V/V$) for a given increase in pressure (Δp) and the compliance as the distensibility multiplied by the arterial volume. Since it is practically impossible to accurately determine $\Delta V/V$ and V noninvasively in man, it is assumed that the increase in volume is caused by distension of the artery, i.e. an increase in cross-sectional area rather than elongation. The assumption that the increase in volume does not result from elongation is reasonable because under normal circumstances lengthening of the artery during systole is not observed on B-mode images (unpublished results). Therefore, the expression for the distensibility coefficient (DC)

$$DC = \frac{\Delta V/V}{\Delta p} \qquad (1)$$

can be rewritten using a first order approximation ($\Delta V \ll V$)

$$DC = \frac{2\pi r . \Delta r/\pi r^2}{\Delta p} \qquad (2)$$

$$= \frac{2\Delta r/r}{\Delta p}$$

$$= \frac{2\Delta d/d}{\Delta p}$$

Similarly the expression for compliance (C)

$$C = DC . V \qquad (3)$$

can be rewritten as cross-sectional compliance (CC)

$$CC = \frac{2\Delta d/d}{\Delta p} . \pi r^2 \qquad (4)$$

$$= \frac{\Delta d/d}{2\Delta p} . \pi d^2.$$

The diameter rather than the radius is used in these equations because the multi-gate pulsed Doppler system measures $\Delta d/d$ (see above). The peak systolic value of $\Delta d/d$, the systolic internal arterial diameter and the pulse pressure (Δp), as determined from brachial artery cuff blood pressure measurements, are used in these calculations.

A critique on this method could be that pulse pressure is measured in the brachial rather the common carotid artery. In this approach it is assumed that the pulse pressure in the brachial artery is representative of that in the common carotid artery. An indication of this assumption is the positive relationship between this pulse pressure and the relative diameter increase of the common carotid artery during systole (Reneman et al, 1986b) Moreover, intra-arterial measurements during surgery have shown that pulse pressure in the carotid and brachial arteries is similar. Besides, it is likely that wall elasticity, an important determinant of pulse pressure, diminishes with age to a similar extent in the brachial and carotid arteries (Reneman et al, 1986b).

Changes in Arterial Wall Properties with Age

It has been known for quite some time that arteries become stiffer at older age. In a study on presumed healthy subjects, varying in age between 20 and 69 years, distensibility and cross-sectional compliance of the common carotid artery, determined as described above, decreased linearly with age; starting in the third age decade already (Reneman et al, 1986b). The decrease in the latter parameter, however, was less pronounced (Fig. 4). This is probably caused by the increase in diameter with age affecting compliance but not distensibility. In this study the average relative arterial diameter increase of the relative elastic common carotid artery decreased by 39% between the third and seventh age decade, a value similar to the one observed in the muscular, relatively stiff femoral artery (35% reduction from under 35 till over 60 years; Mozersky et al, 1972).

It is an interesting observation that along the carotid artery bifurcation inhomogeneities in distensibility have to be appreciated (Reneman et al, 1985). In young presumed healthy volunteers the carotid artery bulb is more distensible than the common

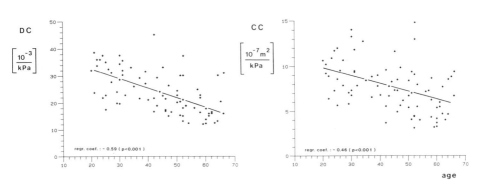

Fig 4. Common carotid artery distensibility coefficient (DC) and cross-sectional compliance (CC) as a function of age. The straight lines represent the linear regressions of DC and CC with age. After Reneman et al, 1986b. With permission.

carotid artery. This difference in distensibility probably cannot be explained by structural differences because the wall of the common carotid artery and the major part of the bulb have a mainly elastic structure (Muratori, 1967). The more pronounced distensibility of the bulb likely results from its thinner wall (Muratori, 1967), so that larger wall tensions are developed at comparable pressures and diameters, which is the case in this situation. The more pronounced distensibility of the carotid artery bulb likely facilitates the functioning of the baroreceptors, which are located mainly in the proximal part of the bulb (Muratori, 1967). In older presumed healthy subjects the distensibility is significantly diminished, as compared with young volunteers, along the whole carotid artery bifurcation but most pronounced in the bulb. The latter might, at least partly, explain the decreased baroreceptor sensitivity at older age (Gribbin et al, 1971). The differences in distensibility at the transition from common to internal carotid artery between older and younger volunteers might explain the differences in flow pattern in the carotid artery bifurcation, as observed between these subjects (see above).

Arterial Wall Properties in Hypertension

In patients with established hypertension arterial wall distensibility is diminished (Simon et al, 1983). In a recent study it was shown that even in young borderline hypertensive subjects (20-35 years) both distensibility and cross-sectional compliance of the common carotid artery are significantly diminished, as compared with age-matched control subjects (Van Merode et al, 1988). The reduced distensibility likely results from a decrease in arterial wall elasticity because the relative increase in common carotid artery diameter during systole is diminished in borderline hypertensives, despite the fact that their pulse pressure is similar to or higher than that in control subjects. The internal arterial diameter is not significantly different between normotensives and borderline hypertensives. Since the stiffer behavior of the common carotid artery in borderline hypertensives is associated with relatively slight changes in blood pressure, the question can be raised as to whether the alterations in arterial wall properties are a result of the elevated arterial blood pressure; these alterations might develop independent of blood pressure elevation. This is supported by the observation that effective antihypertensive treatment is not always associated with improvement of arterial wall distensibility (Simon et al, 1985).

CONCLUSIONS

Although the techniques, which have been developed to assess velocity patterns in and arterial wall properties of arteries in man provide interesting information, they are not ideal yet. Improvements, as far as the spatial resolution of Doppler multi-gate pulsed Doppler systems are concerned, cannot be expected, because it is unlikely that the sample volume of these systems can be made smaller (Reneman et al, 1986b). Therefore, in the near future one has to rely on the detail with which flow patterns in arteries can be recorded with the systems presently available. However, improvements in the non-invasive assessment of arterial wall distensibility can be anticipated, because with a device under development (Hoeks et al, In press) small displacements of the arterial wall can be determined more accurately than with the multi-gate Doppler system in use. This is of special advantage in studies on less distensible muscular arteries, older patients and atherosclerotic vessels. Moreover, the new device, based upon phase tracking of the radio frequency signals generated by the arterial walls, allows the simultaneous recording of displacement of the anterior and posterior wall of an artery, which is problematic with the systems presently available.

Acknowledgement

The authors are indebted to Karin van Brussel, Jos Heemskerk and Rosy Hanssen for their help in preparing the manuscript.

REFERENCES

Gow, B.S., and Hadfield, C.D.,1979, The elasticity of canine and human coronary arteries with reference to postmortem changes, Circ. Res. 45: 588-594.

Gribbin, B., Pickering, T.G., Sleight, P., and Peto, R., 1971, Effect of age and high blood pressure on baroflex sensitivity in man, Circ. Res. 29: 424-431.

Hoeks, A.P.G., Brands, P.J., Smeets, F.A.M., and Reneman, R.S., Assessment of the distensibility of superficial arteries, Ultrasound Med. & Biol. In press.

Hoeks, A.P.G., Peeters, H.P.M., Ruissen, C.J., and Reneman, R.S., 1984,. A novel frequency estimator for sampled Doppler signals, IEEE Trans. Biomed. Eng. BME-31: 212-220.

Hoeks, A.P.G., Reneman, R.S., and Peronneau, P.A., 1981, A multi-gate pulsed Doppler system with serial data processing, IEEE Trans. Sonics Ultrasonics SU-28: 242-247.

Hoeks, A.P.G., Ruissen, C.,J., Hick, P.J.J., and Reneman, R.S., 1985, Transcutaneous detection of relative changes in artery diameter, Ultrasound Med. & Biol. 11: 51-59.

Ku, D.N., and Giddens, D.P., 1983,. Pulsatile flow in a model carotid bifurcation, Arteriosclerosis 3: 31-39.

Laogun, A.A., and Gosling, R.G., 1982, In vivo arterial compliance in man, Clin. Phys. Physiol. Meas. 3: 201-212.

McDonald, D.A., 1974, Blood flow in arteries, Edw. Arnold Publishers Ltd., London.

Mozersky, D.J., Sumner D.S., Hokanson, D.E., and Strandness, D.E., 1972, Transcutaneous measurement of the elastic properties of the human femoral artery, Circulation 46: 948-955.

Muratori, G., 1967, Histological observations on the structure of the carotid sinus in man and mammals. In: Baroreceptors and Hypertension, edited by P. Kezdi, Pergamon Press, New York, 253-265.

Reneman, R.S., and Hoeks, A.P.G., 1977, Continuous wave and pulsed Dopplers flowmeters - A general introduction. In: Echocardiology with Doppler Applications and Real-time Imaging, edited by N. Bom, Martinus Nijhoff, The Hague, 189-205.

Reneman, R.S., Van Merode, T., Hick, P.J.J., and Hoeks, A.P.G., 1985, Flow velocity patterns in and distensibility of the carotid artery bulb in subjects of varying ages, Circulation 71: 500-509.

Reneman, R.S., Van Merode, T., Hick, P.J.J., and Hoeks, A.P.G., 1986a, Cardiovascular applications of multi-gate pulsed Doppler system, Ultrasound Med. & Biol. 12: 357-370.

Reneman, R.S., Van Merode, T., Hick, P.J.J., Muytjens, A.M.M., and Hoeks, A.P.G., 1986b, Age-related changes in carotid artery wall properties in men, Ultrasound Med. & Biol. 12: 465-471.

Reneman, R.S., Van Merode, T., and Hoeks, A.P.G., 1989, Noninvasive assessment of arterial flow patterns and wall properties in humans. NIPS 4: 185-190.

Simon, A.Ch., Laurent, A., Levenson, J.A., Bouthier, J.E., and Safar., M.E., 1983, Estimation of forearm arterial compliance in normal and hypertensive men, Cardiovasc. Res. 17: 331-338.

Simon, A.Ch., Levenson, J., Bouthier, J.D., and Safar, M.E., 1985, Effects of chronic administration of enalapril and propranolol on the large arteries in essential hypertension, J. Cardiovasc. Pharm. 7: 856-861.

Van Merode, T., Hick, P.J.J., Hoeks, A.P.G., Rahn, K.H., and Reneman, R.S., 1988, Carotid artery wall properties in normotensive and borderline hypertensive subjects of various ages, Ultrasound Med. & Biol. 14: 563-569.

ANEURYSMS AS A BIOMECHANICAL

INSTABILITY PROBLEM

Nuri Akkas

Department of Engineering Sciences
Middle East Technical University
Ankara, Turkey

INTRODUCTION

Aneurysms are pathological, localized, blood-filled dilatations of the blood vessels. Their origin may be congenital, traumatic, arteriosclerotic or infectious (Rhoton et al., 1977). The congenital (saccular) aneurysms, which make up over 90% of intracranial aneurysms, are generally found in and about the circle of Willis and especially at bifurcations. Most symptomatic aneurysms range in size from 0.5 to 1.5 cm in diameter. There are also giant aneurysms which expand to 3 cm in diameter or more without rupturing. They are a major cause of stroke-related morbidity and mortality. Saccular aneurysms may eventually rupture or they may expand slowly. Dilatation of the vessel and the eventual rupturing or expansion of the aneurysm bag all involve large deformations of the relevant membrane. In this paper we propose the idea that aneurysm rupture can be considered to be a biomechanical instability problem.

Investigations concerned with the analysis of truly biomechanical models of saccular aneurysms are few (Hung and Botwin, 1975; Perktold, 1987). On the other hand, investigations related to cardiac aneurysms are more in number (Mirsky et al., 1978; Bogen and McMahon, 1979; Radhakrishnan et al., 1980). The question of whether cardiac aneurysms blow out or not has been asked. Indeed, according to Bogen and McMahon (1979), if the mechanism of aneurysm formation and rupture is not elastic stability, then what is it? Degenerative changes in the ventricular wall, including muscle cell necrosis and edema, almost certainly change not only the elastic but the plastic properties of the tissues in the critical post-infarction period. We are of the opinion that the aneurysm wall material should be treated as a nonlinearly elastic material like, for instance, the Mooney-Rivlin material (Mooney, 1940; Treloar, 1975). A realistic biomechanical model for the analysis of the deformation and rupture of cerebral saccular aneurysms should involve large deformations and nonlinear material behavior. Such analyses always bring us to the topic of instability which will be discussed below.

MATERIAL INSTABILITY

In order to deal precisely with the manner in which the deformation of a physical body is related to the forces acting on it, we must express this relation quantitatively, either by equations or by graphs. The simplest relation in mechanics is that of a linear spring which has the characteristic that the force required to deflect it is proportional to the amount of deflection. The amount of force required to produce a unit deflection is called the spring constant of the spring. Those materials which show a linearly elastic behavior are called Hookean. The stress-strain curves of soft biological materials, including the aneurysm membrane, are not linear. Let us now develop a constitutive relation in the form of a graph for the material under consideration. The development to be given below is based on a very simple, logical approach. More sophisticated analyses are available in the literature (Decraemer et al., 1980; Krajcinovic et al., 1987). Soft biological membranes are generally composed of filamentous materials and may be treated as filament-reinforced composites. Our model contains a very large number of filaments. Each filament is treated as a spring which shows a linear elastic behavior until it breaks. The filaments have different initial lengths.

For demonstration purposes, consider the model shown in Fig. 1a. The model membrane consists of six filaments of different lengths and it is stretched by two rigid blocks. The initial nondimensional lengths l_i of the filaments are given in Fig. 1a. The blocks are pulled apart by a force F and the distance between the two rigid blocks is denoted by H, which is initially equal to 1.0 nondimensional unit. The force f versus the stretch ratio $\Gamma = l_f / l_i$ curve for each filament is given in Fig. 1b. Here l_f is the final stretched nondimensional length of the filament. Figure 1b shows that each filament is linearly elastic but it breaks when its final length is double its initial length. The axial force f in each filament can be considered nondimensional also for demonstration purposes and, thus, its maximum value is taken to be 1.0 unit.

The force versus elongation curve for each individual filament is shown in Fig. 1c for the six filaments of different lengths considered. The curve which gives the total force F as a function of H, the distance between the two rigid blocks, for the group of filaments when they are considered to be working as a single unit is obtained simply by superposing the individual curves. The curve obtained from the superposition of the individual curves is shown in Fig. 1c also and it looks like the edge of a saw. This is due to the fact that we modelled the membrane with six filaments only and the variations in length were also very limited. For a realistic model of the membrane, we should include a very large number of filaments and a wide range of lengths. In this case, the saw edge character of the curve will disappear and we will get the smoothed-out curve shown in Fig. 1c as the force-elongation curve of the membrane.

The smoothed-out curve in Fig. 1c has a so-called "limit point" A. Such limit points play important roles in mechanics

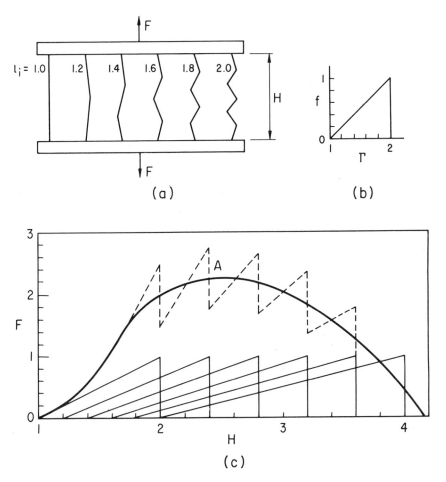

Fig. 1. Material instability

because they correspond to loss of stability of the physical system. Let us now start from the configuration for which H=1.0 corresponding to F=0.0. If we increase the load, we will have an increasing deflection up until the limit point A. When we reach the peak point A, any attempt to further increase the load, however small this increase in load may be, will cause an infinitely large deflection. No equilibrium condition exists above this load level. It is said that the system lost its stability. The limit point A corresponds to this loss of stability. This is called **material instability**. One is referred to (Akkas, 1978; Benedict et al., 1979; Needleman, 1977; Crane, 1973; Crisp and Hart-Smith, 1971) for detailed discussions of this type of instability.

GEOMETRIC INSTABILITY

The analysis of structural instability is a topic of nonlinear mechanics. One type of nonlinearity in mechanics of deformable bodies is material nonlinearity leading to material instability. In the second type, the nonlinearity enters the theory in expressions representing the influence of rotations of structural elements on the behavior of the structure. The geometry of the deformed configuration of the structure enters the problem. Moreover, the kinematic relations, i.e., the strain-displacement relations are also nonlinear because these relations are derived, again, referring to the geometry of the deformed configuration. Consequently, the nonlinearity of the second type is purely geometrical.

A simple illustration of the influence of geometric nonlinearity is provided by a structure consisting of two flexible bars subjected to a concentrated load P at its mid-point Q as shown in Fig. 2a. The bar material is assumed to obey Hooke's law so we cannot have the material instability discussed in the previous section. The two bars are equal in length, cross sectional area and modulus of elasticity, resulting in a structural system symmetric with respect to a vertical line passing through the mid-point Q. As P is increased the mid-point Q will move downward along the line of symmetry. The downward displacement of point Q from its original equilibrium configuration is denoted by y as shown in Fig. 2a. If we plot the applied load P versus the displacement y of point Q, we obtain a curve similar to that shown in Fig. 2b. Each point on the path represents an equilibrium configuration of the structure.

The point A on the equilibrium path of Fig. 2b at which the load P is a relative maximum is called a **limit point**. As the load P starts from zero and is increased, the stiffness of the structure, which corresponds to the slope of the load-displacement curve, decreases. Then, as the maximum point A is reached and passed infinitesimally, the truss snaps into a new position, denoted by B in Fig. 2b, such that the load P subjects the two bars to tension. At point A the structure is said to have lost its stability and this is called **geometric instability** or **snap-through buckling**. As a reference book on geometric instability of many common structural systems and a detailed presentation of general concepts, we will cite Brush and Almroth (1975).

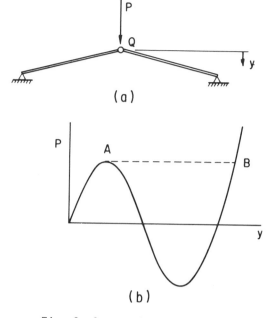

(a)

(b)

Fig. 2. Geometric instability

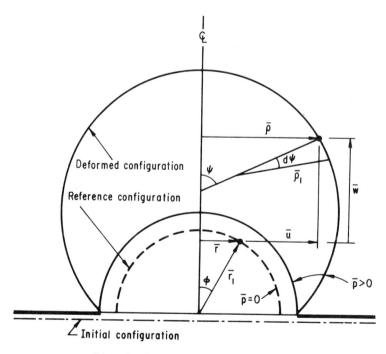

Fig. 3. Geometry of the problem

Referring to Fig. 2b we note that, any further increase in load P will simply increase the displacement of the midpoint Q above the new equilibrium point B. If the load P is now decreased, the truss will come down to the minimum point on the equilibrium path and it will snap back to a point on the initial equilibrium path. One can see that there will be a hysteresis taking place during this back and forth snapping. Here, it should be stressed that geometric instability is not likely to take place for soft biological tissues, except for very extraordinary cases, because soft biological tissues can not sustain any compressive stresses under normal conditions.

MECHANICAL MODEL AND RESULTS

In this section we will present a biomechanical model which describes the aneurysm formation and its eventual loss of stability. Due to the limited space available, we will not give the mathematical equations governing the behavior of the model. One is referred to Engin and Akkas (1983) for a detailed presentation.

The facts that aneurysm formation is a large deformation problem and that the constitutive equation is nonlinear require that the model incorporate both geometric and material nonlinearities. The model utilized in the present work is a pre-stretched circular membrane in its initial configuration. The governing equations are for a membrane made of neo-Hookean material. The reference configuration shown in Fig. 3 has a hemispherical geometry which is defined by $(\bar{r}, \bar{r}_l, \emptyset)$. The equations governing the behavior of the membrane are referred to this configuration. The initial configuration is a prestretched circular membrane of radius larger than that of the reference configuration. If the initially prestretched membrane is inflated it will eventually go into the deformed configuration which is defined by $(\bar{\rho}, \bar{\rho}_l, \Psi)$. The dimensional horizontal and vertical displacement components are denoted by $\bar{u}$ and $\bar{w}$, respectively. $\bar{p}$ denotes the internal dimensional pressure. Equilibrium equations are obtained from a consideration of the equilibrium of forces parallel to the axis of the membrane and in the direction of the radius of a latitude circle. The governing equations turn out to be highly nonlinear and, thus, it is necessary to refer to numerical techniques for their solution. In this work we used the Runge-Kutta method for solution.

From a biomechanics point of view, we will assume that there are at least three parameters which are significant for the growth of the aneurysm sac. These parameters are:

a) Material properties of the aneurysm wall.
b) Geometric properties such as the thickness and radius.
c) The internal fluid pressure acting on the wall.

The effects of these parameters on the growth of the sac are clearly illustrated in Fig. 4 which shows the growth of the nondimensional apex deflection w_0 of the sac as a function of the nondimensional "load" parameter p. As seen in Fig. 4, the

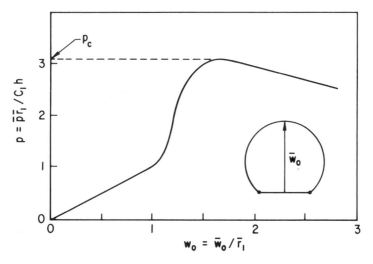

Fig. 4. Nondimensional pressure vs apex deflection

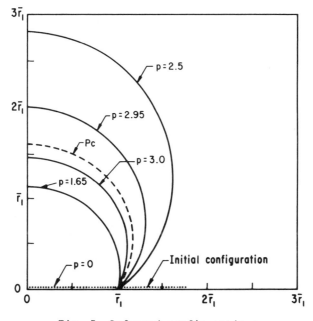

Fig. 5. Deformed configurations

"load" parameter p is nondimensionalized using physical pressure $\bar{p}$ and the elastic stiffness C_l of the aneurysm wall material. Therefore, it is obvious that the growth of the sac under constant physical internal pressure is possible if the strength of the aneurysm wall material is reduced (corresponding to a reduction in C_l) due to structural change and subsequent deterioration of the material properties.

The most significant information one can obtain from Fig. 4 is the existence of a critical value, pc, for the nondimensional "load" parameter, p. As discussed previously, this point is the so-called limit point and it corresponds to the loss of stability. In this case, it is material instability. Once this critical value is reached, the growth of the size of the aneurysm sac does not require any increase in the value of p as shown in Fig. 5 which displays the growth pattern of the sac as a function of p. The configuration at the critical value of p is shown as the dashed curve in the figure. Once the nondimensional "load" parameter is reached and passed infinitesimally the aneurysm will blow out, because the second equilibrium configuration corresponding to pc is too far away, even if it does exist.

SUMMARY AND CONCLUSIONS

In this paper we presented a simple theoretical model elucidating the biomechanics aspects of saccular aneurysm formation and of its eventual blow out. Despite the simplicity of the model, the influence of important biomechanical parameters on the formation of aneurysm is clearly illustrated. Our model indicates that aneurysm blow-out is a material instability problem. If this conclusion is accepted, then we can make the following conjectures:

Aneurysm blowout is not necessarily caused by an increase in the physical internal pressure. It is possible that an increase in the blood pressure may cause the blowout, but this is not the only possibility. Weakening of the wall will have the same effect as the pressure increase. It is reasonable to assume that there may be some chemical agents clustered inside the aneurysm affecting its wall strength. Fragmentation of the internal elastic lamina, a layer of the vessel wall, may be one reason for wall weakening. The degeneration of the elastica and its eventual fragmentation may be speeded up by hypertension; but it is more reasonable to assume a deteriorating wall material behavior rather than an unrealistic increase in the internal pressure.

REFERENCES

Akkas, N. (1978) On the dynamic snap-out instability of inflated nonlinear spherical membranes. Int J Non-linear Mech 13:177-183.
Benedict, R., Wineman, A., and Yang, W.H.(1979) The determination of limiting pressure in simultaneous elongation and inflation of nonlinear elastic tubes. Int J Solids Struct 15:241-249.
Bogen, D.K. and McMahon, T.A. (1979) Do cardiac aneurysms blow out? Biophys J 27:301-316.

Brush, D. O. and Almroth, B. O. (1975) Buckling of Bars, Plates and Shells. McGraw-Hill, New York.

Crane, H.D.(1973) Switching properties in bubbles, balloons, capillaries and alveoli. J Biomech 6:411-422.

Crisp, J.D.C. and Hart-Smith, L.J.(1971) Multilobed inflated membranes: Their stability under finite deformation. Int J Solids Struct 7:843-861.

Decraemer, W.F., Maes, M.A., and Vanhuyse, V.J. (1980) An elastic stress-strain relation for soft biological tissues based on a structural model. J Biomechanics 13:463-468.

Engin, A.E. and Akkas, N.(1983) Etiology and biomechanics of hernial sac formation. J Biomed Engng 5:329-335.

Hung, E.J. and Botwin, M.R. (1975) Mechanics of rupture of cerebral saccular aneurysms. J Biomech 8:385-392.

Krajcinovic, D., Trafimow, J., and Sumarac, D.(1987) Simple constitutive model for a cortical bone. Journal Biomech 20:779-784.

Mirsky, I., McGill, P.L., and Janz, R.F. (1978) Mechanical behavior of ventricular aneurysms. Bull. Math Biol 40:451-464.

Mooney, M. (1940) A theory of large elastic deformation. J Appl Phys 11:582-592.

Needleman, A.(1977) Inflation of spherical rubber balloons. Int J Solids Struct 13:409-421.

Perktold, K. (1987) On the paths of fluid particles in an axisymmetrical aneurysm. J Biomech 20:311-317.

Radhakrishnan, S., Ghista, D.N., and Jayaraman, G. (1980) Mechanical analysis of the development of left ventricular aneurysms. J Biomech 13:1031-1039.

Rhoton, A.L., Jackson, F.E., Gleave, J., and Rumbaugh, C.T. (1977) Congenital and traumatic intracranial aneurysms. CIBA Clinical Symp 29:1-40.

Treloar, L.R.G. (1975) The Physics of Rubber Elasticity, 3rd edn. Clarendon, Oxford.

ENDOTHELIAL CELLS UNDER MECHANICAL AND HUMORAL STRESS

B. Klosterhalfen, H. Rixen, Ch. Mittermayer
C.J. Kirkpatrick and H. Richter

Inst. of Pathology, The Technical
University of Aachen (RWTH), Pauwelsstr.
5100 Aachen, West-Germany

ABSTRACT

Endothelial cells are the interface between blood
components and the vessel wall. Alterations of these
vascular cells provide changing conditions of the vessel
wall surface. Considerations about blood-vessel wall
interactions have to include the pathology of the
endothelium. Complex in vivo and simplified in vitro models
are necessary to advance in our knowledge about the
pathophysiology of these particular cells. Two studies in
our institute, an in vivo as well as an in vitro model, are
described and their results discussed.

Key words: blood-vessel wall interactions, in
vitro/vivo models, endothelial cells, basement membrane,
septic shock, ards

INTRODUCTION

Arteriosclerosis and sepsis are two of the most
frequent diseases found at autopsy in routine pathology.
The endothelial cells of the blood vessels and the heart
play an important part in their pathophysiology. A better
understanding of these pathophysiologic processes seems to
be associated with a better understanding of the pathology
of the endothelial cell layer. In this essay we want to
demonstrate a small part of this pathology to stress the
complexity of the blood - vessel wall interaction **in vivo**.
Mechanical and humoral factors, as well as the relationship
between the endothelial cell layer and basement membrane
components are contents of our studies. Our recent attempt
to simulate in vivo conditions with an **in vitro** and an
animal model is described.

MATERIALS AND METHODS

Cell culture: To study the interactions between the endothelial cell layer and the basement membrane, endothelial cells (EC) were isolated from the bovine cornea (BCEC, Gospodarovicz et al., 1977) and the human umbilical vein (HUVEC, Jaffe et al., 1973) and grown in monolayer culture. Early passage cells were plated on to glass coverslips coated with one basement membrane component (e.g. Collagen I, III, IV , IV-Fragments at 10 μg/ml). Four basic biological phenomena were quantified in standardized assay systems - adhesion, spreading, migration and proliferation. Adhesion was measured by adding an EC suspension (5 x 10^4 cells) to the coated coverslips (12 mm diameter) in tissue culture multiwells. After 2 h the cover-slips were washed and adhesion quantitated by microscopic cell counting in a randomized fashion. The adhesion assay was performed with and without serum supplements. Spreading was quantified using image analysis of cell area in the scanning electron microscope (SEM). Migration was studied using specially constructed metal/silicone inserts for multiwell chambers to enable a defined central area of confluent EC to be established. After removal of the "fences" cells were allowed to migrate for 5 days, followed by fixation, staining and measurement of mean radii to the migrating front using a macroscope. Proliferation was studied over 9 days by constructing time curves for cell number using high and low initial plating densities.

Animal model: Domestic pigs (minimal disease pigs; Deutsche Landrasse) weighing 28-32 kg were used. Prior to the experiments the animals were pretreated with the antibiotic doxycyclin to prevent bronchopneumonia. The animals were anesthetized with Hypnodil R (etomidate) and Pancuronium Organon R (pancuronium) intravenously and subsequently maintained on a mixture of Hypnodil R, Pancuronium Organon R and Dipidolor R (piritramid) for the duration of the study (48 h). The animals were intubated endotracheally and ventilated with an adult pressure ventilator DRÄGER AV 1 with a tidal volume of 10-15 ml/kg, a respiratory rate between 16-20 to keep the arterial PCO_2 at 32-38 mm Hg.

Endotoxaemia was induced with Escherichia coli (W 0111:B4) endotoxin. Experimental animals were given at time 0h 0,5 μg/kg over about 30 minutes. Endotoxin infusion was stopped when the pulmonary pressure was doubled compared with time point 0 h. As soon the pulmonary pressure reached the value of time 0 h, endotoxin infusion was repeated using 0,25 μg/kg over 30 minutes. Control animals were given an equal volume of saline. Blood samples were taken at 0,1,3,7, 11, 15, 24,35,48 h, and thromboxane B2 and prostacyclin (PgI$_2$), metabolites of the cyclooxygenase pathway of arachidonic acid measured with HPLC and RIA.

The experiment was continued for about 48 h and at that time any surviving animals were sacrified by hyperkalemic cardiac arrest using 10 mEq of KCl applied intravenously.

The lungs were immediately removed through a

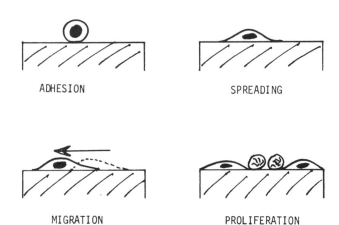

Fig. 1. Quantifiable cell biological steps during
endothelial cell seeding

thoracotomy for morphological study. Representative samples
were also taken from the heart, heart valves, adrenal
gland, kidney, liver stomach, jejunum, ileum, colon,
spleen, lymph nodes and spleen for gross pathological
analysis, i.e. light and electron microscopy.

In the development of our domestic pig model the
objective was to provide a septic-shock like state to
create and produce chronic lung injury and septic
endocarditis comparable with the adult respiratory distress
syndrome of the human lung and endocarditis verrucosa
simplex.

RESULTS

Cell culture: Collagen IV caused a marked increase in
BCEC adhesion (>100%) in the absence of serum. In the
presence of serum (10 and 50% FCS), collagen IV gave
increases in adhesion of the order of 60%. Experiments with
high molecular weight fragments (IV-F) of the tetrameric
collagen IV molecule also markedly increased BCEC adhesion
in the presence or absence of serum, indicating that
adhesion promoting sequences of the collagen IV molecule
are found on the monomeric arm regions. In the absence of
serum, collagen types I and III increased BCEC (but not
HUVEC) adhesion by approximately 45%. With serum, no
adhesion promoting activity of these collagen types could
be established. In the absence of serum, as well as with
10% serum, collagen IV-F caused an increase in spreading of
BCEC. On collagen IV spreading tended to be less than in
non-coated controls, both with and without serum. Spreading
of BCEC on collagen type I was higher than controls in the
absence of serum. However, with serum components present
(10 or 50%) spreading of BCEC on collagen I was markedly
decreased.

Migration studies showed that neither collagen IV or
IV-F showed any marked difference in their promotion of
migratory activity of HUVEC, studied over a 5 day period.
During 9 days observation BCEC proliferation on collagen

IV, either tetrameric or fragmented, did not differ from the glass reference surface.

Animal model: The histopathologic changes seen in the lungs of the pigs infused with endotoxin of E. coli were similar to the early and subchronic changes described in humans.

Pulmonary interstitial edema was common. The endothelial cells showed apical cytoplasmic diverticula, a result of endothelial cell cytoplasmic swelling or vacuole formation. In some specimens endothelial cells of the smallest branches of the pulmonary artery were detached and the basement membrane exposed.

The endothelial cell layer of the heart valves normally provides a continuous covering with anticoagulatory function. Under septic shock-like states these cells undergo a 1.400fold higher cell turnover than under physiologic conditions. The first morphologic alteration of the endothelial cells occurs only a few hours after onset of the septic shock . General microvesicles as a result of an increased pinocytosis were detected in transmission electron micrographs. If the shock state persists the regenerative power of the endothelium decreases and finally is extinguished. A severe edema of the whole heart valve, especially in the spongiosa appears, followed by a detachment of the endothelial cells. Sub-endothelial layers are exposed and small aggregates consisting mainly of platelets are deposited. If the process progresses the clots of platelets start to grow by apposition of additional platelets and other blood components from the blood stream. In the end stage they are visible with the naked eye at autopsy. Numerous wartlike noduli up to 1-2 mm are deposited along the edges of the heart valves, mainly of the left ventricle (aortic and mitral valve).

The thromboxane B2 and prostacyclin levels correlated directly with the endotoxin levels. Both thromboxane B2 and the prostacyclin metabolite 6-keto-PGF1α concentrations begin to increase early in the endotoxin response, but peak thromboxane B2 concentrations precede peak concentrations of the prosta-cyclin metabolite 6-keto-PGF1α.

DISCUSSION

The in vitro study as well as the animal model demonstrate the complexity of the blood-vessel wall interaction. The condition of the endothelium depends on numerous factors such as location, humoral mediators and pathological changes of these vascular cells. One of the essential aspects of an intact blood vessel is the relationship existing between the endothelial cell layer and the underlying componenets of the basement membrane. These cell-matrix interactions take on a particular significance in the case of damage to the endothelium, for example in the course of arteriosclerosis. The need to understand more about these interactions

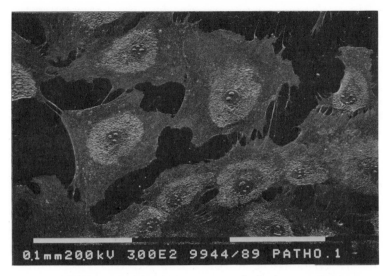

Fig. 2. HUVEC growing on an artificial collagen IV
modified polymer surface

has been highlighted by the recent attempt to "seed" vascular
prothesis with endothelial cells prior to implantation (Shindo
et al., 1987).

Our in vitro experiments indicate that endothelial
cell behavior on collagen matrix components varies
markedly, depending on endothelial cell type, collagen type
and on the presence of serum. Possibly fibronectin, as a
serum component, plays an important role in cell adhesion
(Macarak and Howard, 1983).

The adult respiratory distress syndrome (Ashbaugh et
al., 1967) and septic endocarditis (Mittermayer et al.,
1971a/b; Künzer et al. 1970; Freudenberg et al., 1975;
Scherrer and Schneider, 1978) as a consequence of septic-
shock like states represent two "classical" diseases of
modern critical care unit medicine. An increasing number of
patients nowadays is affected each year. The endothelium
also possesses a central role in their pathophysiology,
these vascular and heart cells offearing to be the actual
target organ. An injury of these vascular cells marks the
onset of these diseases, a destruction probably the so
called "point of no return", where these diseases become
uncontrollable, and in the case of ARDS ending with the
death of the involved patient.

Humoral mediators are of central interest in current
investigations. Fibrin degradation products (FDP) have been
incriminated as possible mediators of endothelial injury
(Dang et al., 1985; Haynes et al., 1980; Haselton, 1983).
Haynes in 1980 found thrombocytopaenia and an increase in
the fibrin/fibrinogen degradation product D as indicator
for DIC during septic shock. FDP D may be a marker or
mediator of injury of the alveolar wall. FDP can induce
interstitial edema, as well as capillary membrane
permeability defects.

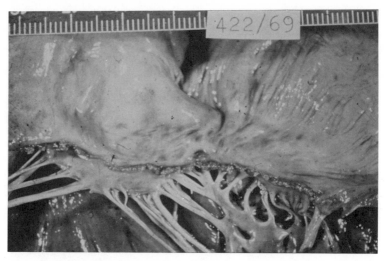

Fig. 3. Endocarditis verrucosa simplex with numerous wartlike noduli along the edges of the mitral valve

They have chemotactic effects for leucocytes, especially neutrophil granulocytes. Neutrophil derivatives like lysosomal constituents (elastase, collagenase, myeloperoxidase etc.) as well as oxygen free radicals and arachidonic acid metabolites are capable of altering endothelial cell membrane structure and function (Westaby, 1988; Harlan, 1981; Sacks et al., 1978; Smedly et al., 1986). Thus neutrophil-mediated injury and DIC/FDP can potentially amplify each other (Rinaldo and Rogers, 1982).

Recent in vitro studies have shown that endotoxin can cause damage of endothelial cells directly. Lipopolysaccharide from E. coli produces a time and concentration-dependent endothelial injury manifested by endothelial cell detachment, prostacyclin production and eventual cell lysis (Brigham and Meyrick, 1985; Harlan et al., 1981; Meyrick et al., 1986; Meyrick, 1986; Weksler et al., 1977). This endothelial injury has been shown to be species- and site-dependent. Endothelial cells from bovine aorta and pulmonary artery are all sensitive to endotoxin, whereas endothelial cells from human pulmonary artery are relatively resistant.

The exact mechanism of endotoxin-induced endothelial cell injury is still unknown. Treatment with inhibitors of protein, prostaglandin synthesis and oxygen radical production have failed to protect the endothelium.

Metabolites of of the arachidonic acid pathway are produced in the lung and can have dramatic effects on lung function (Brigham and Meyrick, 1984). Thromboxanes are potent pulmonary vasoconstrictors (Bowers et al., 1979) and peak thromboxane concentrations in blood samples appear when pulmonary hypertension is most marked. Therefore this

318

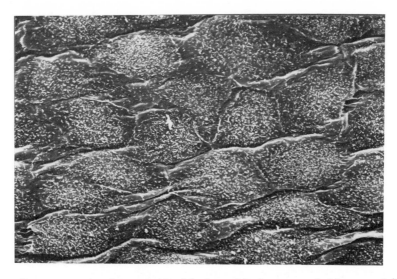

Fig. 4. Intact endothelial cell layer of the aortic valve; domestic pig; SEM micrograph

arachidonic acid metabolite may mediate the initial pulmonary pressure increase. It is possible that thromboxanes do not directly affect the vascular endothelial layer or the vascular permeability, but they are involved in a more complicated metabolic sequence, affecting other humoral factors and inflammatory cells, so that they may be indirectly responsible for an increased permeability effect on endothelial cells.

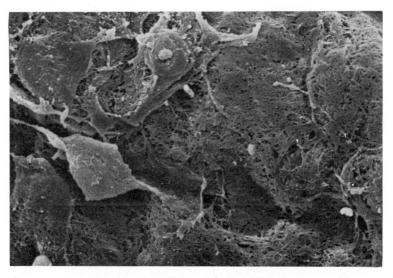

Fig. 5. Exposed subendothelial cell layer after endo-toxinshock of the aortic valve; domestic pig; SEM micrograph

Platelets are generally thought to be the main source of thromboxanes, but in sheep depleted of platelets, large amounts of these mediators can be detected after endotoxin infusion (Snapper et al., 1984). Tahamont et al., (1984) have suggested lung lymphocytes as a possible source of thromboxanes in the lung.

Endothelial cells are the main sources of prostacyclin and injury stimulates the production of this mediator (Ogletree et al., 1986). Prostacyclin is a potent inhibitor of platelet aggregation. Platelets do not adhere to normal vascular endothelium in vivo or to intact endothelial cells grown in monolayer, but do adhere readily to injured endothelium or subendothelial components.

CONCLUSION

It is to be stressed that the endothelial cell layer remains the decisive interface in blood-vessel wall inter-actions. The pathology of these cells remains a particular problem for future investigations. Complex _animal_ (=_in vivo_) and simple _in vitro models_ are both necessary either to study the patho-physiology of several diseases of the endothelium or to examine single problems arising from the complex model.

REFERENCES

Ashbaugh, D.G., Bigelow, D.B., Petty, T.L., Levine, B.E., 1967, Acute respiratory distress in adults
    Lancet, ii:319-323
Bowers, R., Ellis, E., Brigham, K., Oates, J., 1979, Effects of prostaglandin cyclic endoperoxides on the lung circulation of sheep
    J. Clin. Invest., 63:131-137
Brigham, K.L., Meyrick, B., 1985, Endotoxin and lung injury
    Am. Rev. Respir. Dis., 133:913-927
Brigham, K.L., Meyrick, B., 1984, Interactions of granulocytes with the lungs
    Circ. Res., 54:623-625
Dang, C.V., Bell, W.R., Kaiser, D., Wong, A., 1985, Disorganisation of cultured vascular endothelial cell monolayers by fibrinogen fragment D
    Science, 227:1487-1490
Freudenberg, N., Madreiter, H., Mittermayer, C., 1975, Experimental investigations into the pathogenesis of endocarditis due to shock
    Beitr. Path., 155:248-262
Gospodarovicz, D., Mescher, A.R., Birdwell, C.R., 1977, Stimulation of corneal emdothelial cell proliferation in vitro by fibroblast and epidermal growth factors,
    Exp. Eye Res., 25:75-89

Harlan, J.M., Killen, P.D., Harker, L.A., Striker, G.E., 1981, Neutrophil-mediated endothelial injury in vitro - Mechanisms of cell detachment
J. Clin. Invest., 68:1394-1403

Haselton, P.S., 1983, Adult respiratory distress syndrome - a review
Histoapthology, 7:307-332

Haynes, J.G., Hyers, T.M., Giclas, P.E., Franks, J.J., Petty, T.L., 1980, Elevated fibrin(ogen) degradation products in the adult respiratory distress syndrome
Am. Rev. Respir. Dis., 122:841-847

Jaffe, E.A., Nachman, R.L., Becker, C.G., Minick, C.R., 1973, Culture of human endothelial cells derived from umbilical veins
J. Clin. Invest., 52:2745-2756

Künzer, W., Schindera, F., Mittermayer, C., 1970, Fetale Endocarditis und Verbrauchskoagulopathie bei einem Neugeborenen
Dtsch. Med. Wschr., 95:1107-1115

Macarak, E.J., Howard, P.S., 1983, Adhesion of endo-thelial cells to extracellular matrix proteins
J. Cell Pysiol., 116:76-86

Meyrick, B., 1986, Endotoxin-mediated endothelial damage
Fed. Proc., 45:19-24

Meyrick, B., Ryan, U.S., Brigham, K.L., 1986, Direct effects of E. coli endotoxin on structure and permeability of pulmonary endothelial monolayers and endothelial layer of intimal explants
Am. J. Pathol., 122:140-151

Mittermayer, C., Madreiter, H., Schindera, F., Huth, K., 1971, Über Endocarditis verrucosa simplex bei Schock und Verbrauchskoagulopathie
Verh. Dtsch. Ges. Pathol., 55:350-355

Mittermayer, C., Waldthaler, A., Vogel, W., Sandritter, W., 1971, Endocarditis verrucosa simplex/thrombotica bei Verbrauchskoagulopathie (Schock, Leukosen, Karzinome)
Beitr. Path., 143:29-58

Ogletree, M., Begley, C., King, G., Brigham, K., 1986, Influence of steroidal and nonsteroidal antiinflammatory agents on accumulation of arachidonic acid metabolites in plasma and lung lymph after endotoxemia in awake sheep: measurements of prostacyclin and thromboxane metabolites and 12-HETE
Am. Rev. Respir. Dis., 133:55-61

Rinaldo, J.E., Rogers, R.M., 1982, Adult respiratory distress syndrome; changing concepts of lung injury and repair
N. Engl. J. Med., 306:900-909

Sacks, T., Moldow, C.F., Craddock, P.R., Bowers, T.K., Jacobs, H.S., 1978, Oxygen radicals mediate endothelial cell damage by complement -stimulated granulocytes - an in vitro model of immune vascular damage
J. Clin. Invets., 65:1161-1167

Scherrer, P., Schneider, J., 1978, Die thrombotische
     Endocarditis und ihre Beziehung zur
     disseminierten intravasalen Gerinnung
     Verh. Dtsch. Ges. Pathol., 62:238-241
Shindo, S., Takagi, A., Whittemox, A.D., 1987,
     Improved potency of collagen-impregnated grafts
     after in vitro autogeneous endothelial cell
     seeding
     J. Vasc. Surg., 6:325-332
Smedly, L.A., Tonnesen, M.G., Sandhaus, R.A., Haslett,
     C., Guthrie, L.A., Johnston, R.B., Henson,
     P.M., Worthen, G.S., 1986, Neutrophil-mediated
     injury to endothelial cells - Enhacement by
     endotoxin and essential role of neutrophil
     elastase
     J. Clin. Invest., 77:1233-1243
Snapper, J., Hinson, J., Hutchison, A., Lefferts, P.,
     Ogletree, M., Brigham, K.L., 1984, Effects of
     platelet depletion on the unanesthetized
     sheep`s pulmonary response to endotoxemia
     J. Clin. Invest. , 74:1782-1791
Tahamont, M.V., Gee, M.H., Flynn, J.T., 1984,
     Aggregation and thromboxane synthesis and
     release in isolated sheep neutrophils and
     lymphocytes in response to complement
     stimulation
     Prostaglandins Leukotriene Med, 16:181-190
Weksler, B.B., Marcus, A.J., Jaffe, E.A., 1977,
     Synthesis of prostaglandin I2 (prostacyclin) by
     cultured human and bovine endothelial cells
     Proc. Natl. Acad. Sci. USA, 74:3922-3926
Westaby, S., 1988, Mediators in acute lung injury: the
     whole body inflammatory response hypothesis,
     in:"Shock and the adult respiartory distress
     syndrome", W. Kox and D. Bihari, eds., Springer
     -Verlag, Berlin-Heidelberg

ENDOTHELIALIZED PROSTHETIC SUPPORTS AND CELLULAR BIOMECHANICS :

A TEST BENCH FOR EVALUATION OF CELL TEARING

José Sampol, Dominique Arnoux, Régis Rieu, Robert Pelissier, and Claude Mercier

Laboratoire de Techniques Avancées en Hématologie et Biomécanique Vasculaire. Hôp. Conception, Marseille, France

INTRODUCTION

Synthetic vascular substitutes are now successfully used in replacing large vessels.

In vascular prostheses of less than 5 millimeters in diameter, the hemodynamic conditions are often a cause of early occlusion.

In man, as opposed to observations made in various animal species, a complete spontaneous endothelialization of vascular substitutes never occurs. Thus, one of the present lines of research in vascular surgery concerns the implantation, on the internal wall of prostheses, of a continuous and functional endothelial monolayer in order to reduce the thrombogenicity and to improve the long-term patency of vascular prostheses (Graham et al., 1980, Sharefkin et al., 1982, Stanley et al., 1982, Schneider et al., 1988, Örtenwall et al., 1989).

The capacity of prosthetic materials to support adhesion, multiplication and maintenance of functional endothelial cells on their surface is variable (Curtis et al., 1983, Absolom et al., 1988, Seeger and Klingman, 1988, Klein-Soyer et al., 1989, Zilla et al., 1989, Kaehler et al., 1989).

We have undertaken a study of tearing resistance and morphological behaviour of endothelial cells on the surface of different prosthetic materials, under given hemodynamic conditions.

The first experimental phase involved the development of a test bench to evaluate the tearing resistance of an endothelial monolayer cultivated on a plane support, and submitted to physiological flow conditions. The stimulation of a plane flow is interesting for two reasons : firstly, obtaining a confluent cell monolayer is easier on a plane support, secondly, given a known value of instataneous flow, the flow equations admit analytical solutions which, by only calculation, give access to hemodynamic values : speed, pressure and shear wall stress. These analytical solutions therefore allow to validate the measurements and particularly to eliminate from the test bench a number of measuring techniques incompatible with a good conservation of the endothelial cells.

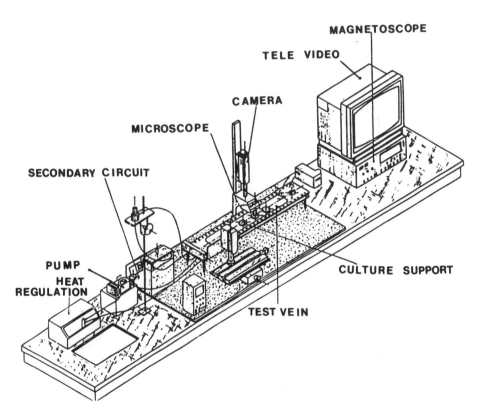

Fig.1.   Schema of the Test Bench

PRESENTATION OF THE TEST BENCH

The test bench (fig. 1) consists of a heat regulated Macrolon test vein of rectangular cross section, with a side ratio of 1 to 10, permitting to consider the flow as plane and to ignore the effects of the vertical sides. The vein includes 3 windows allowing to insert adapted plane culture supports. These supports were designed for an instantaneous adaptation of the cultivated surface to the test bench. A special locking system enables the surface to be correctly positioned in the test vein, avoiding any turbulence and insuring a perfect tightness.

The circulation system consists of a volumetric pump which produces continuous, simple pulsatile or physiological flows. An electromagnetic flowmeter at the beginning of the circuit allows the instantaneous flow evaluation. A tridirectional displacement device permits the presentation of one of the windows to the visualization system, which can receive a fluorescent or normal light microscope. A camera fitted to a video system allows a direct visualization of cell behaviour and an evaluation of the tearing areas.

ENDOTHELIAL CELL CULTURE

Endothelial cells used in our experiments were obtained from human umbilical cord, which represents an easily available source of cells (Jaffe et al., 1973).

The cells were harvested from umbilical vein by collagenase treatment for 15' at 37°C, then cultivated to confluency on 0,2 % gelatin pre-coated plastic dishes. The culture medium consisted of a mixture of M 199 and RPMI 1640, containing 30 % of decomplemented pooled human serum. After 3 to 5 days of culture at 37°C under 5 % Co_2, the confluent cell monolayers were detached by trypsin EDTA. The cells were replated on the various supports to be tested and subcultivated until new confluency, corresponding to a cell density of 10^4 to 10^5 cells/cm^2, cell growth being stopped by contact inhibition when confluency was reached.

Direct cell follow-up can be performed either by optic microscope observation or by supra-vital nuclear labelling with a fluorescent compound. The latter technique allows cell visualization on opaque supports, like Dacron or PTFE vascular prostheses, and cell quantification in the liquid phase by flow cytometry.

The percentage of cell tearing can be determined by counting the cells remaining on the support at the end of the experiment, by reference to an identical endothelialized support not submitted to the test bench. A morphological study of the cells can be performed by optic microscopy after staining or by electron microscopy.

VALIDATION OF THE TEST BENCH

The reliability of the test bench in theoretical determination of shear wall stress has been tested using ultrasonic doppler velocimetry, electromagnetic flow rate, instantaneous pressure transducer and polarographic techniques.

The characteristics of non-stationary plane flow were determined by solving the Navier-Stokes equations, supposing flow to be established and laminary. The analytical solutions obtained gave access to instantaneous velocity profile and to the evolution of shear wall stress with time. The theoretical results were verified (fig. 2), through measurements performed under different flow configurations characterized by Reynolds parameters.

Velocity profiles for sections located at various distances from vertical wall were determined by ultrasonic Doppler velocimetry 15 Mhz with pulse

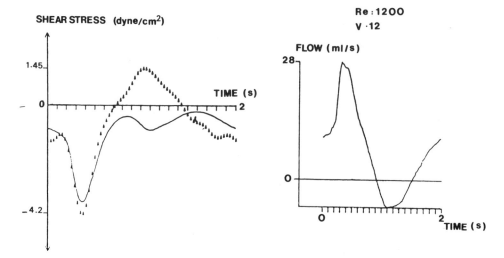

Fig.2. Evolution of Shear Stress

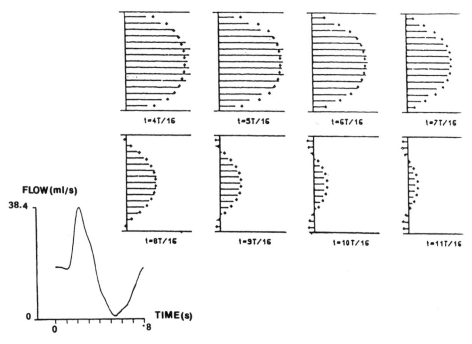

Fig.3. Velocity Profiles
Theoretical +
Experimental -

emission. This non intrusive measuring instrument has the advantage not to disturb the flow.

The comparison between the experimental and theoretical profiles showed that plane flow hypothesis was verified except very close to the wall. The general parabolic shape of the profile corresponded to the theoretical study forecast (fig. 3). The measurement of shear wall stress was performed using a polarographic technique. This method gives excellent results for maximal shear stress values where the tearing risk of endothelial cells is the highest.

Theoretical and experimental results were in concordance, allowing to conclude that the flow was plane. Thus the shear stress could be calculated from the only measurement of the flow.

FIRST EXPERIMENTS AND FUTURE PROSPECTS

Resistance to tearing of endothelial cells cultivated on two types of plastic supports (Macrolon and Primaria®) was tested. These supports were pre-coated with different products in order to improve the adherence of the cell monolayer. We used either gelatin (0.2 %), fibronectin (20 µg/cm^2) or biological fibrin glue (Tissucol®).

The duration of observation varied from 1 to 5 hours.

Whatever the support or the coating tested, the cells submitted to a steady flow at a rate of 13 ml/s, observed for 5 hours, underwent no appreciable tearing. For a physiological flow at average rate of 6,6 ml/s with peaks at 13 ml/s, the percentage of tearing after 5 hours was less than 10 %.

These preliminary experiments have demonstrated the feasibility and reliability of the test bench.

Considering the first results, we can conclude that some improvements of the test bench should be achieved : cell observation should be prolonged over 24 h (or more) to allow a better evaluation of the cell tearing resistance. It will therefore be necessary to provide for an oxygenation system and sterile circuit, and to perfuse a nutritional medium instead of saline buffer.

Future prospects include an extension of the study to other types of culture supports (Dacron, PTFE), to circular prostheses and to adult endothelial cells from different vascular areas (vein or capillary), as well as the use in circulation of complex biological media (plasma or whole blood), which would provide a better appreciation of the interactions between blood cells and endothelium.

REFERENCES

Absolom D.R., Howthorn L.A., and Chang G., 1988, Endothelialization of polymer surfaces. J. Biomed. Mat. Res., 22:271.

Curtis A.S.G., Forrester J.V., Mc Innes C., and Lawrie F., 1983, Adhesion of cells to polystyrene surfaces J. Cell Biol., 97:1500.

Graham L.M., Burkel W.E., Ford J.W., Vinter D.W., Kahn R.M., and Stanley J.C., 1980, Immediate seeding of enzymatically derived endothelium in Dacron vascular grafts : early experimental studies with autologous canine cells. Arch. Surg., 115:1289.

Jaffe E.A., Nachman R.L., Becker C.G., and Minick C.R., 1973, Culture of human endothelial cells derived from umbilical veins. J. Clin. Invest., 52:2745.

Kaehler J., Zilla P., Fasol R., Deutsch M., and Kadletz M., 1989, Precoating substrate and surface configuration determine adherence and spreading of seeded endothelial cells on polytetrafuloroethylene grafts. J. Vasc. Surg., 9:535.

Klein-Soyer C., Hemmendinger S., and Cazenave J.P., 1989, Culture of human vascular endothelial cells on a positively charged polystyrene surface, Primaria : comparison with fibronectin coated tissue culture grade polystyrene. Biomaterials, 10:85.

Örtenwall p., Wadenvik H., and Risberg B., 1989, Reduced platelet deposition on seeded versus unseeded segments of expanded polytetrafluoroethylene grafts : clinical observations after a 6-month follow-up. J. Vasc. Surg., 10:374.

Schneider P.A., Hanson S.R., Price T.M., and Harker L.A., 1988, Preformed confluent endothelial cell monolayers prevent early platelet deposition on vascular prostheses in baboons. J. Vasc. Surg., 8:229.

Seeger J.M., and Klingman N., 1988, Improved in vivo endothelialization of prosthetic grafts by surface modification with fibronectin. J. Vasc. Surg., 8:476.

Sharefkin J.B., Latker C., Smith M., Cruess D., Clagett G.P., and Rich N.M., 1982, Early normalization of platelet survival by endothelial seeding of Dacron arterial prostheses in dogs. Surgery, 92:385.

Stanley J.C., Burkel W.E., Ford J.W., Vinter D.W., Kahn R.H., Whitehouse W.M., and Graham L.M., 1982, Enhanced patency of small-diameter externally supported Dacron iliofemoral grafts seeded with endothelial cells. Surgery, 92:994.

Zilla P., Fasol R., Preiss P., Kadletz M., Deutsch M., Schima H., Tsangaris S., and Groscurth P., 1989, Surgery, 105:515.

EXPERIMENTS AND NUMERICAL ANALYSIS RELATING TO ATHEROGENESIS

J.H. Gerrard

Department of Engineering
University of Manchester, U.K.

ABSTRACT

The arterial wall has constituents which behave as mechano-receptors. It is known, for example, that smooth muscle cells proliferate under the action of applied oscillating stress. There are regions of arterial flow which may be associated with fluctuating wall stresses of elevated magnitude and frequency. These regions are located where there is reversed flow and where the diameter is such that the non-dimensional frequency of the pulse wave lies in the transition region between low and high values, ($\alpha = r \, \omega/\nu$; $2 \leqslant \alpha \leqslant 8$). It is thus possible that this is one of the fluid mechanical contributory factors in the development of atherosclerosis and eventual blockage of an artery.

The main concern is with the formation and development of a ring vortex in sinusoidally oscillating flow. It is found that the principal parameter is the product of peak Reynolds number and the taper semi-angle θ ; the Reynolds number is based on the maximum reversed flow rate and the diameter at the smaller end of the conical tube. At low to moderate values of the frequency parameter the minimum critical value of Reθ for vortex formation was found to be 10. Velocity and wall shear stress determined from numerical analysis were found to execute an oscillation at ten times the fundamental frequency as separation of the flow approached and as the vortex passed down the tube.

The superficial femoral artery has two bifurcations at the distal end and the possibility of a bend at the knee. Models with bends and bifurcations added showed only a slight reduction of the critical value of Reθ. The value of this parameter in the normal femoral artery is about 1.0. Reduction of Reθ towards this value was observed when there was a stenosis in the cylindrical tube at its junction with the narrow end of the tapered tube.

The present evidence shows that it is only marginally possible that a vortex produced by a stenosis as severe as 50% diameter will result in the persistence of the vortex a relatively short distance up the artery. Measurements were made with a sinusoidally oscillating flow. The nonlinearity observed makes necessary the observation of flow with an arterial waveform. The effect of flexing of the thigh muscles in exercise has not been considered.

Keywords: Atherosclerosis/Arterial blood flow/Femoral artery model.

Biomechanical Transport Processes, Edited by F. Mosora *et al.*
Plenum Press, New York, 1990

INTRODUCTION

The occurrence of atherosclerosis in human arteries is often localised
with relatively normal distal and proximal beds. The major sites of
involvement are: (1) The coronary arterial bed: (2) The major
branches of the aortic arch, the innominate, carotid,subclavian arteries
and their major branches: (3) The major branches of the upper abdominal
aorta near to their origin from the aorta and (4) The lower abdominal
aorta and its major branches, principally the aorto-illiac and the superficial femoral arteries.

Many of these sites have in common the size of the artery and the
existence of retrograde flow in part of the cycle. The nondimensional frequency parameter, usually called α, the symbol used by
Womersley (1955), is $d_{/2}$ $\pi f/\nu$ where d is the diameter of the
lumen, f is the pulse rate and ν is the kinematic viscosity of blood.
At rest with f about 1 per second, α lies in the range 2 to 8 when
d lies between 3 and 9mm; at twice this frequency in exercise d lies
between 2 and 6mm. In reversed flow in arteries, which is the
essence of this paper, the frequency associated with the reversed flow
is about three times the above values and the range of d from 2 to 6mm
becomes 1 to 4mm. These arterial sizes correspond to the α range
between low and high frequency oscillating flow at many arterial sites
of blockage. The form of the pulse wave is not known in the exercise
condition.

We have had the idea for some considerable time that if there is a
fluid mechanical contributory factor to the formation of atherosclerotic lesions it could be related to the changes in the flow which
take place in this range of α. Many of the sites of atherogenesis
match the arterial sizes quoted above and also exhibit reversed flow
in part of the cycle. In what follows we examine the flow changes which
occur in rigid tapered tubes in this transition range when there is also
flow reversal. We begin with some elementary fluid mechanics; the result of experiments in rigid conical tubes are presented and the results
compared with a numerical analysis of the same problem.

FLOW IN RIGID CYLINDRICAL TUBES

    The analysis of flow in cylindrical tubes of circular cross section
has been thoroughly treated and the mathematics of this is relatively
simple. Far from the ends of the tube the pressure gradient balances the
fluid acceleration and the skin friction stress at the walls. The
theory is linear and so velocity components determined at different frequencies can be simply added together to find the solution for a composite waveform. The distribution of velocity across the radius of
the tube depends only on the frequency parameter α. When the value of α
is small the flow is quasi-steady and the distribution of velocity
across the diameter (the velocity profile) approximates to the steady
Poiseuille flow which has a parabolic velocity profile. This holds
for α less than about 2. At high α (>8) the velocity profiles
have a double hump with a flat distribution in the centre of the
tube. If the flow is produced by an oscillating piston (or by
the heart) the centre portion moves in phase with the piston and
an oscillating boundary layer next to the walls leads the phase of the
piston motion: The flow at the walls reverses before that at the
centre. The volume flow rate is observed to decrease due to increased
resistance as the frequency increases under the action of a constant
amplitude pressure gradient. The transition from low to high frequency
flow in the α range from about 2 to 8 is typical of oscillating flow in
cylindrical tubes. These flows depend on α only.

FLOW IN TAPERED TUBES

Arteries are tubes of very small taper. In a tapered tube convection
moves the fluid to a region of different frequency parameter (α is
proportional to diameter) and therefore the speed of the flow is
important as well as the frequency. This means that the non-dimensional Reynolds number (= cross sectional mean speed x diameter $/\nu$)
enters as a characteristic parameter as well as α. The equations of
flow are no longer linear even at positions far from the ends and
except in special cases a numerical solution of the equations is in
general the only way to calculate the flow. When the flow is steady
and directed towards the narrow end analytical treatment is possible
as shown by Eagles (1982). When however, the flow is reversed, it is
much more interesting fluid mechanically and also medically.

If the Reynolds number or the taper semi-angle, θ, is large enough the
steady reversed flow in a tapered tube will separate, that is, the
fluid particles at some position down the tube will move away from the
wall and towards this point along the wall from both directions. The
velocity profiles and streamlines of such a flow are shown in Figure 1.
At the separation point the velocity profile approaches zero velocity
at the wall tangentially: The flow near to the wall is in opposite
directions on each side of the separation point and the velocity gradient
perpendicular to the wall decreases to reach zero at separation. At separation the velocity profile is like that of a jet in a unbounded fluid.
Because equal elements of the radius represent much greater area of cross
section near to the walls than they do near to the centre, as the
velocity decreases (approaching separation) near to the wall, that
at the centre increases considerably more since the volume flow rate is
the same at all distances down a rigid tube. Even though in a tapered
tube the mean velocity decreases in proportion to the increase in cross-
sectional area, the centre line velocity has increased considerably.

AN ARTERY MODEL

We have made experiments in a rigid conically tapered tube which models
a human superficial femoral artery except for its distensibility.
Figure 2 shows the geometry and the flow direction sign convention.
All of the measurements reported here were made in this tube of taper
semi-angle 0.025 radian.

When the flows in the model and the artery are the same the two
systems are said to be dynamically similar. To achieve dynamical
similarity the non-dimensional paramters which characterise the flow
must be the same. These parameters are the Reynolds number, Re; the
taper semi-angle.θ; the frequency paramenter α and the ratio,λ, of
the oscillation amplitude to the mean flow. We have been able to
show that in the lower range of α the peak reversed flow Reynolds number
is the relevant Reynolds number independently of λ. Theoretical work
(e.g. by Eagles (1982) and by the author's colleague G.Y. Buss) shows
that in steady flow the product Reθ is the significant parameter
rather than Re and θ independently. Our numerical analysis shows that
this is still the case in oscillating flow. The nondimensional parameters which have to be matched to model the artery flow are therefore
only Reθ and α at the lower values of α. Dynamical similarity can
thus be achieved in a model with different taper than the artery.

EXPERIMENTS AND NUMERICAL ANALYSIS

Flow visualisation was achieved by injecting dye into the cylindrical
tube at the narrow end of the tapered tube whilst the flow was in the

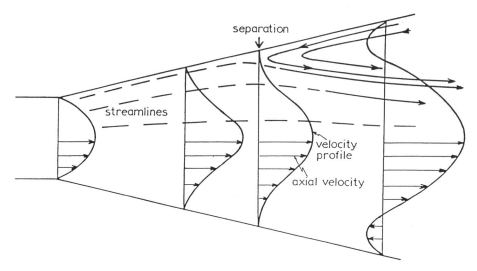

Fig.1. Velocity profiles and streamlines in a steady flow
in an expanding tube in which there is flow separation.

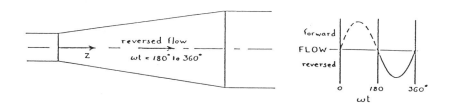

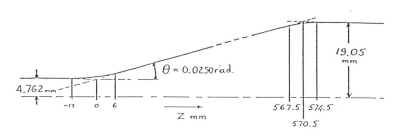

Fig.2. Conical section: Dimensions and sign convention.

forward direction and after the piston was stopped at the $\omega t = 180^\circ$
position. After the disturbances due to dye injection had dissipated
the flow was started from rest in the reversed flow direction. In
the experiments conducted in this geometry the following cycles were
found to be the same as the first at low values of α. At higher values
later cycles were found to be different from the first. The manner in
which the flow develops depends principally on the Reynolds number. When
separation occurs it is initiated well down the tube but once separation
occurs the point of separation moves rapidly towards the narrow end.
Reattachment of the flow moves out of the tapered tube at the wide end.

The experiments reveal that the flow has a series of regimes which
depend on the characteristic values which are the peak reversed flow
Reynolds number, Re, and α at the narrow end of the tube. Figure 3
shows the regimes of the flow which at low α values depend principally
on the Reynolds number, but there is a change when α is greater than
about 3. At low Re the flow is attached: as Re increases the flow
separates and forms a jet: at still higher Re the jet exhibits a ring
vortex at its head; at higher Re, the vortex and jet become turbulent:
at the highest Re the transition to turbulence of the jet entrains enough
fluid to suck the jet back onto the wall (turbulent reattachment takes
place) and the whole reversed flow becomes undirectional and turbulent.
The dependence on α is weak but the value of α must lie in the transition
range of 2 to 8 in the whole tube for the vortex production to occur. The
transition to turbulence takes different forms depending on the Reynolds
number. At low Reynolds number, the head of the vortex goes turbulent
first: large scale eddying motions are seen and the motion of the vortex
departs from the axial direction. At higher Reynolds number the jet goes
turbulent before the vortex. At α values greater than 3 it was observed
that later cycles behaved differently than the first which implies that
the flow is no longer quasi-steady. The laminar and turbulent vortex
formation curves coalesce at α greater than 6. It is also known from the
work of Schneck (1977) that separation becomes independent of α at high α.
Figure 4 shows a late stage in the development of a flow in which the
jet goes turbulent: this was called flow 5 and is indicated on figure 3.
In this flow turbulence starts in the jet following the vortex and is
seen as waves in the dye at the bottom of the jet. Below the photograph is the computed flow pattern which is in fair agreement with the
photograph as far as the speed and position of the vortex and the turbulent waves are concerned. We are able to compute the phase and position
of vortex appearance. The method used is the solution of the equations of
motion by an explicit finite difference method, published by the author
(Gerrard,1971)

THE MEDICAL APPLICATION

The application of this work to atherosclerosis is based on the
currently held view that the profileration of smooth muscle cells in
the arterial wall is implicated in atherogeneses and that this process
is accelerated by oscillating stresses. Figure 5 shows the oscillations of wall shear stress at three positions in the tapered tube for
what we have called flow A. At those positions not reached by the
vortex (e.g. z = 360mm) the wall shear oscillates with the fundamental
frequency. At those positions which the vortex passes one or more peaks
of wall stress occur and then occupy a time of the order of one tenth
of the fundamental period. When only the jet is present, at the
observation point on the wall of the tube there is also a brief peak
of shear stress.

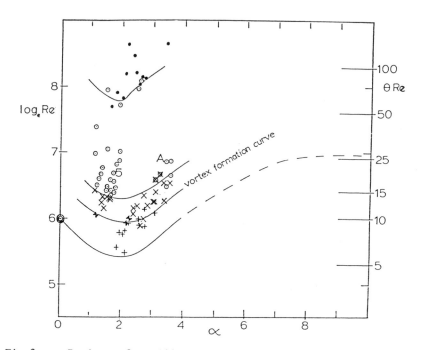

Fig.3.   Regimes of oscillatory reversed flow.

◎ steady flow separation, ⬥ separation but
no vortex.
+ no vortex (method did not indicate separation).
✕ laminar vortex, ⊙ vortex going turbulent.
• turbulent reattachment.

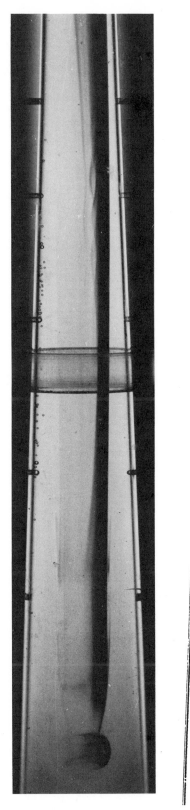
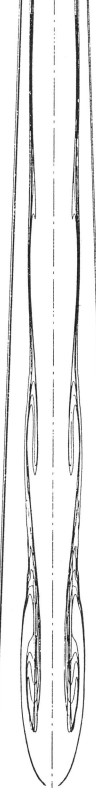

Fig.4. Flow visualisation of the vortex in the tapered tube by dye injected into the cylindrical end tube and comparison with a flow pattern produced by numerical analysis

$\alpha = 1.85$, $Re = 392$, $\omega t = 279^\circ$

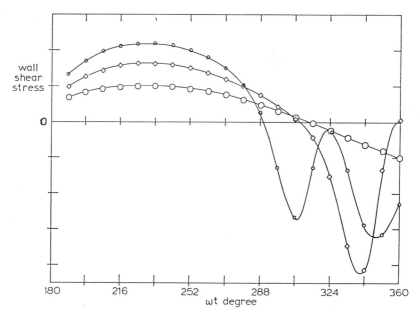

Fig.5.   Computed wall shear on an arbitrary linear scale
as a function of time.   Distance from the distal
end of the conical tube:
°, 180mm; ◊, 240mm; o, 360mm.

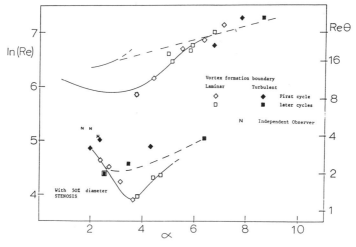

Fig.6.   Laminar and turbulent vortex formation at higher
values of α with and without 50% diameter stenosis.

We have seen that when the $Re\theta$ product is high enough the reversed flow in the artery model exhibits a jet and vortex formation. Separation was found to occur at a minimum $Re\theta$ of about 6. The taper of the human sperficial femoral artery has $\theta = 0.0034$ radian and when the subject is at rest the maximum reversed flow Reynolds number is about 300 at the narrow end of the artery at the popliteal junction and this corresponds to $Re\theta=1.0$, six times too small for separation to occur.

Having noticed that disturbances reduced the critical value of $Re\theta$, we made observations with specifically disturbed flows. The femoral artery bifurcates twice at the knee where there also may be a bend. The effects of a bifurcation and of a bend were found to reduce the critical $Re\theta$ for vortex formation from about 10 to 7 and we will assume that the separation value of $Re\theta$ is reduced from 6 to about 4. The effect of a stenosas was to produce further reductions. Figure 6 shows the vortex formation curves obtained when an axisymmetric 50% diameter stenosis was positioned just distal to the tapered tube. A ring vortex is produced at the stenosis (θ is very large). The points shown in the figure indicate conditions under which the vortex just persists and travels up the tapered tube. The smallest value of $Re\theta$ for this condition is seen to be about unity. At smaller values the vortex dies out. The penetration of the vortex into the taper is much less than at the higher Re presented above.

We conclude that the production of atherosclerosis in the superficial femoral artery by means of a vortex (or jet) travelling up the artery is not a likely possibility. There are however, other factors which require further investigation before a definite conclusion can be reached. The unsteady flows at the higher values of α are non-linear and unsteady and so examination of the response to the arterial waveform rather than a sinusoidal oscillation should be investigated. It is not known what effect other factors have on the flow and its waveform. These may include the effect of muscle flexure and of sitting with the legs crossed. Blockage is found to begin about 1/3 of the way up the artery from the knee.

ACKNOWLEDGEMENTS

Two colleagues Dr. G.Y. Buss and Dr. C.N. Savvides were responsible for the whole of the preparatory work on this project and their devoted efforts are acknowledged. Discussions with David Charlesworth of the Department of Surgery have been invaluable in the progress of this work. Dr. Buss and Dr. Savvides were jointly financed and supervised by the two departments.

REFERENCES

Eagles P.M., 1982, Steady flow in locally exponential tubes.
    Proc. Roy. Soc. Ser.A, 383: 231-145.

Gerrard J.H., 1971, The stability of unsteady axisymmetric incompressible pipe flow close to a piston: I Numerical analysis.
    J. Fluid Mech. 50: 625-644.

Sir James Lighthill 1978. Waves in fluids. C.U. Press.

Schneck D.J., 1977, Pulsatile flow in a channel of small exponential divergence: III Unsteady flow separation. ASME Ser.J. J. Fluids Eng. 99: 333-338.

Womersley J.R., 1955, Oscillatory motion of a viscous liquid in a thin walled elastic tube: I The linear approximation approximation for long waves, Phil. Mag. 46: 199-221.

INVESTIGATIONS CONCERNING LOCALIZATION OF INDUCED

THROMBUS FORMATION AT ARTERIOLAR BRANCHINGS

Dirk Seiffge, Eberhard Kremer, Volker Laux and Peter Reifert

Pharmacology, Hoechst AG Werk Kalle-Albert
6200 Wiesbaden 1, FRG

ABSTRACT

The areas of predilection for atherosclerosis and thrombosis in the vascular system are characterized by curvature, branching, bifurcation or embanking. We have investigated induced thrombus formation at arteriolar bifurcations with migrating stagnation points and local vortex flow. Thrombogenic endothelial lesions were produced by argon laser injury or by photochemical reactions (using FITC-dextran 70). The following items have been evaluated: local red blood cell velocity, geometry of the vessel, localization of the first thrombus growth. The first appearance of thrombus formation at presumably reattachment points at bifurcations in the microcirculation could be directly demonstrated by intravital microscopy.

Key words: Onset of thrombosis, atherosclerosis, vessel branching, flow dynamics

INTRODUCTION

Blood flow at vessel segments with complicated geometry are the objects of fluid-dynamically orientated investigations concerning atherogenesis and thrombosis (Brech and Bellhouse, 1973; Caro, 1981). The development of atherosclerosis in large arteries is determined by two pathogenetic factors: the arterial wall and the contents of the vessel, e.g. platelets, white and red blood cells, cholesterol levels (Schettler et al., 1983). These factors are linked to intimal lipid accumulations and development of fibrous changes in the intima of the arterial wall. In addition, blood is a complex fluid, comprising liquid and flexible cells. Hemodynamic factors like shear stress, high and low velocity gradients, pressure gradients and mass transfer variations playing a major role in atherogenesis are not yet understood (Stehbens, 1975; Texon, 1980), but the areas of predilection for atherosclerosis and thrombosis in the vascular system are characterized by curvature, branching, bifurcation or embanking, where laminar flow is altered and vortices, stagnation points and reverse flows are found (Marshall and Hess, 1970).

Biomechanical Transport Processes, Edited by F. Mosora *et al.*
Plenum Press, New York, 1990

Local hemodynamic conditions appear to play a major role in the localization of atherosclerotic plaques and thrombus formation (Liepsch, 1981; Nerem and Cornhill, 1980). Atherosclerotic symptoms are intensified when blood rheology is altered. There is a considerable number of investigations concerning blood flow at arterial bifurcations. Most of those studies were performed in artifical systems utilizing flow visualization techniques (Liepsch, 1986; Liepsch, 1988). The non-Newtonian flow behaviour of blood has never been studied at bifurcations of the microcirculation, where flow is also believed to be disturbed and secondary flows are generated (Krings, 1984; Naumann, 1975).

The aim of our study was to investigate the localization of the area of the onset of thrombus formation. We have induced thrombus formation at arteriolar bifurcations with migrating stagnation points and local vortex flow. We assume that there are sufficient similarities in flow conditions - but not flow situations - between micro- and macrocirculation.

MATERIALS AND METHODS

Studies were carried out in male Sprague Dawley rats with a body weight of about 250 g. The animals were fed commercial pellets (Altromin 1214) and allowed to drink water ad libidum. The animals, pretreated subcutaneously with atropine sulfate at a dose of 0.1 mg/kg body weight, were anesthetized with ketamine hydrochloride (100 mg/kg i.p.) and xylaxine (4.0 mg/kg i.p.). The mesentery, exposed by a hypogastric incision, was superfused with degazed paraffin liquid (37°C) and spread carefully flat on a specially constructed object stage which was mounted on the microscope table. Investigations were performed in arteriolar bifurcations or junctions of the same diameter in the fat-free ileo-caecal portion of the mesentery.

Laser Thrombosis

The 3 W Argon laser beam (Messrs. Spectra Physics, Darmstadt, FRG) was coaxially inserted into the inverted light beam path of a microscope (ICM 405, LD-Epiplan 40/0.60, Messrs. Zeiss, Oberkochen, FRG) by means of a ray adaption and adjusting device (Seiffge and Kremer, 1986). The wave length used was 514.5 nm with a capacity above the objective of 30.0 mW in the study. The exposure time for a single laser shot was 1/30 sec. The number of laser lesions necessary to induce a defined thrombus was counted.

Photochemical Thrombosis

A similar microscope was used for the photochemically induced thrombus formation (Herrmann, 1983). The ICM was connected to induce fluorescence with a high-pressure mercury lamp XBO-75. A condenser with long working distance and a numerical aperture of 0.6 was used to transilluminate the field of observation. Light intensity was adjusted by interposing a neutral grey filter with 50 % absorbance. For observations of fluorescent objects we used the Ploemopack system which utilizes the objective for both illumination and observation. An appropriate set of

340

filters (BP 450 - 490 nm, FT 510 nm, LP 520 nm Zeiss, Oberkochen, FRG) separated light at different wavelengths for excitation and observation. Thirty minutes after surgery, 0.3 ml of a 10 % FITC-dextran 70 solution were injected into the femoral artery. Only preparations with brisk flow in all vessels and without obvious leakage were used. In transillumination an arteriole with a diameter between 15 - 40 μm was centered in the field of observation and then FITC was excited by switching over to epi-illumination. Using stopclocks, we determined the time elapsing between onset of the noxious stimulus and appearance of the first fluorescent platelet aggregate adhering to the vessel wall. To have time to select the next vessel, the maximal observation time was limited to 6 min.

Routinely, all investigations were controlled by red blood cell velocity measurements (Seiffge and Kremer 1986). For statistical evaluation the x^2 test was used whereby the arithmetic mean was expressed as x and standard deviation as SD.

RESULTS

Thrombogenic endothelial lesions were produced by argon laser injury directly on the angular point of arteriolar branching. Results were compared with control values of the same animal obtained in branchless vessels. As shown in Table 1a and 1b there are increased number of laser lesions necessary to induce thrombus formation in branchless vessels compared with bifurcations. At junctions 40 % and at bifurcations 54 % lower number of laser lesions were necessary to induce the same thrombus compared with branchless vessels. An observation has to be noted, which is not included in the results. Thrombi at bifurcations or junctions were more stable and were rapidly growing at its edges into the diverging vessels.

Table 1a.   Laser-induced thrombus formation at arteriolar branching:
Results of vessel geometry and red blood cell velocity.

vessel segment	n	vessel diameter μm	branching angle ^{o}	red blood cell velocity mms^{-1}
rat mesenterium				
branchless	26	15.6	---	6.0
junction				
total	18	`21.7	84°	12.1
antegrad	10	21.1	60°	7.5
retrograd	4	20.3	89°	20.6
orthograd	4	23.6	128°	16.1
bifurcation	10	19.9	69°	5.2

Table 1b. Laser-induced thrombus formation at arteriolar branching:
Extent of induced thrombosis.

vessel segment	n	No. of laser lessions		%
		x	SD	
rat mesenterium				
branchless	26	2.42	1.24	--
junction				
total	18	1.44	0.62	40
antegrad	10	1.30	0.48	46
retrograd	4	2.00	1.00	17
orthograd	4	1.40	0.55	42
bifurcation	10	1.10	0.32	54

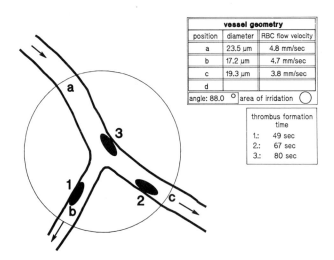

vessel geometry		
position	diameter	RBC flow velocity
a	23.5 µm	4.8 mm/sec
b	17.2 µm	4.7 mm/sec
c	19.3 µm	3.8 mm/sec
d		

angle: 88.0 ° area of irridation ○

thrombus formation time	
1.:	49 sec
2.:	67 sec
3.:	80 sec

Fig. 1. Thrombus formation at arteriolar bifurcations induced by photochemical
reaction. The circle describes the area of irradiation and arrows the
flow direction. The first thrombus (1) appeared after 49 seconds in
the branching vessel (b).

Therefore we have investigated thrombus formation at bifurcations induced by photochemical reaction. In this method the thrombogenic endothelial lesions covers the entire area of the vessel junction excited by the fluorescence dye. We measured the first location of thrombus growth (Fig. 1).

We measured vessel diameter, red blood cell velocity, branching angle and thrombus formation time at arteriolar bifurcations or junctions and compared them with data obtained at branchless vessel segments (Table 2). In contrast to results obtained with the laser (Table 1) we measured a decrease in the extent of thrombus formation expressed as an increase in thrombus formation time, at arteriolar branchings compared with branchless vessels.

Table 2.   Vessel geometry (d), rbc-velocity (V) and thrombus formation time induced by photochemical reaction at arteriolar bifurcations or junctions.

vessel segment	n	vessel diameter (μm)	rbc-velocity (mm/s)	branching angle	thrombus formation time (s)
bifurcation	27	d_a 22.1 d_b 17.4 d_c 18.5	V_a 4.0 V_b 3.4 V_c 3.2	57°	69.7
junction	16	d_a 20.8 d_b 18.2 d_c 16.0	V_a 3.9 V_b 3.4 V_c 3.4	74°	44.2
branchless	24	d_a 18.8 d_b -- d_c --	V_a 3.6 V_b -- V_c --	--	25.0

To quantify the measured data of the onset of thrombus formation at arteriolar bifurcations or junctions we have performed a nomenclature for the areas at vessel branchings (Fig. 2).

On the basis of the above mentioned nomenclature we have calculated the frequency concerning the areas of the onset of thrombus formation at arteriolar bifurcations or junctions. In our test model the onset of thrombus formation occurs at bifurcations in nearly 90 % and at junctions in more then 80 % of the cases within the branching vessels (Tab. 3).

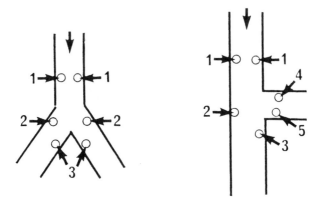

Fig. 2.   Nomenclature concerning the areas of the onset of thrombus
            formation at arteriolar bifurcations or junctions.

Table 3.   Areas of the onset of photochemically-induced thrombus formation at
            arteriolar bifurcations and junctions.

area		No./cases	frequency (%)
bifurcation	1	3/27	11
	2	12/27	44.5
	3	12/27	44.5
junction	1	3/16	18.75
	2	4/16	25
	3	3/16	18.75
	4	3/16	18.75
	5	3/16	18.75

DISCUSSION

With the applied methods of induced localized endothelial injury we could demonstrate that thrombus formation at arteriolar bifurcations or junctions is increased compared with branchless vessel segments. It is possible to induce thrombus formation at the stagnation point of the flow divider, where the kinetic of the flowing blood is directly transformed into normal pressure. Also, the tangential stresses and hence the local shear stresses acting on the blood components are higher than average. It is notable that the growth of a stable thrombus at stagnation points takes place at its edges within the branching vessels, where areas of recirculation are presumed. Basically, our observations show that a thrombus will grow at the stagnation point of the flow divider.

With the second investigation with diffuse injury we could demonstrate that first thrombus formation will not appear at the flow divider, but at areas of complicated flow within the branching vessel segments. It remains to be open whether those areas are reattachment points or areas of recirculation and local vortex flow.

The relevance of investigations in the microcirculation concerning events primarily taking place in the macrocirculation remains to be proven. It is not the intention of the presently reported experiments to answer these questions. Important differences between micro- and macrocirculation are known, e.g. nutritive requirements, reactions to growth- and cell migration factors, vessel wall permeability, and the Reynolds numbers. But we also know some important common features like the monolayer of endothelial cells, the Weibel-Palade bodies, the production of factor VIII - antigen, prostacyclin production, and the endothelium-derived-relaxing-factor (EDRF). But for the appearance of areas with complicated flow it is very important to mention that there is also a pulsatile flow present in the microcirculation. From our findings we assume that thrombus formation at branching vessels in the microcirculation takes place primarily at areas, where high and low shear forces are in close proximity, e.g. the area between the stagnation or reattachment point and the area of recirculation.

Another interesting result of the reported data is that thrombus formation at the stagnation point of the flow divider following localized laser injury is increased compared with branchless vessel segments. But the onset of thrombus formation in our experiments using a photochemical reaction as a thrombogenic stimulant never takes place at the flow divider. Under these circumstances thrombosis occurs within a shorter period of time in branchless vessels compared with branching segments of the microvascular system. But in contrast to the time related appearance of thrombosis, the onset of thrombus formation primarily takes place at the branching vessel segments. This discrepancy between the time dependent reaction towards thrombogenic stimuli and the location of the first thrombus growth seems to be of interest and importance.

REFERENCES

Brech, R.; Bellhouse, B.J., 1973, Flow in branching vessels.
    Cardiovasc. Res. 7: 593.

Caro, C.G., 1981, Arterial fluid mechanics and atherogenesis, in: Recent advances incardiovascular disease", Vol. II, Supplement: 6.

Herrmann, K.H., 1983, Platelet aggregation induced in hamster cheek pouch by a photochemical precess with excited fluorescein isothiocyanate-dextran. Microvasc. Res. 26: 238.

Krings, W.B., 1984, Das Verhalten menschlicher Blutplättchen in der Rohr- und verzweigten Kanalströmung (Durchmesser kleiner als 0,1 mm). Med. Diss., Aachen

Liepsch, D.W.; Moravec, St. and Zimmer, R., 1981, Einfluß der Hämodynamik auf Gefäßveränderungen (Einfluß der Blutströmung in Arterien auf die Bildung von Thrombosen und Stenose). Technik 26: 115.

Liepsch, D.W., 1986, Flow in tubes and arteries - a comparison. Biorheology 23: 395.

Liepsch, D.W., 1988, Effect of blood flow parameters on flow patterns at arterial bifurcations - studies in models. in: Proceedings, International Symposium on Biofluid Mechanics: 47.

Marshall, M.; Hess, H., 1978, New findings concerning pathogenesis and nonsurgical treatment of peripheral arterial diseases. Vasa 7: 49.

Naumann, A., 1975, Strömung in natürlichen und künstlichen Organen und Gefäßen. Klin. Wschr. 53: 1007.

Nerem, R.M.; Cornhill, J.F., 1980, The role of fluid mechanics in atherogenesis. ASME J. Biomech. Eng. 102: 181.

Schettler, G.; Nerem, R.M.; Schmid-Schönbein, H.; Mohrl, H.C.and C. Diem, 1983, in: Fluid dynamics as a localizing factor for atherosclerosis. Springer, Berlin.

Seiffge, D.; Kremer, E., 1986, Influence of ADP, blood flow velocity, and vessel diameter on laser-induced thrombus formation. Thromb. Res. 42: 331.

Stehbens, W.E., 1975, The role of hemodynamics in the pathogenesis of atherosclerosis. Progressive Cardiovascular Diseases 18: 89.

Texon, M., 1980, Hemodynamic basis of atherosclerosis. Hemisphere, Washington.

THE INFLUENCE OF VASCULAR SMOOTH MUSCLE ON

THE DEVELOPMENT OF POST-STENOTIC DILATATION

S.E. Greenwald[*], U. Kukongviriyapan[¶] and B.S. Gow[¶]

[*]Department of Morbid Anatomy, London Hospital
London E1 1BB, England and [¶]Department of Physiology
University of Sydney, Sydney NSW, Australia

SUMMARY

A segment of the thoracic aorta distal to a stenotic ring in
four rabbits was frozen with liquid nitrogen in order to determine
whether post-stenotic dilatation (PSD) would develop in the
absence of viable vascular smooth muscle (VSM). Four animals were
similarly prepared except that the rings did not constrict the
aorta. Another four served as controls, having constricting rings
but no aortic freezing. PSD developed as expected in the non-
frozen stenosed animals (60% diameter increase in 8 weeks).
Freezing was associated with a rapid dilatation of the vessel when
compared to the proximal segment (56% diameter increase) which
regressed to normal dimensions during the next 8 weeks. In spite
of marked fibrosis of the frozen areas their static elastic moduli
did not differ significantly from that of their dilated counterparts, although in all cases the segments distal to the
constriction had significantly greater modulus values than those
proximal to it. The dynamic elastic modulus of the frozen segments
was significantly greater than that of the PSD region and the
untreated areas proximal to the stenosis site.
We conclude that destruction of VSM by freezing prevents the
development of PSD and that this may be due in part to the
increased dynamic stiffness of the frozen vessels.

INTRODUCTION

The causes of post-stenotic dilatation (PSD), a region of an
artery downstream from a partial occlusion or stenosis in which
the lumen diameter is increased, are not well understood. It has
been observed distal to valve coarctations in the aorta and
pulmonary arteries (Wood, 1956), to occlusive lesions in the renal
(Roach and Macdonald, 1970) and femoral arteries (DeBakey et al,
1961); and has been induced at various sites in experimental
animals by applying an external constriction to the vessel.
It is generally agreed that the formation of PSD is associated with turbulence of sufficient magnitude to produce an audible
thrill and bruit at the skin surface (Bruns et al 1959; Roach,
1963a; Miller et al, 1980), although it is probable that other

forms of disturbed flow are involved in its formation. (Giddens et al, 1976; Khalifa and Giddens,1978,1981). The phenomenon has been described in rubber tubes (Holman et al, 1954; Bruns et al, 1959) and excised arteries with inactive smooth muscle (Roach, 1963b; Roach and Harvey, 1964) which implies that passive mechanisms such as fatigue may weaken the wall sufficiently for the swelling to occur.However in some reports no dilatation has been observed (Foreman and Hutchison, 1970). Furthermore estimates of the time required for PSD to develop vary from hours (Roach and Harvey, 1964) to weeks (Gow et al, 1984; Zarins et al, 1986). It has also been shown that vibration alone containing frequencies similar to those generated by turbulent flow downstream from a stenosis can produce a dilatation (Gow et al, 1984; Roach, 1963b; Vito et al, 1975). However, these experiments have been confined to rubber tubes or excised vessels; and attempts to reproduce this observation in-vivo (Gow et al, 1984) have failed. The results of elasticity measurements on the dilated region have also been contradictory. In a number of studies Roach (1963b; Roach and Harvey,1964) the dilated region has been found to be more distensible than normal; while others have observed a decrease (Stefanadis et al, 1988).

Removal of the stenosis in living experimental animals results in regression of the dilatation within 6-24 h, whilst in-vitro the regression may be complete within 8h (Roach, 1970; Gow et al, 1984). Some of the observations described above suggest that PSD is a passive process; whilst others imply that it only occurs in viable blood vessels. For instance, if the regression of PSD can occur within a few hours and its formation in in-vitro has not been consistently observed, and may therefore depend on the state of VSM activity, how can a passive remodeling process be responsible? Similarly, why does vibration alone in-vivo, which imparts as much energy to the vessel wall as turbulent flow, not result in dilatation? Since these questions suggest that the formation of PSD is an active process which may require the mediation of vascular smooth muscle (VSM) and or endothelial cells we describe here an attempt to determine whether destruction of VSM prevents the development of PSD.

METHODS

Twelve adult male New Zealand rabbits were anaesthetised with pentobarbital sodium (30 mg/kg i.v.), the descending thoracic aorta was exposed through a left thoracotomy and mobilised between ribs 5 and 8. Smooth muscle viability was confirmed by topical application of 30 µl noradrenaline (30 µg/ml) while measuring vessel diameter with an electrical caliper.(Gow, 1966) In 4 animals an area 1 cm long at the distal end of the segment between intercostal arteries 7 and 8 was rapidly frozen by clamping between two hollow brass blocks soldered to a pair of tongs through which liquid nitrogen was drawn for 15s. The vessel was then thawed by drawing water at room temperature through the tongs for 5s. This treatment resulted in an immediate dilatation of the frozen area which was complete within 3 minutes and subsequent loss of contractile response to noradrenaline. A split nylon ring (2.6 mm I.D., 2 mm long) was positioned proximal to the frozen area. Stainless steel markers were glued to opposite sides of the vessel proximal to the ring and distal, at the mid point of the frozen length. The animals were allowed to recover and the diameter of the vessel was measured 3 times weekly by radiography, (60kV, 2.3mAs). A second group of 4 animals was treated in the same way except that a larger ring (causing no constriction) was fitted to the aorta. The third group was fitted with a stenotic ring but the distal segment of the vessel was not frozen.

Eight weeks after the initial operation the animals were again anaesthetised and the aorta was exposed as before. The previously frozen areas remained unresponsive to noradrenaline showing that freezing resulted in permanent incapacity of the muscle. A nylon cannula was passed into the descending aorta via the right common carotid and positioned 5-10 mm proximal to the stenosis site. A second cannula was passed into the right femoral artery and positioned the same distance distal to the stenosis. Dynamic elastic modulus of the segments proximal and distal to the ring was calculated from pressures measured through each of the two cannulae. Diameter was measured with the caliper positioned in turn above the tip of each cannula. Next the aorta between ribs 3 and 11 was cannulated and inflated through a syringe connected to a pressure transducer. Diameter was measured at fixed pressures in the range 10 to 160 mmHg and these data were used to calculate a static elastic modulus. Following the elasticity measurements the segment was perfused at a pressure of 80 mmHg with Karnovsky's fixative (pH 7.4) and processed for microscopy. Finally the animals were killed with an intravenous injection of 2ml saturated KCl solution.

The bulk elastic properties of the vessel were expressed as a circumferential incremental elastic modulus:

$$E_{inc} = 1.5 \ E_p \ D_i^2/(D_o^2 - D_i^2) \ \text{(Gow and Taylor, 1968)}$$

where D_o, the outside diameter of the vessel, is measured by the caliper. The inside diameter, D_i, is calculated from the cross sectional area of the wall obtained by measurement of the microscopic sections. E_p, a measure of functional stiffness is given by:

$$E_p = \Delta P . D_o/\Delta D_o.$$

For the dynamic modulus ΔP is the pulse pressure and ΔD, the corresponding diameter change. For the static case $\Delta P/\Delta D$ is replaced by dP/dD and calculated by differentiating a third order polynomial fitted to the measured pressure/ diameter curve.

The significance between groups of differences in vessel dimensions and elasticity were tested by one way analysis of variance.

RESULTS

Radiographic measurements showed that the immediate and localised dilatation observed in both groups with frozen aortas gradually regressed (figure 1). Thus the ratio of the diameter of distal segment to that of the proximal segment (D_{dist}/D_{prox}) fell from an average value of 1.56 ± 0.21(SD) on the second post-operative day to a value of 1.05 ± 0.02 by day 56; whereas the non-frozen group showed the expected progression of PSD,

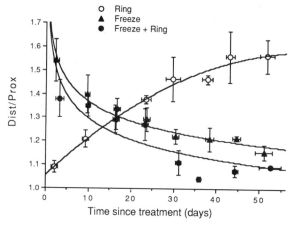

Figure 1. Freezing of the aorta was associated with a rapid
dilatation followed by a regression towards normal
diameter (solid symbols). The unfrozen segments (open
symbols) dilated as expected. Mean values ± SD.

D_{dist}/D_{prox} increasing from 1.0 ± 0.11 at day 2 to 1.6 ± 0.22 by
day 56. Immediately after freezing the smooth muscle cells (s in
figure 2B) showed acute degenerative changes, disintegrating and
fragmenting (arrows). Eight weeks after freezing (figure 2C) no
viable smooth muscle cells could be identified and remnants of
degenerated cells had largely disappeared. There was a slight
increase in the number of collagen fibrils (c) and possible
fusion and straightening of elastic lamellae (e).

Table 1 Dimensions and static elasticity of aortic segments
distal and proximal to the stenosis site, measured at a
pressure of 90 mmHg (12.0 kPa). (Mean values for each
group ± SEM)

Group No.	Relative Diameter	Wall (mm) Thickness	E_p (Nm^{-2}) (x 10^4)	E_{inc} (Nm^{-2}) (x 10^5)
DISTAL				
PSD 3	1.37±0.07	0.14±0.003	5.81±2.12	8.11±2.69
F no C 4	1.24±0.04	0.15±0.009	10.83±1.99	14.79±3.22
F + C 3	1.24±0.03	0.15±0.010	10.40±4.60	12.55±4.25
PROXIMAL				
PSD 3	1.23±0.01	0.16±0.003	4.35±0.10	4.55±0.08
F no C 4	1.49±0.08	0.15±0.003	3.42±0.39	4.30±0.49
F + C 3	1.48±0.26	0.18±0.025	4.11±0.15	4.48±0.27

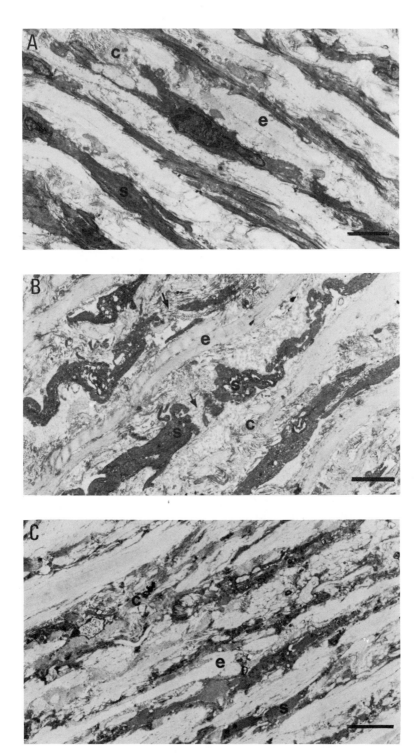

Figure 2. Transmission electron-micrograph of rabbit thoracic aorta. A. normal vessel, B. immediately after freezing, C. 8 weeks after freezing. Magnification: x5000. Bar indicates 3 μm.

Table 2. Dynamic elastic modulus of aortic segments distal and proximal to the stenosis site (Mean values for each group ± SEM)

Group	No	Pressure (mmHg)	E_p (Nm^{-2}) (x 10^4)	E_{inc}(Nm^{-2}) (x 10^5)
DISTAL				
PSD	2	78±2.5	7.6±0.73	11.6±0.14
F no C	2	85±9.0	35.4±5.00	53.2±9.30
F + C	2	75±7.5	35.7±4.45	45.1±9.92
PROXIMAL				
PSD	2	75±5.0	6.92±0.53	5.79±0.55
F no C	2	84±3.1	8.13±1.88	10.80±1.84
F + C	2	82±9.2	9.19±2.22	9.71±1.90

Figure 3 (upper panel) shows, as expected, that when the vessels were inflated the region distal to the stenosis in the unfrozen animals dilated more than the two groups with frozen vessels.In the proximal segment (figure 3, lower panel) the unfrozen stenosed vessels appeared to distend significantly less than those of the two frozen groups.

At physiological pressure the wall thickness of the distal segments were consistently but not significantly less than those of the proximal (table 1). There were no significant differences between the groups.

The distal segments of the two frozen groups were consistently stiffer (both functional (E_p) and structural stiffness, E_{inc}) than those of the PSD animals although because of large standard errors (table 1) these differences were not significant ($p > 0.1$). In the proximal segment values for all groups were similar. The distal segments of each group were significantly stiffer than their proximal counterparts ($p < 0.01$ for E_p; $p < 0.005$ for E_{inc}); whilst the wall thickness was significantly less ($p < 0.05$).

The dynamic modulus values of the distal segments in both frozen groups were significantly higher than those of the PSD group. Again in the proximal segment no significant differences were found, (table 2). Similarly, the distal segments in each group were significantly stiffer than the corresponding proximal areas, ($p < 0.02$ for E_p; $p < 0.05$ for E_{inc}). However, it should be noted that data were obtained for only two animals in each group and that although the mean pressures for which modulus values were calculated were similar, they were not identical. Consequently any conclusions based on these dynamic data must remain, at least until further results are available, tentative.

DISCUSSION

The relationship between disturbed flow distal to a stenosis and the consequent dilatation of the vessel in this region is not well understood. The magnitude of the dilatation and the time taken for it to develop depend strongly on the experimental conditions.

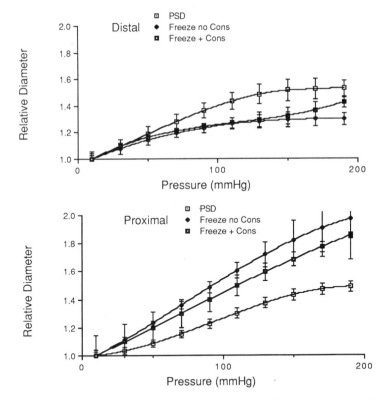

Figure 3. In the distal segment (upper panel) the ratio of radius
at various pressures to the radius measured at 10 mmHg
was greater in the unfrozen than in the 2 frozen
groups. The opposite effect was seen in the proximal
segment (lower panel)

In-vitro it has been detected in as little as three hours
after the start of turbulent flow (Roach and Harvey, 1964). In
living dogs it reaches a maximum (approximately a 30% increase in
diameter) within 10 to 14 days (Roach, 1963a), in this study on the
rabbit a larger dilatation ($\approx$ 60%) is attained after 50 days and
in baboons the diameter may continue increasing for 6 months
(Zarins et al, 1986). It is therefore possible that PSD may evolve
by several different processes in different species, in-vitro and
in-vivo. Thus general conclusions about the mechanism of its
progression should perhaps be treated with caution.

In 1959 it was suggested by Bruns and others that vibration
due to turbulent flow is associated with structural fatigue of the
vessel wall. Hutchison (1974) has shown that turbulent flow
through stenoses contains significant energy in frequencies at
which the vessel wall was observed to resonate and they have
suggested that this enhanced energy transfer to the wall may
accelerate the fatigue process. Further evidence for the fatigue
hypothesis has been adduced by Roach (1963b) and Roach and Harvey
(1964) who have shown that PSD can occur in excised arteries and

that vibrating excised vessels alone in the absence of flow also leads to dilatation (Boughner and Roach, 1971). It has also been shown that the dilated region is more distensible than the area proximal to the stenosis and has suggested that this increased distensibility is a consequence of the fatigue (Roach, 1963b; Roach and Harvey, 1964).

However in spite of this evidence it is clear that a number of observations cannot easily be explained by a passive process such as material fatigue. Firstly, in a complex multi-component material such as the vascular media undergoing a large increase in diameter in a period as short as 10 days, changes in the structure of the vessel would be expected. Although degenerative changes which result in elastin fragmentation, an increase in fibrous tissue and collagenase activity are seen after several months (Trillo and Haust, 1975; Zarins et al, 1986), few significant histological or chemical changes have been observed in the initial period during which the dilatation has occurred. The observed increase in the size of the fenestrations in the internal elastic lamina of the distal region when compared to the proximal, are presumably a consequence of the dilatation alone since their area density does not change (Potter and Roach, 1983).

Secondly, the changes in elasticity are not straightforward. Thus although Roach (1963a) has shown that the dilated region is more distensible than normal, a close look at the results of this study reveals that at 100 mmHg the dilated region was not significantly more distensible than the same part of the vessel in control animals. The present study confirms this finding, as does a recent report on the distensibility of the dilated ascending aorta due to aortic valve stenosis (Stefanadis et al, 1988).

Thirdly, vibrating the aorta in living animals (Gow et al., 1984) in the absence of abnormal shear stress due to disturbed flow has repeatedly failed to produce any measurable dilatation.

Fourthly, dilatation due to turbulence or vibration in-vitro does not always develop (Foreman and Hutchison, 1970).

Finally, it is unlikely that a repair process which must involve some remodeling could occur in the 6-24h period in which PSD can regress when a stenosis is removed. Furthermore a repair process would be expected to result in a permanent alteration of the vessel's structure and mechanical properties. It should be noted that in man, PSD of long duration resulting in the formation of fibrous tissue or degenerative changes leading to aneurysm, will not usually regress when the stenosis is removed.

It seems then that under some conditions at least the development and regression of PSD is an active process requiring the mediation of VSM.

Changes in the metabolic activity of VSM leading to PSD and long-term remodeling of the vessel wall are probably mediated by endothelial cells. Their orientation in and around PSD reflects the observed flow patterns (Legg and Gow, 1982) and possible mechanisms relating disturbed flow to endothelial cell damage and consequent medial changes have been discussed by Schmid-Schonbein and Wurzinger (1986). The phenomenon of flow-dependant vasodilatation (see for example Griffith et al, 1989) suggests another route by which abnormal shear stress may chronically activate VSM leading to long lasting changes in the vessel wall.

The results presented here show that freezing the aorta for 15s destroys VSM in a few minutes. Since the VSM in the frozen regions remains absent eight weeks after its destruction, the gradual return to normal diameter of the vessels in the two frozen groups is difficult to explain. Contraction of the fibrotic material investing the frozen area seems unlikely because, in two animals, on removing the fibrosed adventitia, no increase in the internal diameter of the vessel (measured radiographically) was seen. Furthermore, if PSD had occurred in the absence of muscle dilatation, it would probably have occured before the fibrotic changes had progressed.

Since the increase in static elastic modulus of the frozen regions was not statistically significant one cannot conclude that the failure of PSD to occur after freezing is attributable to a stiffer wall. However, the increased dynamic modulus of the frozen vessels when compared to the dilated ones may be a factor. This observation was surprising since the viscous component of the dynamic modulus is expected to be reduced in the segments that no longer contained VSM. We would emphasise however that in this initial study dynamic elasticity data were only obtained from 2 animals in each group.

In conclusion, we have observed that, 1) freezing a segment of the aorta leads to a rapid destruction of VSM while leaving scleroprotein largely unaffected. 2) The resulting removal of all muscle tone leads to an initial dilatation which regresses during the following 8 weeks, and that the presence of a proximal constriction does not alter the time course of this regression. 3) The development of PSD is inhibited when VSM is destroyed, although increasing stiffness in the frozen area may also contribute to the long-term return to normal diameter.

ACKNOWLEDGEMENT

This work was supported in part by a Royal Society Study Visit Grant to SEG.

REFERENCES

Boughner, D.R. and Roach, M.R.,1971, Effect of low frequency vibration on the arterial wall, Circ Res, 29: 136.

Bruns, D.L., Connolly, J.E., Holman, E., and Stofer, R.C., 1959, Experimental observations on post stenotic dilation, J Thor Cardiovasc Surg, 38: 662.

De Bakey, M.E., Crawford, E.S., Morris, G.C. Jr. and Colley, D.A., 1961, Surgical considerations of occlusive disease of the innominate, carotid, subclavian and vertebral arteries, Ann Surg,154: 698.

Foreman, J.E.K. and Hutchison, K.J., 1970, Arterial wall vibration distal to stenoses in isolated arteries of dog and man, Circ Res, 26: 583.

Giddens, D.P., Mabon, R.F. and Cassanova, R.A.,1976, Measurement of disordered flows distal to subtotal vascular stenoses in the thoracic aortas of dogs, Circ Res, 39: 112.

Gow, B.S. and Taylor, M.G., 1968, Measurement of the viscoelastic properties of arteries in the living dog, Circ Res, 23: 111.

Gow, B.S., 1960, An electrical caliper for the measurement of pulsatile arterial diameter in vivo, J Appl Physiol, 21: 1122.

Gow, B.S., Devenish-Mears, S.E., Crosby, D.G. and Legg, M.J., 1984, The role of vascular smooth muscle in post-stenotic dilatation, in: "The Peripheral Circulation" Hunyor, S., Ludbrook, J. and Mcgrath M., eds. Elsevier.

Griffith, T.M., Edwards, D.H., Davies, R.L. and Henderson, A.H., 1989, The role of EDRF in flow distribution: a microangiographic study of the rabbit isolated ear, Microvasc Res, 37: 162.

Holman, E., 1954, The obscure physiology of poststenotic dilatation: its relation to the development of aneurysms, J Thor Surg, 28: 109.

Hutchison, K.J., 1974, Effect of variation of transmural pressure on the frequency response of isolated segments of canine carotid arteries, Circ Res, 35: 742.

Legg, M.J. and Gow, B.S., 1982, Scanning electron microscopy of endothelium around an experimental stenosis in the rabbit aorta using a new casting material, Atherosclerosis, 42: 299.

Khalifa, A.M.A. and Giddens, D.P., 1978, Analysis of disorder in pulsatile flows with application to poststenotic blood velocity measurement in dogs, J Biomech, 11: 129.

Khalifa, A.M.A. and Giddens, D.P.,1981, Characterization and evolution of poststenotic flow disturbances J Biomech, 14: 279.

Miller, A., Lees, R.S., Kistler, J.P. and Abbot, M.D., 1980, Spectral analysis of arterial bruits (Phonoangiography): Experimental validation, Circulation, 61: 515

Potter, R.F. and Roach, M.R., 1983, Are enlarged fenestrations in the internal elastic lamina of the rabbit thoracic aorta associated with poststenotic dilatation?, Can J Physiol Pharmacol, 61: 101.

Roach, M.R. and Harvey, K., 1964, Experimental investigation of poststenotic dilatation in isolated arteries, Can J Physiol Pharmacol, 42: 53.

Roach, M.R. and MacDonald, A.C., 1970, Poststenotic dilatation in renal arteries. Preliminary report, Invest Radiol, 5: 311.

Roach, M.R., 1970, Reversibility of postenotic dilatation in the femoral arteries of dogs, Circ Res, 27: 985.

Roach, M.R.,1963a, Changes in arterial distensibility as a cause of poststenotic dilatation, Am J Cardiol, 12: 802.

Roach, M.R., 1963b, An experimental study of the production and time course of poststenotic dilatation in the femoral and carotid arteries of adult dogs, Circ Res, 13: 537.

Schmid-Schonbein, H. and Wurzinger, L.J., 1986, Transport phenomena in pulsating post-stenotic vortex flow in arteries, Nouv Rev Fr Haematol, 28: 257.

Stefanadis, C., Wooley, C.F., Bush, C.A., Kolibash, A.J. and Boudoulas, H., 1988, Aortic distensibility in post-stenotic dilatation: The effect of co-existing coronary artery disease, J Cardiol, 18: 189.

Trillo, A. and Haust, M.D., 1975, Arterial elastic tissue and collagen in experimental poststenotic dilatation in dogs, Exp Mol Pathol, 23: 473.

Vito, R., Tso, W.K. and Schwartz, C.J., 1975, Poststenotic dilatation: arterial wall mechanics in response to vibration, Can J Physiol Pharmacol, 53: 998.

Wood, P. cited in Roach, 1963b.

Zarins, C.K., Runyon-Hass, A., Zatina, M.A., Chien-Tai Lu and Glagov, S., 1986, Increased collagenase activity in early aneurysmal dilatation, J Vasc Surg, 3: 238.

INSTRUMENTATION AND HEMODYNAMIC INVESTIGATIONS

BLOOD RHEOLOGY IN EXTRACORPOREAL CIRCULATION

J. Martins-Silva, R. Lima, L. Cardoso, J. Cravino, M. Nunes,
A. Lemos, T. Quintão, A. Nobre, J. Cruz, M. Dantas and
C. Saldanha

Institute of Biochemistry, Fac. Medicine of Lisbon, and
Department of Cardiothoracic Surgery, Hospital Santa Maria
Lisbon, Portugal

ABSTRACT

Sixteen patients undergoing cardiopulmonary bypass for coronary artery surgery were evaluated for hemodynamic, hemorheologic and metabolic responses as a function of extracorporeal circulation flow, continuous or pulsatile. There were no major differences between the pulsatile and non-pulsatile groups, although statistically significant time course responses have been observed in both groups.

Key words - cardiac surgery, hemodynamics, pulsatile and nonpulsatile flow, hemorheologic factors.

INTRODUCTION

In surgical interventions requiring extracorporeal circulation, dysfunctions of some vital organs can occur, related to diminished peripheral perfusion and local ischemia (Doberneck et al., 1962; Rosky and Rodman, 1966; Kirklin and Kirklin, 1981). The modality of flux (pulsatile and continuous) used during cardiopulmonary bypass (CPB), and alterations of blood composition and rheologic properties, may be causes of the afore-mentioned disturbances (Osborn et al., 1982; Ekeström et al., 1983; Kamada et al., 1987).

In most clinical (Trinkle et al., 1969; Chiu et al., 1984) and experimental (Trinkle et al., 1969; Shephard and Kirklin, 1969) studies, pulsatile flux is considered to be closer to physiologic conditions, since it probably decreases peripheral vascular resistance, and increases oxygen consumption. The advantages of pulsatile flux are however still a matter of discussion, as result of contraditory reports regarding perfusion of different organs (Frater et al., 1980; Kono et al., 1983; Chiu et al., 1984).

The present study was undertaken to evaluate possible advantages of pulsatile over nonpulsatile flow in a group of sixteen patients with coronary occlusive disease scheduled for multiple coronary artery grafts under cardiopulmonary bypass. Blood rheology and tissue oxygenation repercussions were also investigated in both types of perfusion.

Biomechanical Transport Processes, Edited by F. Mosora *et al.*
Plenum Press, New York, 1990

PATIENTS AND METHODS

Case Selection

Sixteen patients (males) who underwent coronary artery by-pass graft, were selected for study and arbitrarily divided into two comparable groups. Informed consent was previously obtained from each patient.

All the patients requiring associated cardiac procedures were excluded from the study. None of the patients had a history of significant previous illness other than coronary disease.

Conventional non-pulsatile cardio-pulmonary by-pass (CPB) was used in one group (8 patients), whereas pulsatile flow was utilized in the other group (8 patients). The two groups were similar for age (non-pulsatile 52.5±5.1; pulsatile 54.6±4.8 years old), weight (non-pulsatile 80.3±9.0; pulsatile 73.2±7.3 kg), body surface (non-pulsatile 1.90±0.10; pulsatile 1.82±0.09 min), by-pass time (non-pulsatile 185.3±18.8; pulsatile 182.8±40.7 min) and aortic clamping time (non-pulsatile 126.5±19.8; pulsatile 109.6±25.3 min). The number of grafts was also comparable for both methods (non-pulsatile 4.0±1.1; pulsatile 3.8±0.7), and all the patients had an ejection fraction above 50%.

Hemodynamic Measurements

On the day before operation each patient had electrocardiographic leads placed, and percutaneous catheters inserted for measurements of hemodynamic values and blood sampling. A thermal-dilution Swan-Ganz catheter was introduced through the internal jugular vein into the pulmonary artery for measurement of central venous pressure (CVP), pulmonary artery pressure (PA), pulmonary capillary wedge pressure (PCWP) and cardiac output (CO). The cardiac output was determined by the thermal-dilution technique, and the left atrium pressure (LAP) was infered from the PCWF.

A radial pressure line was inserted into a radial artery for monitoring the systemic (systolic, diastolic and mean) arterial pressures (SAP, DAP, MAP). All pressures were monitored continuously with a Siemens system (Sirecust 404.1 monitor). A thermodilution catheter was placed into the coronary sinus through the right femural vein for measurement of coronary sinus blood flow (CSO) on a Webster Laboratories Flowmeter (CF-300 model), and blood sampling. Total systemic vascular resistance (SVR) was calculated from the following formula:

$$SVR\ (dynes.sec.\ cm^{-5}) = \frac{MAP\ (mmHg) - CVP\ (mmHg)}{CO\ (L/min)} \times 80$$

where 80 is a correction factor.

An indirect value of myocardial oxygen consumption was calculated by the triple index (systolic blood pressure x heart rate x left atrium pressure).

Anesthesia and Medication

A standard anesthestic technique was used in all patients. On the day before surgery, and two hours before catheter placement, all patients received premedication including hydroxizine (50mg) and lorazepam (1mg) orally, buprenorphine S.L., adhesive tape nitroglycerin, and oxygen 35%.

Premedication on the day of surgery consisted of hydroxizine (50mg) and lorazepam (1mg) both administered orally 120 minutes before arrival in the operating room. Morphine sulfate (0.1mg/Kg weight) and scopolamine (0.005mg/Kg) were administered intramuscularly, along with inspired 35 percent oxygen and cutaneous nitroglycerin, 60 minutes before induction of anesthesia. Anesthesia was induced with diazepam (0.5mg/Kg i.v.) and fentanyl (10µg/Kg, i.v.), and the patients were intubated after muscle relaxation with pancuronium (0.1mg/Kg i.v.), ventilated with a Servo ventilator Model 900, using a mixture of O_2 and air, maintaining the arterial oxygen pressure (PaO_2) between 100-150mmHg and the arterial carbon dioxide pressure ($PaCO_2$) between 35-40mmHg. During CPB, PaO_2 levels were increased to 150-200mmHg whereas $PaCO_2$ was kept at 35-40mmHg.

Anesthesia was maintained with diazepam, midazolan, fentanyl and pancuronium bromide in doses as required; after release of aortic clamp, if ventricular fibrillation occured xylocaine (1mg/Kg i.v.) was adminis-tered. No other drugs were administered during the study period.

Cardiopulmonary Bypass

The heart was approached by median sternotomy, with cannulation for bypass through the ascending aorta and both venae cavae via right atrium. Systemic body temperature was reduced to 28°C during cardiac arrest and bypass. Distal anatomoses were performed first, and heart protection was achieved by perfusion with cold cardioplegic solution (Plegisol) and topical hypothermia.

The Stockert cardiopulmonary bypass pump and a bubble oxygenator (Bentley BN-10) were used in all patients. In cases were the pulsatile flow was instituted, a rate of 70 pulses/minute was maintained during the period of aortic clamping, and after that, pulsatile flow was triggered by ECC. In both types of flow, circulatory support with the pump was used for 15 minutes after releasing the partial aortic cross clamp used to accomplish the proximal anastonosis. The system was primed with Ringer's lactate solution (1000mL). Arterial flow rates were set to 2.2-2.5 L/min/m2) and adjusted as necessary to maintain mean arterial pressure above 60mmHg during the bypass period.

The patients were heparinized (300 U/Kg) just before the beginning of CPB, and received a solution of protamine sulfate i.v., in the propor-tion of about 1/3 to the heparin administered for heparin neutralization. Protamine sulfate dose was repeated whenever necessary.

Blood Parameters

Arterial and venous blood samples were taken through the radial line and centrally placed catheters into coronary sinus and right heart. Blood samples in K3-EDTA (1.5mg/ml) were obtained to determine hemato-crit, hemoglobin, plasma viscosity, haptoglobin and hemopexin levels. Plasma fibrinogen, plasminogen, alpha 2-antiplasmin and antitrombin III concentrations were measured from blood samples collected in 3.8% sodium citrate buffer. Blood anticoagulated by addition of heparin (15 IU/ml) was utilized for determination of red cell aggregation and filterability indices, erythrocyte osmotic fragility, lactate and pyruvate concentra-tions, and pH.

Blood hemoglobin and hematocrit were measured by standard techni-ques. Arterial and venous blood lactate and pyruvate determinations were assessed in supernatants from deproteinized (perchloric acid) blood samples, according to the enzymatic methods of Gutmann and Wahlefeld (1963), and Czok and Lamprecht (1974) respectively. Erythrocyte filtera-bility measurements were carried out in a filterometer (Myrenne MF4) as described by Teitel and Mussler (1981), using 8% erythrocyte suspensions in phosphate buffered solution (PBS: 300 mosm/l; pH 7.4, with 0.25% human albumin). Polycarbonate filters (Nuclepore Corporation) of the same batch with pores of 5 µm of diameter were used. The increase in the filtration index signifies a loss of red cell deformability. Red cell aggregometer was quantified by a Myrenne aggregation, as described by Schmid-Schönbein et al. (1982). High aggregation readings indicate stronger rouleaux formation.

Plasma was separated from anticoagulated specimens after centifuga-
tion as soon as possible. Plasma viscosity was measured at 37°C in a
Coulter-Hakness capillary viscometer (Harkness, 1963), and according to
the recommendations of the International Committee for Standardization
in Haematology (1986). Fibrinogen, plasminogen, alpha-2-antiplasmin,
antitrombin III, hemopexin and haptoglobin were also determined in
plasma samples collected under the same technical conditions. Hemopexin
and haptoglobin were quantitated by radial immunodiffusion with M-Parti-
gen plates from Behring. Antithrombin III, plasminogen, alpha-2-anti-
plasmin were determined by Berichrom test, and fibrinogen was measured
in a coagulometer with a turbo-densitometric detection device (Behring
Fibritimer).

A colorimetric method was employed for determination of erythrocyte
osmotic fragility, the results being expressed as degree of hemolysis.
The values of NaCl concentration that cause 50% of red cell lysis are
refered as degree of hemolysis (Fievet et al., 1971).

Except for red cell osmotic fragility and filtration measurements,
which are measured from fixed volumes of packed red cells, the preinduc-
tion hematocrit level was used to make corrections for the hemodilution
effect on all other laboratory determinations made in each patient,
according to the formula:

$$\text{Corrected concentration of each blood parameter} = \frac{\text{Obtained level of each blood parameter x Preinduction hematocrit}}{\text{Hematocrit at time of sampling}}$$

Frequency and Timing of Assessment

Arterial and coronary sinus blood samples were drawn for pyruvate
and lactate measurements at the following times (Fig. 1): day before
surgery (time 1), immediately after ECC (T7) and 30 minutes later (T8).
Venous blood samples were collected for: hemoglobin, hematocrit and
plasma viscosity measurements at all periods (T1 to T8); red cell filte-

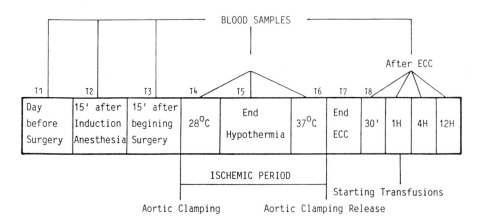

Fig. 1. Experimental protocol, showing the time sequence of blood collection and ischemia-reperfusion phases (ECC = extracorporeal circulation). Numbers T1 to T8 represent the sequence of pre and post
surgery events. Blood sample collection was extended from the
day before surgery (time 1) up to time 8, before starting blood
transfusions.

rability and aggregation evaluation at the day before surgery (T1), 15 minutes after induction of anesthesia (T2), 15 minutes after begining surgery (before ECC, T3), and immediately after ECC (T7); fibrinogen, plasminogen, alpha 2-antiplasmin and antithrombin III determinations at the day before (T1), 15 minutes after induction of anesthesia (T2), end of ECC (T7), and 30 minutes later (T8); osmotic fragility, hemoglobin and haptoglobin levels 15 minutes after begining surgery (T3) and after ECC (T7).

Statistics

All results are expressed as mean $\pm$ standard deviation (m$\pm$SD). The analysis was done using repeated-measures of variance model (MANOVA) after the descriptive analysis had been performed. Statistical performance was assumed when the probability value was 0.05 or less.

RESULTS

All of the patients survived the operation and presented no evidence of major complications. The hemodynamic response from the beginning of hypothermia through the post-cardiopulmonary bypass periods is shown in Table I. All patients were normotensive before operation.

However, the pattern of arterial pressure measurements (SAP, MAP, DAP) differed between the two main periods: during ECC and following ECC. Following the onset of cardiopulmonary bypass all blood pressure values decreased in both groups to about half of control range. As could be expected, the systolic blood pressure recovered from the lowest levels during ECC until 12h after operation. Patients of the pulsatile group had significantly (p<0.05) higher arterial pressure values after ECC than those from the other group. Meanwhile, no significant differences between both groups were reflected on systemic vascular resistance, that decreased progressively since the ending of ECC until the later stages of observations. During the period of bypass, cardiac output changed significantly (p<0.05) in the nonpulsatile group, having been stable in the pulsatile cases. After bypass the cardiac output rose rapidly and was indistinguishable between both groups. The pulmonary capillary wedge pressure and the central venous pressure after ECC were significantly higher (respectively p<0.05, and p<0.01) in the pulsatile group than in the nonpulsatile patients. Also the triple index was higher (p<0.01) in patients that have been submitted to pulsatile perfusion, than in the other group.

Due to perfusion with solutions in pre-ECC and machine priming at the beging of ECC, all laboratory data (except red cell osmotic fragility and erythrocyte filterability) recorded during and after ECC showed clear variations. Generally a decline was noticed after the ECC onset, caused by hemodilution. In order to evaluate the effect of hemodilution on plasma concentration of each parameter, a correction was necessary with reference to the packed cell volume (as described in Methods). The actual variation of all laboratory data before and during ECC is shown in Tables 2 and 3. The apparent decline induced by hemodilution, as exemplified by plasma viscosity time course (Fig. 2), is no longer observed. Note that after correction, plasma viscosity levels actually increased significantly (p<0.001) before and during ECC, in both groups of patients. Figure 3 shows the effect of hemodilution in hematocrit levels.

With one exception, there was no statistically significant difference between the pulsatile and non-pulsatile cases regarding hemorheological (Table 2) and further laboratory parameters (Table 3), throughout the period of study. Only alpha-2- antiplasmin was significantly different (p<0.01) between the two groups by the time ECC was ended.

Although no distinct changes in blood parameters studied have been

Table 1. Averaged (mean and SD) hemodynamic adaptation to pulsatile and non-pulsatile flow.

		During ECC			End ECC	After ECC			
		28°C	End-Hyp.	37°C		30'	1h	4h	12h
SAP (torr)	nonpulsatile	54.1 (12.3)[x]	65.6 (7.1)	53.3 (7.7)	86.8 (11.4)	83.8 (11.0)	84.5 (8.9)	104.2 (15.0)	108.0 (11.7)[x]
	pulsatile	61.1 (19.5)	85.4 (14.6)	64.7 (19.2)	99.9 (14.2)	95.1 (12.8)[x]	101.5 (15.3)[x]	115.0 (19.0)[x]	122.6 (14.5)[x]
DAP (torr)	nonpulsatile	45.6 (8.8)	55.1 (7.1)	44.1 (7.9)	52.6 (9.0)	54.9 (7.6)	56.9 (9.1)	61.7 (12.4)	55.6 (10.5)
	pulsatile	47.9 (17.0)	62.1 (11.1)	51.0 (14.5)	64.9 (15.2)	60.8 (16.2)[x]	67.5 (14.5)[x]	68.6 (9.4)[x]	59.6 (8.5)[x]
MAP (torr)	nonpulsatile	55.5 (16.6)	63.4 (8.5)	49.4 (7.4)	63.1 (8.9)	63.8 (8.9)	64.8 (4.9)	71.9 (12.2)	67.4 (7.9)
	pulsatile	54.0 (16.7)	71.6 (10.3)	59.6 (15.3)	71.3 (14.8)	70.3 (15.3)[x]	75.9 (13.3)[x]	81.5 (17.1)[x]	77.1 (7.1)[x]
CO (L/min)	nonpulsatile	3.23 (0.81)[x]	2.62 (0.31)[x]	4.07 (0.20)[x]	3.57 (1.49)[x]	3.26 (1.36)	4.01 (1.49)	5.18 (1.12)	7.96 (1.64)
	pulsatile	3.40 (0.63)	3.33 (0.63)	3.78 (0.94)	3.67 (1.39)	3.51 (1.62)	3.99 (2.37)	5.58 (2.31)	6.65 (2.16)
SVR (dynes.sec/cm-5)	nonpulsatile	1548 (924)	1963 (403)	976 (183)	1577 (664)	1680 (501)	1329 (500)	1080 (621)	654 (247)
	pulsatile	1307 (468)	1786 (477)	1456 (853)	1651 (703)	1662 (630)	1741 (746)	1191 (562)	900 (351)
CVP (torr)	nonpulsatile	—	—	—	2.50 (1.19)	2.12 (0.99)	4.62 (1.68)	10.25 (3.28)	8.57 (1.81)
	pulsatile	—	—	—	5.12 (4.12)	6.37 (4.47)[xx]	6.62 (3.37)[xx]	11.12 (3.27)[xx]	10.00 (3.05)[xx]
PCWP (torr)	nonpulsatile	—	—	—	4.37 (2.13)	4.37 (2.72)	5.87 (2.23)	10.12 (2.85)	9.85 (2.11)
	pulsatile	—	—	—	7.00 (3.51)	9.00 (3.25)[x]	9.37 (2.72)[x]	11.12 (3.27)[xx]	13.42 (4.86)[x]
SAPxHR x PCWP (torr./beats/min)	nonpulsatile	—	—	—	37,286 (2,294)	33,930 (1,760)	46,236 (1,870)	82,608 (1,483)	86,225 (1,555)
	pulsatile	—	—	—	61,069 (2,756)[xx]	73,829 (2,601)[xx]	80,904 (2,478)[xx]	87,476 (2,322)[xx]	95,204 (1,267)[xx]

Abbreviations – ECC: Extra-corporeal circulation; End-Hyp.: End of hypothermia; SAP: Systolic arterial pressure; DAP: Diastolic arterial pressure; MAP: Mean arterial pressure; CO: Cardiac output; SVR: Systemic vascular resistance; CVP: Central venous pressure; PCWP: Pulmonary capillary wedge pressure; HR: Heart rate.

Significance levels were determined from 2 (groups) x4 (examinations) repetead-measures analysis of variance.

[x] $p<0.05$ and [xx] $p<0.01$, pulsatile group versus nonpulsatile group main effect.

detected for each period between the group of patients with pulsatile flow and the other with steady perfusion, a clear time course variation was also evident for hematocrit and fibrinogen levels (p<0.005), pyruvate and lactate concentrations either in arterial or coronary sinus samples (p<0.010), red cell aggregation (p<0.05), alpha-2 antiplasmin (p<0.05) antithrombin III (p<0.01), and hemopexin (p<0.001) values.

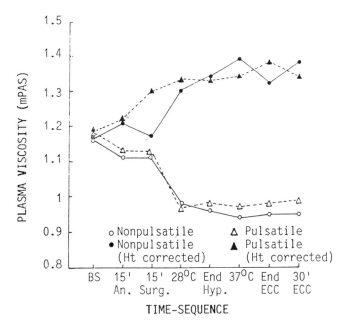

Fig. 2. Time course of mean plasma viscosity values as measured and compared with and without correction of the effect of hemodilution in two groups of patients, ascribed to pulsatile or non--pulsatile cardiopulmonar bypass.
(Legend - BS: before surgery; 15' An: 15 minutes after induction of anesthesia; 15'S: 15 minutes after onset of surgery; 28°C: hypothermia; E.Hyp: Ending of hypothermia; 37°C: rewarming period; E-ECC: Ending of extracorporeal circulation; 30'-ECC: 30 minutes after ending ECC).

Fibrinogen, as well as plasma viscosity, pyruvate and lactate levels rose significantly, whereas hematocrit values decreased from the start of the observation period. Plasma viscosity variation was significantly affected (p<0.01) by fibrinogen levels, also with no apparent effect on red cell aggregation index. Conversely, the further three plasma proteins that showed significant variations with time, showed consistent decreasing after the operation (Table 3).

DISCUSSION

The data obtained in the present randomized study demonstrated that hemodynamics and laboratory measurements clearly change during and after cardiopulmonary bypass (Tables 1 to 3). Nevertheless, the variations observed were not statistically different between pulsatile and non-pulsatile groups. The exception was an apparently early decrease in plasma alpha-2-antiplasmin showed by the non-pulsatile group at the end of total cardiopulmonar bypass.

The hemodynamic results here observed are generally supported by

Table 2. Time course of the main hemorheologic determinants before, during and immediately after extracorporeal circulation with pulsatile or non-pulsatile perfusion (see text for details).

	Before surgery		15' after onset surgery	During ECC			End ECC	30' after ECC
	Day before	15' after anhestesia		28°C	End-Hyp	37°C		
Hemoglobin (g/dL)								
nonpulsatile	12.8 (1.5)	13.7 (2.6)	12.5 (1.4)	12.9 (1.9)	12.6 (1.4)	12.5 (1.5)	12.9 (1.7)	12.6 (1.8)
pulsatile	12.9 (1.7)	12.8 (2.5)	13.6 (4.9)	12.7 (3.0)	12.6 (3.0)	12.8 (3.0)	13.1 (3.5)	13.0 (3.1)
Hematocrit (%)								
nonpulsatile	41.0 (5.9)	37.6 (7.2)	39.1 (6.4)	31.2 (5.1)	29.6 (6.4)	28.1 (5.6)	29.9 (6.2)	28.7 (6.5)
pulsatile	40.1 (7.6)	37.2 (4.9)	36.0 (8.3)	29.6 (5.2)	29.1 (3.1)	28.8 (3.7)	28.4 (2.8)	29.3 (3.5)
Plasma viscosity (mPas)								
nonpulsatile	1.16 (0.04)	1.21 (0.11)	1.17 (0.11)	1.30 (0.12)	1.34 (0.08)	1.39 (0.08)	1.32 (0.14)	1.38 (0.19)
pulsatile	1.18 (0.04)	1.22 (0.14)	1.30 (0.38)	1.33 (0.21)	1.33 (0.16)	1.34 (0.15)	1.38 (0.24)	1.34 (0.16)
Red cell aggregation								
nonpulsatile	19.5 (2.7)	18.9 (6.8)	19.3 (7.1)	—	—	—	14.7 (4.9)	—
pulsatile	20.0 (3.8)	19.6 (4.8)	22.4 (3.8)	—	—	—	17.9 (6.2)	—
Red cell filterability								
nonpulsatile	26 (16)	21 (6)	28 (22)	—	—	—	44 (36)	—
pulsatile	21 (10)	19 (5)	19 (6)	—	—	—	69 (93)	—
Fibrinogen (mg/dL)								
nonpulsatile	337.4 (33.7)	399.0 (75.4)	—	—	—	—	408.7 (32.8)	382.7 (42.1)
pulsatile	354.1 (97.1)	411.9 (95.9)	—	—	—	—	467.7 (204.6)	432.1 (213.1)

Table 3. Time course of laboratory data in pulsatile and non-pulsatile groups (for details see the text).

	Before ECC			End ECC	30' after ECC
	Day before	15' after anesthesia	15' after onset of surgery		
Venous blood					
Plasminogen (%)					
nonpulsatile	88.9 (16.9)	88.7 (20.9)	—	90.6 (19.0)	90.4 (28.9)
pulsatile	80.0 (31.5)	88.9 (34.2)	—	88.2 (41.7)	80.2 (37.1)
Alpha-2-antiplasmin (%)					
nonpulsatile	80.6 (7.6)	78.2 (18.5)	—	61.2 (22.6)[x]	60.7 (23.6)
pulsatile	85.6 (24.1)	82.8 (28.7)	—	82.8 (28.2)	66.1 (30.0)
Antithrombin III (%)					
nonpulsatile	96.0 (21.7)	90.7 (25.8)	—	74.0 (22.7)	73.2 (27.5)
pulsatile	94.8 (18.6)	89.0 (17.8)	—	83.9 (11.7)	73.4 (19.1)
Haptoglobin (mg/dL)					
nonpulsatile	—	—	300.2 (82.3)	301.4 (110.9)	—
pulsatile	—	—	219.8 (104.5)	225.3 (124.2)	—
Hemopexin (mg/dL)					
nonpulsatile	—	—	74.5 (14.1)	55.6 (17.2)	—
pulsatile	—	—	96.4 (30.2)	68.3 (23.7)	—
Osmotic fragility (50%)					
nonpulsatile	—	—	40.5 (17)	41.7 (23)	—
pulsatile	—	—	41.3 (21)	40.9 (22)	—
Arterial blood					
Pyruvate (μmoles/mL)					
nonpulsatile	0.07 (0.02)	—	—	0.35 (0.25)	0.34 (0.25)
pulsatile	0.07 (0.02)	—	—	0.25 (0.20)	0.21 (0.17)
Lactate (μmoles/mL)					
nonpulsatile	0.81 (0.24)	—	—	3.89 (2.06)	3.71 (1.37)
pulsatile	0.72 (0.22)	—	—	3.53 (1.87)	3.43 (1.37)
Lactate/pyruvate					
nonpulsatile	11.4 (2.9)	—	—	16.5 (10.2)	18.9 (16.6)
pulsatile	11.3 (3.6)	—	—	28.1 (27.4)	35.9 (44.9)
Coronary simes blood					
Pyruvate (μmoles/mL)					
nonpulsatile	0.05 (0.02)	—	—	0.23 (0.18)	0.30 (0.25)
pulsatile	0.04 (0.02)	—	—	0.18 (0.12)	0 (0.12)
Lactate (μmoles/mL)					
nonpulsatile	0.55 (0.21)	—	—	3.95 (2.63)	2.75 (0.64)
pulsatile	0.56 (0.20)	—	—	2.99 (1.79)	2.75 (1.68)
Lactate/pyruvate					
nonpulsatile	11.3 (4.6)	—	—	32.6 (40.4)	19.3 (18.9)
pulsatile	13.8 (5.5)	—	—	36.2 (47.2)	35.6 (42.5)

Significance levels were determined from 2 (groups) x 4 (examination) repetead-measures analysis of variance

[x] p< 0.05, pulsatile group versus nonpulsatile group main effect.

other studies (Frater et al., 1980) although recognizable differences in time sequence have been described (Taylor et al., 1979; Chiu et al., 1984). The most remarkable contrast was detected with systemic vascular resistance, that was reported to increase during non-pulsatile cardiopulmonar bypass (Taylor et al., 1979; Chiu et al., 1984), and also on pulsatile perfused patients (Landymore et al., 1979), although other studies refer a significant increase exclusively after the bypass in pulsatile and non-pulsatile group (Frater et al., 1980). Conversely our results have indicated that a consistent fall in SVR in sucessive post-bypass periods is associated with clear rise of cardiac output, with no significant differences between pulsatile and non-pulsatile groups (Table 1). In the two cases, the time course of SVR and CO observations might be an acceptable hemodynamic advantage in cardiac operation of patients with co-existing severe myocardial ischemia.

Besides differences in patient population, also variations in operation techniques and anesthetics, pump flow rate and perfusion pressure utilized, type and total volumes of pressure used may be potential determinants of the conflicting results published.

Higher values of central venous pressure and left arterial pressure, as detected in pulsatile treated patients after ECC, and particularly in the following 60 minutes (Table 1), are apparently not associated with increased fluid replacements required during bypass, regarding the similar hematocrit levels observed in pulsatile and non-pulsatile groups (Table 2). Alternatively, both hemodynamic findings could represent a significant advantage of pulsatile perfusion, since patency of large and end-arterioles (Dunn et al., 1974), and higher bypass flow rates (Trinkle et al., 1969) could be obtained. Meanwhile, myocardial oxygen consumption was also significantly increased in the pulsatile perfused group, as shown by the triple index (Table 1). The calculated increase in myocardial oxygen consumption may reflect a higher myocardial work under pulsatile flow (Katz et al., 1989).

The absence of a significant difference in the lactate/pyruvate

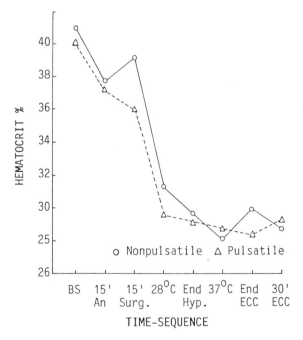

Fig. 3. Mean hematocrit variation during the observation period of two group of patients, under pulsatile or non-pulsatile cardiopulmonary bypass. (See explanation of abbreviation used in Fig. 2).

ratio between non-pulsatile and pulsatile patients (Table 3), suggests that whatever the perfusion type used in ECC, similar effects are induced in myocardial oxygenation. Additionally, we may propose that the increase in myocardial oxygen consumption detected in the pulsatile group is associated with a redistribution of coronary blood flow. This might be supported by the increased blood pressure levels that co-exist with pulsatile flow.

In the two groups of patients the comparable increases in blood lactate and pyruvate concentrations during ECC were statistically significant when compared with the mean values before CPB. The arterial concentrations of both metabolites were approximately at the levels determinated simultaneously in coronary sinus, and exhibited a peak at the end of ECC (Table 3). An increased peak lactate concentration could indicate an enhanced release due to improved blood flow, and increased lactate formation through the predominance of anaerobic metabolism (Karlsson, 1986).

As blood levels obtained from coronary sinus are usually much lower than the arterial concentration of lactate and pyruvate under physiological conditions, the high range observed in the two groups of patients may indicate transient myocardial underperfusion and hypoxia. This situation may be also partly attributed to the low hematocrit levels reached by our cases (Fig. 3), as secondary effects of intentional hemodilution. The onset of tissue oxygen supply dependence is a recognized consequence of critical reduction of systemic delivery of oxygen to tissues, as seen for instance during experimental hypovolemia or hypothermia (Schumacker et al., 1987), and hemodilution (Kiel et al., 1989).

The increased cardiac output and lower peripheral resistances after ECC, as described (Table 1), may be adaptive responses to maintain oxygen flux sufficiently to meet oxygen demands of tissues under hypoxia (Messmer et al., 1972). By lowering blood viscosity, the decrease in systemic hematocrit that follows hemodilution may also lead to higher cardiac output, as reflected in this study.

Whole blood viscosity is determined by the rheological behavior of all blood components, and is also one of the two major determinants of flow resistance (Chien and Lipowsky, 1982). However the hemodynamic repercussiosn of blood viscosity are affected by autoregulatory changes in vascular hindrance, as well as vessel wall co-existing pathology. In the absence of primary disease of the vascular system, compensatory vasodilation becomes a natural response to higher blood viscosity. Conversely, if the vascular hindrance is high, increased blood viscosity would lead to a reduced blood flow and lower oxygen supply to depending tissues. Comparing with the large vascular beds, flow conditions in the microcirculation of different organs are dependent not only on the vascular geometry and functional behavior of local angio-architectural arrangement but, with relevance, on the rheological properties of the blood cells, plasma composition and properties and blood-vascular endothelium interactions (Schmid-Schönbein, 1981, Chien, 1985). Although hematocrit is a major determinant of whole blood viscosity (Chien and Lipowky, 1982) in the microcirculation, increased plasma viscosity and reduction of red cell deformability in small capillaries, as well as increased red cell aggregation in post-capillary venules becomes increasingly relevant (Schmid-Schönbein, 1981; Gaehtgens 1987).

In the present study, plasma viscosity levels increased significantly, accompanied by a less evident elevation of fibrinogen concentration (Table 2). These results support the notion that increased levels of plasma fibrinogen are the principal cause of higher plasma viscosity in certain hematological and vascular diseases (Leblond, 1987) or acute phase reactions (Dowton and Colten, 1988). Pre and peri-operative trauma could explain the increase of fibrinogen levels in both groups of patients, probably initiated by a factor that increased fibrinogen synthesis.

In general, macromolecules causing increased plasma viscosity may also contribute to increased red cell aggregation. This phenomenon is a non-specific index of disease involving an acute phase elevation in plasma proteins (Chien and Sung, 1987).

This interdependence was absent in our patients, probably caused by more-complex changes and opposite effects. However, the variations observed reveal that sluggish blood could be a significant finding before surgery (Table 2).

The variations in the magnitude of red cell filterability in our patients could suggest an impairement of erythrocyte rigidity caused by the operation, thus becoming a significant rheologic factor potentially interfering with the delivery of oxygen to tissues (Chien and Lipowsky, 1982; Chien, 1985). Destruction or mechanical lesions of blood cell components in contact with the extra-corporeal circuit (Osborn et al., 1962), in addition to the apparent reduction of erythrocyte deformability in surgery with CPB (Ekeström et al., 1983; Kamada et al., 1987), may well be potential reasons of diminished peripheral perfusion and ischemia.

Accordingly, with the red cell osmotic fragility test, the erythrocytes did not become more susceptible to lysis after CPB (Table 3). Although other studies have indicated increased hemolysis during perfusion (Osborn et al., 1962; Landymore et al., 1979), we have also no reasons to accept that the diminution of red cell deformability, or further effects caused by CPB on red cell properties, were involved in higher cell destruction. The present study clearly shows that hemoglobin concentration (with volume correction) was maintained in stable values during the observation period. Also no variation was observed in haptoglobin levels; decreased levels of this alpha-2 plasma protein would occur whether intracelular or extravascular hemolysis exist. However, the clear decrease of hemopexin levels after ECC (Table 2) do not completely exclude that small fractions of hemoglobin might be released from erythrocytes into plasma upon cell damage (Bunn, 1972).

Besides thrombocytopenia and platelet dysfunction (Harker et al., 1980, Czer et al., 1987) other alterations of hemostasis may oppear after CBP surgery, such as occasional hemorahages (Rosky and Rodman, 1966; Kirklin and Kirklin, 1981; Czer et al., 1987), increased fibrinolysis (Kalter et al., 1979; Stibbe et al., 1984) and decreased activity of some coagulation factors (Kalter et al., 1979; Knobl et al., 1987). These dysfunctions were not observed in our cases, although the time--course of antithrombin III in both group and alpha-2-antiplasmin in the non-pulsatile group indicate a transient tendency to opposing coagulable states, as probably side-effects of post-surgery and/or heparin rebound (Buller and Ten Cate, 1989; Williams, 1989).

CONCLUSIONS

The present investigations in two groups of patients with similar clinical profile affected by severe coronary artery disease failed to demonstrate differences between pulsatile and non-pulsatile bypass, either in circulatory response or blood levels of metabolic, hemorheologic and coagulation factors. The exception was lowering of alpha-2-antiplasmin after ECC in the non-pulsatile group. Meanwhile, the increases on plasma viscosity and fibrinogen are expected responses to surgical trauma. The elevation of lactate/pyruvate ratio after surgery may indicate that tissue oxygen supply, and particularly heart oxygenation, could be insufficient to meet oxygen requirements. The effect of hemodilution in lowering oxygen delivery to tissues should be stressed. Hemodynamic adaptation, by decreasing systemic peripheral resistance, and increasing cardiac output and blood pressure, could provide a theoretical advantage in terms of tissue oxygenation. The role of blood rheological alterations, particularly the increase of plasma viscosity and the role of lower red cell filterability in blood slugging remains to be evaluated in future study.

ACKNOWLEDGEMENTS

This study was partly supported by a grant from INIC-MbL2. A preliminary presentation of part of this paper was submitted to the VIIth International Congress of Biorheology at Nancy (France), 18-23 June 1989.

REFERENCES

Buller, H. R., and Ten Cate, J. W., 1989, Acquired antithrombin III deficiency: laboratory diagnosis, incidence, clinical implications, and treatment with antithrombin III concentrate, Am. J. Med., 87 (suppl. 3B):44S

Bunn, H. F., 1972, Erythrocyte destruction and hemoglobin catabolism, Sem. Hematol., 9:3.

Chien, S., and Lipowsky, H. H., 1982, Correlation of hemodynamics in macrocirculation and microcirculation, Int. J. Microcirc.: Clin. Exp., 1:351.

Chien, S., 1985, Role of blood cells in microcirculatory regulation. Microvasc. Res., 29:129.

Chien, S., and Sung, L. A., 1987, Physicochemical basis and clinical implications of red cell aggregation, Clin. Hemorheology, 7:71.

Chiu, I-S., Chu, S-H., and Hung C-R., 1984, Pulsatile flow during routine cardiopulmonar bypass. J. Cardiovasc. Surg., 25:530.

Czer, L. S. C., Bateman, T. M., Gray, R. J., Raymond, M., Stewart, M.E., Lee, S., Goldfinger, D., and Chaux, A., 1987, Treatment of severe platelet dysfunction and hemorrhage after cardiopulmonar bypass: reduction in blood product usage with desmopressin, Cardiology, 9:1139.

Czok, R., and Lamprecht, W., 1974, Pyruvate, phosphoenolpyruvate and D-glycerate-2-phosphate, "Methods of Enzymatic Analysis", Acad. press, Inc., New York.

Doberneck, R. C., Reiser, M. P., and Lillehei, C. W., 1962, Acute renal failure after open-heart surgery utilizing extracorporeal circulation and total body perfusion. Analysis of one thousand patients, J. Thorac. Cardiovasc. Surg., 43:441.

Dowton, S. B., and Colten, H. R., 1988, Acute phase reactants in inflammation and infection, Sem. Hematol., 25:84.

Dunn, J., Kirsh, M. M., Harness, J., Carroll, M., Straker, J., and Sloan, H., 1974, Hemodynamic, metabolic, and hematologic effects of pulsatile cardiopulmonary bypass, J. Thorac., Cardiovasc. Surg., 68:138.

Ekeström, S., Koul, B. L., and Sonuenfeld, T., 1983, Decreased red cell deformability following open-heart surgery, Scand. J. Thorac. Cardiovasc. Surg., 17:41.

Fievet, C. Y., Gigandet, M. P., and Ansel, H. C., 1971, Hemolysis of erythrocytes by primary pharmacologic agents, Am. J. Hosp. Pharm., 28:961.

Frater, R. W. M., Wakayama, S., Oka, Y., Becker, R. M., Desai, P., Oyama, T., and Blaufox, M. D., 1980, Pulsatile cardiopulmonary bypass: failure to influence hemodynamis or hormones, Circulation, 62 (suppl. 1):1.

Gaehtgens P., 1987, Blood rheology and blood flow in the circulation current knowledge and concepts, Rev. Port. Hemorreologia, 1:5.

Gutmann, I., and Wahledeld, A. W., 1963, L-(+)-lactate. Determination with lactate dehydrogenase and NAD, "Methods of Enzymatic Analysis", Acad Press Inc, New York.

Harker, L. A., Malpass, T. W., Branson, H. E., Hessel, E. A., and Slichter, S. J., 1980, Mechanism of abnormal bleeding in patients undergoing cardiopulmonar bypass: acquired transient platelet dysfunction associated with selective alpha-granule release, Blood, 56:824.

Harkness, J., 1963, A new instrument for measurement of plasma viscosity, Lancet, 2:280.

International Committee for Standardization in Hematology (Expert Panel on Blood Rheology), 1986, Guidelines for measurement of blood viscosity and erythrocyte deformability, Clin. Hemorheology, 6:439.

Kamada, T., McMillan, D. E., Sternlieb, J. J., Bjork, V. O., and Otsuji, S., 1987, Erythrocyte crenation induced by free fatty-acids in patients undergoing extracorporeal circulation, Lancet, 2:818.

Kalter, R. D., Saul, C. M., Wetstein, L., Soriano, C., and Reiss, R. F., 1979, Cardiopulmonary bypass: associated hemostatic abnormalities. J. Thorac. Cardiovasc. Surg. 77:427.

Karlsson, J., 1986, Muscle exercise, energy metabolism and blood lactate, Adv. Cardiol., 35:35.

Katz, L. A., Swain, J. A., Portman, M. A., and Balaban, R. S., 1989, Relation between phosphate metabolites and oxygen consumption of heart in vivo, Am. J. Physiol., 256:H265.

Kiel, J. W., Riedel, G. L., and Shepherd, A. P., 1989, Effects of hemodilution on gastric and intestinal oxygenation, Am. J. Physiol., 256:H171.

Kirklin, J. K., and Kirklin, J. W., 1981, Management of the cardiovascular subsystem after cardiac surgery, Ann. Thorac. Surg., 32:311.

Knöbl, P. N., Zilla, P., Fasol, R., Müller, M. M., and Vukovich, T. C., 1987, The protein C system in patients undergoing cardiopulmonary bypass, J. Thorac. Cardiovasc. Surg., 94:600.

Kono, K., Philbin, D. M., Coggins, C. H., Slater, E. E., Triantafillou, A., Levine, F. H., and Buckley, M. J., 1983, Adrenocortical hormone levels during cardiopulmonary bypass with and without pulsatile flow. J. Thorac. Cardiovasc. Surg., 85:129.

Leblond, P. F., 1987, Hemorheology and blood diseases, "Clinical Hemorheology", Martinus Nijhoff Publishers, Dordrecht-Boston-Lancaster.

Messmer, K., Sunder-Plassmann, L., Klöverkorn, W. P., and Holper, K., 1972, Circulatory significance of hemodilution; rheological changes and limitations. Adv. Microcirc., 4:1.

Osborn, J. J., Cohn, K., Hait, M., Russi, M., Salel, A., Harkins, G., and Gerbode, F., 1962, Hemolysis during perfusion. J. Thorac. Cardiovasc. Surg., 43:459.

Rosky, L.P., and Rodman, T. R., 1966, Medical aspects of open-heart surgery, N. Engl. J. Med., 274:833.

Schmid-Schönbein, H., 1981, Interaction of vasomotion and blood rheology in haemodynamics, "Clinical Aspects of Blood Viscosity and Cell Deformability", Springer-Verlag, Berlin-Heidelberg-New York.

Schmid-Schönbein, H., Volger, E., Teitel, P., Kieswetter, H., Dauer, V., and Heilman, L., 1982, New hemorheological techniques for the routine laboratory, Clin. Hemorheology, 2:93.

Schumacker, P. T., Rowland, J., Saltz, S., Nelson, D. P., and Wood, L. D. H., 1987, Effects of hyperthermia and hypothermia on oxygen extraction by tissues during hypovolemia. J. Appl. Physiol., 63:1246.

Shephard, R. B., and Kirklin, J. W., 1969, Relation of pulsatile flow to oxygen consumption and other variables during cardiopulmonary bypass, J. Thorac. Cardiovasc. Surg., 58:694.

Stibbe, J., Kluft, C., Brommer, E. J. P., Gomes, M., deJong, D., and Nauta, J., 1984, Enhanced fibrinolytic activity during cardiopulmonary bypass in open-heart surgery in man is caused by extrinsic (tissue-type) plasminogen activador. <u>Eur. J. Clin. Invest.</u>, 14:375.

Taylor, K. M., Bain, W. H., Russell, M., Brannan, J. J., and Morton, I. J., 1979, Peripheral vascular resistance and angiotensine II levels during pulsatile and non-pulsatile cardiopulmonary bypass. <u>Thorax</u>, 34:594.

Teitel, P., and Mussler, K., 1981, "Microcirculation and Ischaemic Vascular Disease - Advances in Diagnosis and Therapy", (Proc. Congr. Munich), Abbott Laboratories.

Trinkle, J. K., Helton, N. E., Wood, R. E., and Bryant, L. R., 1969, Metabolic comparison of a new pulsatile pump and a roller pump for cardiopulmonary bypass. <u>J. Thorac. Cardiovasc. Surg.</u> 58:562.

Williams, E., 1989, Plasma alpha-2-antiplasmin activity: role in the evaluation and management of fibrinolytic states and other bluding disorders. <u>Arch. Int. Med.</u>, 149:1769.

A HIGH RESOLUTION PERSONAL COMPUTER CONTROLLED LASER DOPPLER VELOCIMETER

Robert M. Heethaar and Evert-Jan Nijhof

Dept. of Medical and Physiological Physics
University Hospital, Utrecht, The Netherlands

ABSTRACT

For modelling tissue and organ perfusion blood cell and flow characteristics, like viscosity, blood cell transport and interaction are important parameters. To study these parameters in-vitro a laser Doppler velocity- and concentration meter was developed using a standard personal computer. The high spatial resolution ($7*7*20$ μm) of the system allowed high precision measurements of velocity and concentration of blood cells in artificial vascular structures.

KEYWORDS

Laser Doppler velocimeter/ microscopic probing volume/ personal computer/ signal analysis/ blood flow velocity/ in vitro studies/ local blood viscosity/ local hematocrit/ platelet transport/ particle distribution.

INTRODUCTION

To understand principles of tissue and organ perfusion in normal and pathological situations two phase models, or models based on porous media theories are introduced (Huyghe et al, 1983, 1989). To describe blood flow (patterns) through such complex vascular structures the rheological characteristics of blood (e.g. viscosity) play an important role. Viscosity of the blood depends on hematocrit, deformability of the red cells and local shear rate. Therefore viscosity and hematocrit in a complex vascular structure may change over the vessel diameter as red cells tend to migrate to the axis of the flow (Aarts et al, 1988) and from site to site in the structure. Modelling tissue perfusion needs the input of these relevant experimental data.

A second important field in physiology and clinical research is the process of atherogenesis. Laboratory studies have pointed out that platelet transport to the vessel wall is an essential step in the formation of a thrombus and may play a role in atherogenesis. This platelet transport is mediated by hematocryt, size and deformability of the red blood cell (Aarts et al, 1984).

Biomechanical Transport Processes, Edited by F. Mosora *et al.*
Plenum Press, New York, 1990

Both topics mentioned above are under investigation in our hospital. To obtain experimental data about blood cell transport through vascular structures in-vitro models are developed. Because we are working with red blood cells and platelets no hydrodynamic scaling can be applied. This implies that we have to work with measuring equipment with high spatial resolution. The laser-Doppler system described in this paper provides us with such a system and enables us to measure local velocity and concentration of blood cells in a measuring volume of only 7*7*20 μm.

METHODS

In figure 1 a schematic representation of the laser Doppler velocimeter is shown. A 2mW He-Ne laser beam (wavelength λ = 633 nm) is focussed by the first lens on the diffraction grating G. This grating is specially designed and manufactured by the Dutch Space Laboratory. About 70% of the energy of the laser bundle is recovered in the two first order beams. The main bundle is blocked by a bundle stop. The two first order beams are focussed with two lenses in a probe volume in the fluid flow. The angle of intersection (θ) of the first order beams is about 34°. The fringe distance (d) in the probe volume is given by the formula d = λ/(2sin(θ/2)) and amounts to 2 μm. The dimensions of the probe volume depend on the positioning of the lenses and their focussing distances, but can be made as small as 6*6*20 μm in air. Lens L_3, designed to be used in optical disk drives, has a short focussing distance of 4 mm. In the fluid the probe volume is enlarged and shifted from the laser to the detector due to refraction at the transitions between the tube and fluid. Special care is taken so that the waists of the laser bundles are at their point of intersection, resulting in a probe volume as small as possible. If a cell traverses the probe volume the laser light is scattered. Part of the scattered light is collected by lens L_5 and focussed upon a photodiode. The electrical signal of the photodiode is fed into a computer by a 1 MHz A/D converter card inserted in one of the extension slots of a standard PC or AT compatible computer. This card is has been designed in our laboratory. Software control of the card allows adjustment of the sampling frequency up to 1 MHz. The digitized data is stored into the RAM memory of the computer via DMA access. In a standard PC with 640 kB of memory 448 kB can be reserved for the data to be digitized. With the highest sample frequency this leads to about 0,5 second of real time data collection.

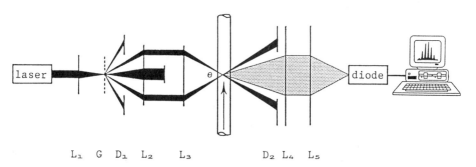

L_1 G D_1 L_2 L_3 D_2 L_4 L_5

Fig. 1. Schematic representation of the laser Doppler velocimeter.
L_1-L_5: lenses, G: diffraction grating, D_1-D_2: bundle stops.

Data analysis is performed with an extensive menu-driven software package. Single bursts can be analyzed in different ways:

- the fast Hartley transform (Bracewell, 1983), yielding the spectra of single Doppler bursts. From the main frequency the velocity of the particle is derived.
- zero-crossing detection, allowing a relatively fast method to calculate particle velocities.
- burst counting. Detection of proper Doppler bursts at a specific location in a certain time interval in combination with the measured velocities is a measure for the relative local particle or cell concentration. Integration of the local concentrations over the cross section of the vessel diameter yields the total concentration which can be measured by a particle (Coulter) counter.

With translation tables in x- and y-direction the vascular model can be positioned relative to the probe volume. Concentration and velocity profiles can be made at different cross sections of the tubular flow.

RESULTS

In figure 2 a typical example is shown of velocity measurements in a model of the basilar artery. Two vessels (internal diameters 3.4 mm) with developed parabolic profiles merge, leading to a parabolic profile in the common branche (internal diameter (R) 3.9 mm). The angle between the merging vessels is 51°. These results are used to test the mathematical models which are developed to describe the flow in the basilar artery and circle of Willis (Hillen, 1986).

In figure 3 a sample is shown of a distribution of micro spheres in perfusate, flowing through a glass tube of 1 mm inner diameter (R) under laminar flow conditions (Re ≈ 65). The open circles indicate the measured velocities. Each datapoint is the average of 5 measurements. The experimental values match the calculated parabolic flow profile (solid line). The mean velocity, equal to half the maximum velocity deviates less than 3% from volumetric velocity measurements. The closed circles indicate the concentration measurements of the spheres. The mean concentration ($3 * 10^5/\mu l$) is indicated by the dashed line.

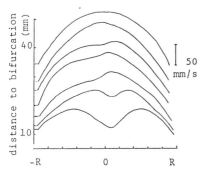

Fig. 2. Velocity profiles in the common branche of a model of the basilar artery.

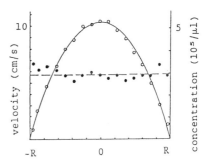

Fig. 3. Concentration and velocity profiles in a suspension with micro spheres.

DISCUSSION

The laser Doppler system described above is designed to measure blood cell and particle concentrations and velocities. Its high spatial resolution makes it suitable to perform studies in small artificial transparent vessels and in the proximity of the walls of the vascular models used. Data acquisition is performed by a special designed AD converter card that can be software controlled. The sampling frequency can be as high as 1 MHz in a standard personal computer running under MS-DOS. Velocities can be measured up to 200 mm/s, sufficient to study flow characteristics in vessels modelling small arteries and veins. With the highest sampling rate and a standard memory of 640 kB data acquisition can be performed over a period of 0.5 s. With additional memory longer sampling times are possible. If the highest velocities are lower than 200 mm/s the sampling rate can be lowered accordingly, allowing longer sampling times. If studies are performed with a pulsatile flow with frequencies up to the physiological range a sampling time of one second is sufficient to acquire the data of one simulated cardiac cycle.

Concentration measurements can be done with particle densities up to $3 * 10^5$ /μl, in the order of physiological values. To enable measurements with red blood cells these cells are transferred into light transparent ghost cells. Further study is required to measure the flow behaviour of these cells relative to the more flexible red blood cells.

The digitized data are analyzed with an extended menu driven software package. Velocities and concentration can be obtained relatively fast. A mathematical processor speeds up the fast Hartley transform. The zero crossing method is an even faster method to obtain velocity data. As the individual Doppler bursts are available for analysis it is in principle possible to try to detect the nature of the scattering particle from its scattering profile. Further research is needed to establish this fact.

In summary with the laser Doppler system described above a flexible tool is created to study off-line flow characteristics in vitro.

ACKNOWLEDGEMENT

The authors would like to thank mr. W.S.J. Uijttewaal, G. Streekstra, mrs L.M. Groeneveld and prof.dr. J.J. Sixma for their participation in this study and their valuable suggestions.

REFERENCES

Aarts, P.A.M.M., Heethaar, R.M. Sixma, J.J., 1984, Red blood cell deformability influences platelet-vessel wall interaction in flowing blood, Blood 64,6: 1228

Aarts, P.A.M.M., Van den Broek, S.A., Prins, G.W., Kuiken, G.D., Sixma, J.J., Heethaar, R.M., 1988, Blood platelets are concentrated near the wall and red blood cells, in the centre of flowing blood. Arteriosclerosis, 8(6): 819

Bracewell, R.N., 1983 Discrete Hartley transform, J. Opt. Soc. Am. 73: 1832

Goldsmith, H.L., Turitto, V.T., 1986, Rheological aspects of thrombosis and Hemostasis: Basic principles and applications. Thrombosis Hemostasis, 55:415

Hillen, B., Hoogstraten, H.W., Post, L., 1986, A mathematical model of the flow in the circle of Willis. J. Biomechanics, 19: 187

Huyghe, J.M., Grootenboer, H.J., Van Campen, D.H., Heethaar, R.M., 1983, Intramyocardial blood rheology and heart muscle mechanics. Proc. ASME Symp. on Biomech. 56: 241

Huyghe, J.M., Oomens, C.W., Van Campen, D.H., Heethaar, R.M., 1989, Low Reynolds number steady state flow through a branching network of rigid vessels: I. A mixture theory, Biorheology, 26: 55

INTRA OPERATIVE FLOWMETRY FOR VASCULAR SURGERY

Patrice Bergeron*, V. Pugliesi**, R. Rieu** and P. Garcia**

*Fondation SAINT-JOSEPH - 26, Bd de Louvain
13008 Marseille - FRANCE
** Institut Mécanique des Fluides - U.M 34 - 1, rue Honnorat
13003 Marseille - FRANCE

INTRODUCTION

Current methods of in vivo flowmetry lack reliability. Previously, we described a new computer-assisted ultrasound flowmeter (Bergeron et al, 1988). In vitro and in vivo test showed errors with this system to be less than 8 %. Design improvements have been made on the original system. Its size has been reduced and measurement time has been shortened to 30 seconds. This paper discusses the clinical applications of this second generation system in vascular surgery.

MATERIAL, METHODS AND RESULTS

The principles of measurements as well as the results of in vitro and in vivo tests were previously reported. The dimensions of the second generation devices now are only 50 x 30 x 30 cm^3. Measurement time has been reduced to 30 seconds. Further reduction could only be made at the expense of reliability since this is the minimum time needed to eliminate the respiration wave which is the main source of error (Bergeron et al, 1988).

Two models are under development. The simpler one which gives only mean flow rate is already available. A more sophisticated one, displaying spatio-temporal velocity profiles at eleven points across the vessel section using a multi-window pulsed Doppler velocimeter is to be ready soon.

CASE REPORTS

Case 1 : A 56-year-old man underwent an aortobifemoral bypass graft for severe claudication caused by atherosclerotic lesions involving the aorta and iliac arteries. Flowmetry was done in the body and each limb of the dacron graft. Flow distribution was assessed from distal run off. Flow was 155 ml/mn on the

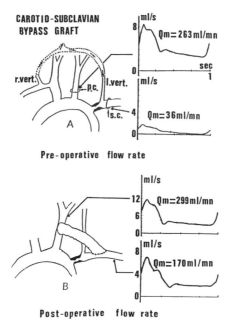

Figure 1 . Intraoperative flowmetry before and after an carotido-subclavian bypass. Redo. Flowmetry on the new bypass graft

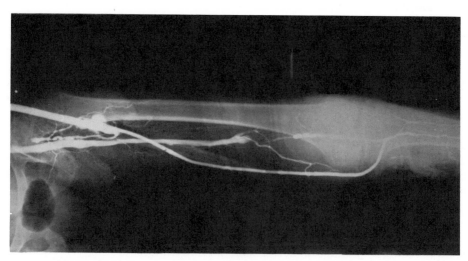

Figure 2 . Intraoperative   arteriography   of   an   in-situ   femoropopliteal saphenous   vein bypass.   Importance of fistulas with the deep venous return

right and less than 112 ml/mn on the left. Significant occlusion was found in the left superficial femoral artery.

Case 2 : a 50-year-old man presented a textbook case of steal syndrom due to a tight stenosis of the left sub-clavian artery proximal to the vertebral artery. Intra operative flowmetry was performed on the common carotid and subclavian arteries before and after placement of a bypass graft between the two vessels. Results (Fig. 1) showed that carotid flow distal to the graft was not affected by the increased subclavian flow and that the steal syndrome was corrected.

Case 3 : a 70-year-old woman was operated on for an occluded superficial artery causing rest pain. An in-situ saphenous vein graft was performed. Intra-operative arteriography (Fig. 2) disclosed distal opacification and several collaterals causing fistulas between the vein graft and deep venous return. Flowmetry was performed on the vein graft proximal to the collaterals. Flow proximal to the collaterals was 200 ml/mn and only 6 ml/mn distal to the collaterals. After division of all the collaterals, the distal flow rate increased to 50 ml/mn. Follow-up examination at two years demonstrated patency.

Case 4 : a 65-year-old man with rest pain underwent an in-situ femoropopliteal vein graft. Intra operative arterial X-ray was normal but successive flowmetry measurements showed a progressive drop in the flow. A defect in the vein graft was suspected and it was replaced by a Goretex graft (Fig. 3). Gross examination of the vein showed vein graft thrombosis due to a long tear caused by the valvulotome.

DISCUSSION

The main methods of intra operative investigations for vascular surgery are arteriography, duplex scanner, angioscopy and flowmetry.

Arteriography, the current gold standard, is routinely performed in our institution. *Standard X-ray* provide only morphologic and static data. By allowing detailed analysis correction of technical errors, arteriography can greatly enhance patency (Courbier et al., 1977).

Duplex scanner has has been proposed as an alternative to X-rays. Although it does provide better visualization of fine details, screening area is limited. Therefore, this technique appears useful only as an adjunct to X ray controls.

Angioscopy (White 1989 ; Mehigan and Schell, 1989) is a recent method that is becoming widespread. It seems to be very useful for intraluminal arterial reconstruction, checking anastomoses, devalvulation of vein grafts and analysis of artery walls. Further study is needed to fully evaluate its role and indications.

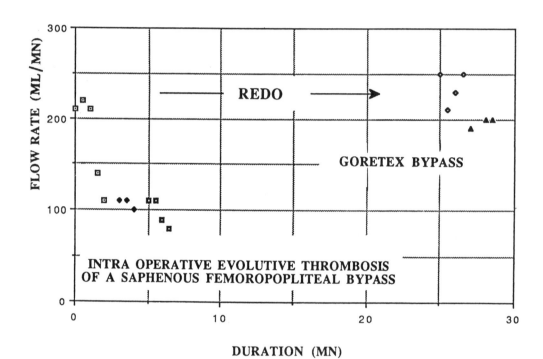

Figure 3. Intraoperative flowmetry showing the occuring thrombosis of the venous bypass

Flowmetry, once widely used, was all too frequently abandonned because of lack of reliability. The simple and accurate system described herein calls for a re-evaluation the usefulness of flowmetry as a means of obtaining much needed dynamic data for arterial reconstruction.

Up to now, we have not used flowmetry because of the long measurement time. However with the second generation device, we now practice this investigation regularly. This report describes a few of the applications for flowmetry and further experience in a large series should reveal further uses.

Indeed flowmetry has great potential of development. Moreover, with the development of a more sophisticated model reproducing a hemodynamic view of the arterial section, it will become to detect turbulent or low flow zones and compare this data with X ray findings.

Case 1 shows that flowmetry can be used to determine outflow from a bypass. Correlating this information with late patency could be valuable as a prognostic factor.

Case 2 involves surgical correction of a steal syndrom. Carotido-subclavian bypass is one of the surgical techniques reversing flow in the vertebral artery. Diversion of blood from the carotid artery to the subclavian artery does not affect the brain. Good clinical results are corroborated by flowmetry showing an increase in the flow in the common carotid artery.

Case 3 suggests that flowmetry could be useful in controversial situations involving diversion via collaterals of in-situ vein bypasses. If, by shunting blood to the deep venous return, collaterals deprive distal arteries, division is indicated. However if, by maintaining high flow in the graft, collaterals prevent the rapid graft thrombosis due to very low flow in the distal run off, division should be avoided.

Case 4 shows the limitations of intra operative angiography in some patients. Some defects compromising graft patency cannot be visualized on X-rays. Successive hemodynamic measurements showing a progressive drop in flow rate allows early diagnosis of thrombosis. Thus though flowmetry cannot supplant arteriography for intraoperative investigation, it could be an important adjunct and a good indication for evaluating distal bypass.

CONCLUSION

In vivo ultrasound flowmetry can be a reliable technique with a mean error less than 10 %. Ultimate conclusions as to the usefulness of this technique must await widespread clinical application in large series of patients.

In vascular surgery the intraoperative applications now seem clear. Flowmetry can be used effectively in many areas mainly to determine diversion of blood via collaterals and to assess flow for predicting patency.

REFERENCES

Bergeron, P., Jausseran, J. M., Reggi, M., Courbier, R., Pugliesi, V., Rieu, R., Garcia, P., Pelissier, R., 1988, Fiabilité et perspectives de la débimétrie per opératoire. Une technique ultrasonore informatisée originale, J. Mal. Vasc., 13:20-26

Courbier, R., Jausseran, J. M., Reggi, M., 1977, Detecting complications of direct arterial surgery - the role of intraoperative arteriography, Arch. Surg., 112:1115-1118

White, G. H., 1989, Angioscopy to monitor arteria thromboembolectomy, in "Endovascular Surgery", W. S. Moore and S. S. Ahn, ed., W.B Saunders Company

Mehigan, J. T., Schell, W. W., 1989, Angioscopic control of in situ saphenous vein arterial bypass, in "Endovascular Surgery", W. S. Moore and S. S. Ahn, ed., W.B Saunders Company

IN VITRO ANALYSIS OF A MODEL OF INTRACARDIAC JET:

A LASER DOPPLER ANEMOMETRIC STUDY

B. Diebold, A. Delouche, E. Abergel, and P. Péronneau

Unité INSERM 256
Hopital Broussais
Paris, France

ABSTRACT

The intracardiac jets are mostly turbulent jets and their precise description is needed for supporting the development of imaging techniques. So far, most of the reported data have been obtained in steady flow conditions. The present study was designed to verify the currently accepted description, in a model of intracardiac jet, using laser Doppler anemometry both in steady and in pulsatile flow conditions. Our results qualitatively agree with previous results but give another constant factor and suggest that, in pulsatile conditions, other parameters should be taken into account.

KEY WORDS : free turbulent jet, laser Doppler anemometry

INTRODUCTION

Intracardiac flows and especially intracardiac jets can be studied using a variety of new imaging technology: Doppler ultrasound, digital angiography, magnetic resonance imaging. These tools have specific capabilities, limitations and potentialities which can only be precisely evaluated by means of a plain knowledge of the investigated phenomena: a precise description of blood velocities within turbulent jets would be of interest both for the interpretation of currently available flow images and for the orientation of future advances in cardiac imaging.

Most of the valvular lesions correspond to the presence of an abnormal orifice connecting two chambers submitted to different pressures. This orifice is abnormally narrowed in the case of valvular stenosis and abnormally present in the case of valvular regurgitation or intracardiac shunt. These conditions generate a pulsatile turbulent jet which is more or less confined within the receiving chamber.

In vivo measurements of the initial velocity within the jets have been obtained using continuous wave Doppler ultrasound. They have allowed the description of a relationship between the jet velocity and the pressure difference which is extensively used for the clinical evaluation of the pressure drop at the level of aortic (Hegrenaes, 1985), mitral (Hatle, 1978), pulmonic (Oliveira Lima, 1983) valvular stenoses, of prosthetic valves (Wilkins, 1986) and of the pulmonary artery pressure (Yock, 1984). More recently several studies conducted on patients using color flow mapping have been reported. They attempted to establish a relationship between the spatial extension of the jet and the severity of aortic (Perry, 1987), mitral (Miyatake, 1986, Helmecke, 1987) or tricuspid (Suzuki, 1986) regurgitation. Unfortunately, they are affected by the limitations of color flow imaging (Péronneau, 1986) : velocity ambiguity, signal-to-noise ratio, artifacts due to wall motion and mean velocity estimators.

On the other hand, Davies (Davies, 1972) and Moore (Moore, 1977) have reported the following description of a turbulent free jet in steady flow conditions. When the origin of a turbulent jet is laminar, it generates a central laminar core where the velocity is uniform. This core is roughly conical. It is surrounded and progressively invaded by a diverging cone made of structured vortices. When studied at increasing distances from the orifice, these vortices exhibit dimensions growing until the tip of the core. More downstream, they are destroyed into turbulence. In addition, Davies proposed the following numerical relationships:

(1)
$$L = 6.4 \cdot D,$$

where "L" is the length of the laminar core and "D" the diameter of the orifice. This formula suggests that, when the flow at the orifice is laminar, the length of the core is independent of the initial velocity.

(2)
$$V = Vo \cdot L / X$$

where "V" is the mean velocity measured on the axis of the jet downstream of the tip of the core and within turbulence, "X" being the distance apart from the plane of the orifice and "Vo" the velocity within the orifice. The combination of (1) and (2) gives the following formulas:

(3)
$$V(x) / Vo = 6.4 \, D / X$$

(4)
$$D = [X \, V] / [6.4 \, Vo]$$

(5)
$$Q = \prod D^2 \, Vo / 4$$

388

(6) $$Q = [\Pi X^2 V^2] / [4\ 6.4\ Vo]$$

The present study was designed to verify these relationships on a model intracardiac jet, using laser Doppler anemometry.

MATERIAL AND METHOD

Hydraulic Model

The hydraulic model has already been described (Diebold, 1987, Diebold, 1989). The pulsatile flow was obtained using an oscillating pump activated by a tiltable disc, whose angulation determined the ejected volume. The pulse frequency was adjusted between 50 and 100 cycles per minute by using the rotation speed of the disc. This pump was connected to the discharge tank through an inflow ball valve and to the study circuit through an outflow ball valve. The ejected fluid flowed through a divergent-convergent structure in order to dissipate the upstream disturbances, before entering the selected nozzle. A series of circular orifices was investigated (diameters: 3.5, 4.4, 5.8, 7.1, 8.2, 9.0, 9.8, 11.3 mm). This design generated a flat velocity profile at the origin of the jet. The studied nozzle was connected to the tube within which the measurements were performed. The use of a laser Doppler anemometer allowed to overcome the limitations due to the ultrasonic technique that we previously used (Diebold, 1989) and, thus, to use of a 50 mm diameter tube. The initial peak velocity was adjusted in the region of 4 m.s^{-1}. This tube was connected to adjustable downstream resistance and compliance which allowed, together with upstream resistance and compliance, the adjustment of the velocity wave form. The fluid was a mixture of water, glycerol, corn starch and salt. A 30% concentration of glycerol provided a 0.04 P viscosity at room temperature, a <1 % corn starch introduced moving targets for the ultrasound measurements and the salt allowed the use of an electromagnetic velocimeter for continuously monitoring the velocity waveform.

Velocity Measurements

The velocity measurements were performed using an laser Doppler ane-mometer (DANTEC, He-Ne 15 mW). The optical system was mounted on a mechanical device allowing to displace the sample volume within the equa-torial plane and to electronically monitor its position with respect to the center of the orifice. The two beams were oriented along an equatorial plane, in order to measure the longitudinal velocity on the axis of the jet every 2 mm. The velocity curves were digitized (730 values / cycle) and stored using a microcomputer (PDP 11/02) together with their position in the plane of investigation. In the case of pulsatile flow, a synchronization signal delivered by the pump was used, the cycle was divided in 730 consecutive and adjacent periods and the instanta-neous values meaned over 10 cycles. This set of data allowed an off-line reconstruction of a series of 730 quantitated longitudinal velocity profiles.

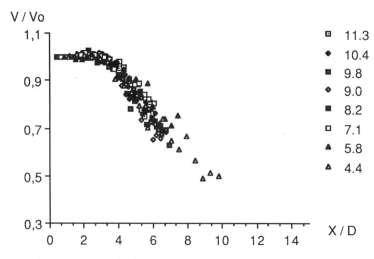

Figure 1. Steady flow : Normalized velocity profiles.

RESULTS

Steady Flow

Eight orifices were studied in steady flow conditions with an initial velocity in the region of 4 m.s[-1]. The raw data are presented on figure 1 with, on the abscissa, the distance from the orifice (X) divided by the diameter of the orifice (D) and, on the ordinate, the velocity (V) divided by the initial velocity (Vo). The points are clearly distributed along a single curve with a plateau corresponding to the central core within which the velocity is stable and with a progressive velocity decay.

The length of the plateau (L) was compared to the diameter of the orifice and the results are showed on figure 2. The relationship between L and D is very closed (r = .99) and linear. The ordinate at the origin does not differ significantly differ from 0, but the slope is of about 3.8.

The relationship between V/Vo and D/X is presented on figure 3. It appears closed (r= .89), linear and its ordinate at the origin does not differ significantly differ from 0. Its slope is, again, in the region of 3.8.

Thus, the results qualitatively agree both with formula (1) and (3) but the value of the constant clearly differs from the 6.4 value reported by Davies.

Pulsatile Flow

Nine orifices were studied in pulsatile flow conditions with a peak initial velocity in the region of 4 m.s[-1]. The raw data, measured at peak velocity, are presented on figure 4 with, on the abscissa, the distance from the orifice (X) divided by the diameter of the orifice (D) and, on the ordinate, the velocity (V) divided by the initial velocity (Vo). Like in steady flow conditions, the points are

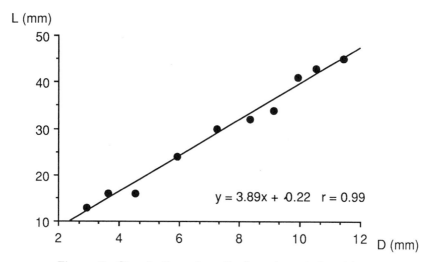

Figure 2. Steady flow : Length diameter relationship.

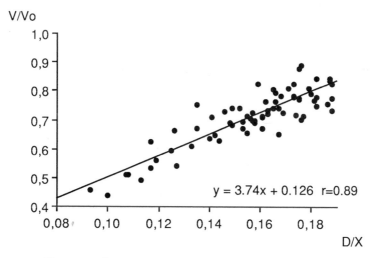

Figure 3. Steady flow : Normalized velocity decay.

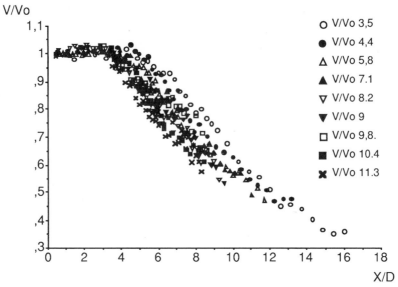

Figure 4. Pulsatile flow : Normalized velocity profiles.

distributed along a single curve with a plateau corresponding to the central core within which the velocity is stable and with a progressive velocity decay. Nevertheless, a larger degree of scatter is noticed in the region of the velocity decay with the smaller orifices in the upper and right part of the distribution and the larger orifices in the lower and left part.

The length of the plateau (L) was compared to the diameter of the orifice and the results are showed on figure 5. The relationship between L and D is very closed (r = .99) and linear. The slope is in the region 3.8, but the ordinate at the origin significantly differs from 0.

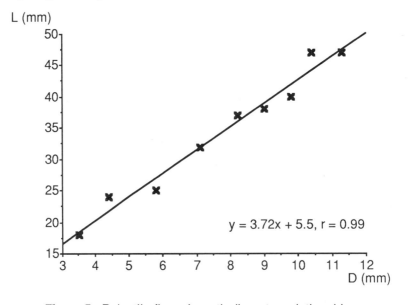

$$y = 3.72x + 5.5, r = 0.99$$

Figure 5. Pulsatile flow : Length diameter relationship.

The relationship between V/Vo and D/X is presented on figure 6. It appears weaker than in steady flow conditions but still roughly linear. The ordinate at the origin is significant and the slope clearly below 3.8.

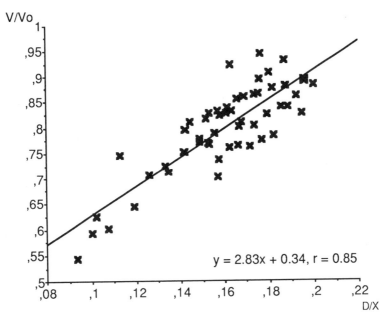

Figure 6. Pulsatile flow : Normalized velocity decay

Thus, the qualitative description of turbulent jet still apply to pulsatile jets, at peak velocity, but the formula clearly miss some phenomena which would need additional studies.

Clinical Implications

In spite of the above mentioned limitations, we have tested the value of formula (4) and (6) for predicting the diameter of the orifice and the actual volumic flow at peak velocity. The comparison between the actual volumic flow at peak velocity ("Q") and the corresponding predicted flow ("Qcalc") on figure 7.

The two relationships appear closed and suggest that the measurement of the initial velocity and of the velocity at a known distance from the orifice could be used for assessing some valvular regurgitations.

DISCUSSION

Previously Reported Data

So far, only a few number of studies has been devoted to in vitro models of intracardiac jets. Wong (Wong, et al. 1987) compared the spatial extension of

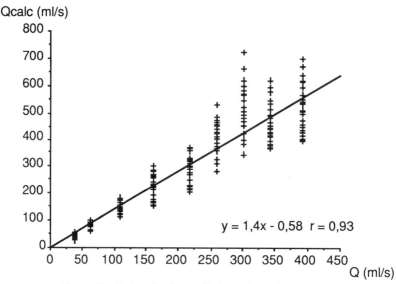

Figure 7. Pulsatile flow : Calculation of volumic flow.

the jet image to the actual flow rate using an echoDoppler flow imager providing a maximal measurable velocity of .92 m.s[-1] while initial velocities were up to 3 m.s[-1]. Using the same apparatus, Switzer (Switzer et al., 1987) investigated various parameters determining this spatial extension and described the predominant influence of the velocity at the origin of the jet and of the pressure difference driving the jet. Due to the limitations of their ultrasound velocimeters, these two groups were unable to provide a quantitative description of velocity fields within the jet: in most of the cases, the initial velocity far exceeded the maximal measurable velocity. The same comments apply to the data reported by Teague (Teague et al., 1984).

Indeed, precise measurements of velocities fields within physiologic or pathologic intracardiac jets require the use of a methodology providing both a good spatial resolution and a capability for measuring high velocities. Some studies have been conducted using hot film or laser Doppler anemometry.

Clark (Clark, 1976) reported measurements obtained in a model of aortic stenosis using hot film anemometry. They were mainly concentrated on the analysis of turbulences. He clearly demonstrated the laminarity of the flow at the origin of the jet, in spite of peak Reynolds numbers up to 18 000. These results apply to other cardiac lesions since the same values (in terms of Reynolds number) are found at the origin of the jet of aortic or mitral regurgitation. Some studies have been performed under conditions simulating peripheral vascular disease (Casanova, 1978, Ahmed, 1983), they have been conducted with relatively low Reynolds numbers (<2500) and, thus, do not apply for intracardiac jets.

Two studies performed in steady flow conditions, with high Reynolds numbers have been reported (Lu, 1983, Yoganathan, 1986). Lu and coworkers have studied velocity fluctuations and pressure spectra to analyze the mechanism of vascular murmurs in the presence of aortic stenosis. They measured mean velocity profiles at nine positions immediately downstream from an orifice (diameter: 0.83 cm) included in a cylindrical testing chamber (diameter: 2.5 cm). They found, in this configuration of confined turbulent jet (Reynolds number: 50 000), a laminar core having a length equal to about 5 times the orifice diameter. Yoganathan and coworkers investigated a model of pulmonic stenosis in steady flow conditions. This particular geometrical situation induced the development of a jet impinging against the inner wall of the left branch of the pulmonary artery.

Recently, Cape (Cape et al., 1989) have studied a configuration similar to our model using pulsed Doppler ultrasound velocimetry. Velocity values were visually extracted from spectrum tracings. They have hypothesized that the formula (6) is verified and have tested its value for predicting the actual flow. For that purpose they have studied only two orifices.

Contribution of Our Data

In steady flow conditions, our data demonstrate, like in previously reported studies, the existence of a core within which the velocity is stable and equal to the initial velocity. The length of this core is found equal to about 3.8 times the orifice diameter. This value is clearly inferior to the theoretical one but it appears consistent since it is found both when studying the length of the plateau and when analyzing the velocity decay. This particular point would need additional studies but it is not surprising since in the data reported by Moore, the constant was also below 6.4. The hyperbolic description of the decay fits with our data but the scatter is relatively large. This could be accounted for by the lower accuracy of the velocity measurements, in the presence of turbulence.

In pulsatile conditions, the presence of a plateau at peak velocity is clearly demonstrated. Its length is still related to the orifice diameter, but the ordinate at the origin cannot be neglected and suggest that other parameters should be included. Similarly, the hyperbolic description of the velocity decay appears too simplistic.

In spite of these limitations, the potential clinical implications are promising since they could be used on patients with presently on going developments of the transoesophageal echocardiography.

Validity of the Results

The jet was deemed to be free since the ratio of the orifice diameter to the tube diameter is below 0.23. Moreover, the theoretical velocity calculated using the formula proposed by Davies ($V(X,Y) = V(0,0) . 6.4 . D . 10^{-40(Y/X)^2} /X$, where X is the distance from the orifice, on the axis of the jet, and Y the distance from the axis of the jet) in the region of the tip of the laminar core and at 50 mm apart from the axis, corresponding to the position of the tube wall, is below 10^{-5} m.s^{-1}. This suggests that this jet can be considered as insensitive to the presence of the tube. As concerns the validity of the hydraulic model, the values simulate aortic and, to some extend, mitral or tricuspid regurgitation. Aortic regurgitant jets have an initial peak velocity (Masuyama, 1986) of about 4 ms^{-1} and appear in the left ventricular outflow tract whose diameter ranges from 10 to 30 mm, depending on the size of the patient and on the dilatation of the aortic annulus. The peak velocity of mitral regurgitation is of about 5 ms^{-1} (Hatle, 1985). It appears in the left atrium whose diameter often exceeds 30 mm in adult patients with mitral valve lesion. Tricuspid insufficiencies have an initial peak velocity ranging roughly from 2 to 5 ms^{-1}, depending on the systolic pulmonary artery pressure (Yock, 1984) and appear within the right atrium whose diameter often exceeds 30 mm in adult patients with severe tricuspid regurgitation. The volume of these regurgitations is ranged from 0 and 8 l.mn^{-1} or more, depending on the size of the patient and on the severity of the lesion. Consequently, the investigated values correspond to mitral or tricuspid regurgitation. The form of the velocity wave is similar to this of a systolic jet (aortic or pulmonic stenosis, mitral and tricuspid regurgitation). Diastolic jets (aortic or pulmonic regurgitation, mitral and tricuspid stenosis) are slightly different since the acceleration and the deceleration are more rapid and the velocity is more stable during middiastole. These differences could lead to a different behavior of velocity fields during early- and late diastole but would not be able to modify the hydraulic events during middiastole since our results demonstrate that in the presence of moderate acceleration or deceleration, the central core of pulsatile jets is similar to this found in steady flow conditions. In fact, the Strouhal number calculated at the orifice is very low.

All the orifices included in this study were circular whereas pathologic orifices have often different shapes, some of them being very irregular. Similarly, it would be necessary to study the influence of the orientation of the jet. Indeed, the present configuration produced central jets whereas, in patients, the jets can impact against a wall, in particular, in mitral regurgitation at the bottom of the left atrium, or adhere to a wall, as it is the case with some aortic or mitral regurgitant jets. These various configurations can significantly modify (Deshpande, 1982, Solzbach, 1987) the hydraulic events within the central core.

Finally, our experiments were all achieved within rigid tubes while heart chambers are compliant. The influence of the distensibility of the chambers can probably be neglected in all the configurations where the velocity is below the threshold value. On the other hand, when the jet is no longer free, the damping effect of elastic walls should be taken into account. Thus, it would have to be studied, despite the technical problems that have to be solved for conducting such an experiment.

REFERENCES

Ahmed S.A., Giddens D.P. 1983 Velocity measurements in steady flow through axisymmetric stenoses at moderate Reynolds numbers. J. Biomechanics 16, 505-516.

Ask P., Loyd D., Wranne B., 1986, Regurgitant flow through heart valves: a hydraulic model applicable to ultrasound Doppler measurements. Med. & Biol. Eng. & Comput. 24, 643-646.

Cape E.G., Skoufis E.G., Weyman A.E., Yoganathan A.P., Levine R.A. ,1989, A new method for noninvasive quantification of valvular regurgitation based on conservation of momentum, in vitro validation. Circulation 79, 1343-53.

Casanova R.A., Giddens D.P. , 1978, Disorder distal to modeled stenoses in steady and pulsatile flow. J. Biomechanics 11, 441-453.

Clark C., 1976, : Turbulent velocity measurements in a model of aortic stenosis. J. Biomechanics 9, 677-687.

Davies J.T., 1972, Turbulent free jet in Turbulence phenomena, pp.69-73, Academic press, New York.

Deshpande M.D., Vaishnav R.N. , 1982, Submerged laminar jet impingement on a plane. J. Fluid Mech. 114, 213-236.

Diebold B., Touati R., Delouche A., Guglielmi J-P., Forman J., Guermonprez J-L., Péronneau P. , 1987, Doppler imaging of regurgitant jet in aortic insufficiency: experimental validation and preliminary clinical evaluation. Eur. Heart J. 8 (Supp C), 45-52.

Hatle L., Brubakk A., Tromsdal A., Angelsen B. , 1978, Non-invasive asses sment of pressure drop in mitral stenosis by Doppler ultrasound. Br. Heart J. 40, 131-140.

Hatle L., Angelsen B., 1985, Mitral regurgitation in Doppler Ultrasound in Cardiology, pp. 176-188, Lea & Febiger ed., Philadelphia.

Hegrenaes L., Hatle L., 1985, Aortic stenosis in adults, non- invasive estima tion of pressure differences by continuous wave Doppler echocardio graphy. Br. Heart J. 54, 396-404.

Helmcke F., Nanda N.C., Hsiung M.C., Soto B., Adey C.K., Goyal R.G., Gatewood R.P., 1987, Color Doppler assessment of mitral regurgitation with orthogonal planes. Circulation 75, 175-183.

Lu P.C., Hui C.N., Hwang N.H.C., 1983, A model investigation of the velocity and pressure spectra in vascular murmurs. J. Biomechanics 16, 923-931.

Masuyama T., Kodama K., Kitabatake A. et al.,1986, Non-invasive evaluation of aortic regurgitation by continuous wave Doppler echocardiography. Circulation 73, 460-466.

Miyatake K., Izumi S., Okamoto M., Kinoshita N., Asonuma H., Nakagawa H., Yamamoto K., Takamiya M., Sakakibura H., Nimura Y. , 1986, Semiquantitative grading of severity of mitral regurgitation by real-time two-dimensional Doppler flow imaging technique. J. Am. Coll. Cardiol. 7, 82-88.

Moore C.J. , 1977, The role of shear-layer instability waves in jet exhaust noise. J. Fluid Mech. 80, 321-367.

Nowicki A., Karlowicz P., Piechocki M., Secombi W. , 1985, Method for the measurement of the maximum Doppler frequency. Ultrasound in Med & Biol 11, 479-486.

Oliveira Lima C., Sahn D.J., Valdes-Cruz L.M., Goldberg S.J., Vargas Barron J., Allen H.D., Grenadier E. , 1983, Noninvasive prediction of transvalvular pressure gradient in patients with pulmonary stenosis by quantitative two-dimensional echocardiographic Doppler studies. Circulation 67, 866-871.

Péronneau P., Diebold B., Guglielmi J-P., Lanusel O., Bele R., Souquet J., 1986, Structure and performances of momo- and bidimensional pulsed Doppler systems in Color Doppler flow imaging, pp 3-18, J.Roelandt ed., Martinus Nijhoff Publishers, Dordrecht.

Perry G.J., Helmcke F., Nanda N.C., Byard C., Soto B. , 1987, Evaluation of aortic insufficiency by Doppler color flow mapping. J. Am. Coll. Cardiol. 9, 952-959.

Solzbach U., Wollschla~ger H., Zeiher A., Just H.,1987, Effect of stenosis geometry on flow behavior across stenotic models. Med. & Biol. Eng. & Comput. 25, 543-550

Stein P.D., Walburn F.J., Blick E.F., 1980, Damping effect of distensible tubes on turbulent flow : implications in the cardiovascular system. Biorheology 17, 275-281.

Suzuki Y., Kambara H., Kadota K., Tamaki S., Yamazato A., Nohara R., Osakada G., Kawai C. , 1986, Detection and evaluation of tricuspid regurgitation using a real-time, two-dimensional, color-coded, Doppler imaging system: comparison with contrast two-dimensional echocardiography and right ventriculography. Am J. Cardiol. 57, 811-815.

Switzer D.F., Yoganathan A.P., Nanda N.C., Woo Y-R., Rigway A.J., 1987,

Calibration of color Doppler flow mapping during extreme hemodynamic conditions in vitro: a foundation for a reliable quantitative grading system for aortic incompetence. Circulation 75, 837-846.

Teague S.M., von Ramm O.T., Kisslo J.A., 1984, Pulsed Doppler spectral analysis of bounded fluid jets. Ultrasound in Med. & Biol. 16, 435-441.

Wilkins G.T., Gillam L.D., Kritzer G.L., Levine R.A., Palacios I.F., Weyman A.E., 1986, Validation of continuous wave Doppler echocardiographic measu rements of mitral and tricuspid prosthetic valve gradients: a simultaneous Doppler-catheter study. Circulation 74, 786-795.

Wong M., Ramirez M.L., Alejos R., 1987, Quantitative flow from two-dimen sional Doppler color imaging of jets: feasibility and limitations from an in vitro study. J Cardiovasc. Ultrason. 6, 3-8.

Yock P., Popp R.L., 1984, Non-invasive estimation of right ventricular systolic pressure by Doppler ultrasound in patients with tricuspid regurgitation. Circulation 70, 657-662.

Yoganathan A.P., Ball J., Woo Y-R., Philpot E.F., Sung H-W., Franch R.H., Sahn D.J., 1986, Steady flow measurements in a pulmonary artery model with varying degrees of pulmonic stenosis. J. Biomechanics 19, 129-146.

NATO ARW and European Mechanics Colloquium : "Bio-mechanical Transport Processes"

October 9-13, 1989

IESC Cargèse, Corsica

1. K.AFFELD, Berlin, B.D.R. 2. N. AKKAS, Ankara, Turkey 3. Ch. BAQUEY, *Scientific and Organizing Committee,* Bordeaux , France 4. D. BARTHES-BIESEL, Compiègne, France 5. P.BERGERON, Marseille, France 6. A.F.BERTELSEN, Oslo, Norway 7. C.G.CARO, *Scientific and Organizing Committee,* London, U.K. 8. F.CASSOT , Marseille, France 9. P.CORIERI, Rhode-St-Genese, Belgique 10. A.M.DAMAS, Porto, Portugal 11. A.DROCHON, Compiègne, France 12. J.H.GERRARD, Manchester, U.K. 13. M.GRAHAM, London, U.K. 14. R.GREBE, Aachen, B.D.R. .15. S.GREENWALD, London, U.K. 16. G.GUIFFANT, Paris , France 17. M.-F. HANSELER, *General Secretary of IESC,* Cargèse, Corsica, France 18. R.M.HEETHAAR, Utrecht, The Netherlands 19. P.E.HYDON Cambridge, U.K.. 20. J.M. HUYGHE, Maastricht, The Netherlands 21. O.E.JENSEN, Cambridge, U.K. 22. Y.Y.KISLYAKOV, Leningrad, U.R.S.S. 23. R.I.KITNEY, London, U.K . 24. B.KLOSTERHALFEN, Aachen, B.D.R. 25. E.KRAUSE, *Scientific and Organizing Committee,* Aachen, B.D.R. 26. D.LERCHE, Berlin, D.D.R. 27. M.J.LEVER, London, U.K. 28. D.LIEPSCH, München , B.D.R. 29. J.MARTINS-SILVA Lisboa , Portugal. 30. F.MOSORA,*Colloquium Chairperson,* Liège, Belgique 31. C.ODDOU, Paris, France 32. H.-D.PAPENFUSS , Bochum , B.D.R. 33. R.PELISSIER, *Scientific and Organizing Committee,* Marseille, France 34. P.PERONNEAU, Paris, France 35. Th.POCHET , Liège, Belgique 36. R.RIEU, Marseille, France 37. M.SAHLOUL, Liège, Belgique 38. H.SCHMID-SCHÖNBEIN, Aachen, B.D.R. 39. R.C.SCHROTER, London, U.K. 40. M.THIRIET, Paris, France 41. S. TSANGARIS, Athens, Greece 42. A.A.VAN STEENHOVEN Eindhoven, The Netherlands 43. N.WESTERHOF, Amsterdam, The Netherlands 44. T.M.WICK, Atlanta, Georgia, U.S.A. 45. T.YAMAGUCHI, Suita, Osaka, Japan.

PARTICIPANTS

AFFELD K., Professor, Universitätklinikum Rudolf Virchow, Standort Charlottenburg Medizintzchnik-Forschungslabor, Spandauer Damm 130, 1000 Berlin 19, B.D.R.

AKKAS N., Professor, Bahçelievler, P.O.Box 70, 06502 Ankara, Turkey.

BAQUEY Ch., Dr., Univ. de Bordeaux II, U306 I.N.S.E.R.M., Rue Léo Saignat 146, 33076 Bordeaux Cedex, France.

BARTHES-BIESEL D., Professor, Université de Technologie de Compiègne, Département de Génie Biologique, B.P. 649, 60206 Compiègne, France.

BERGERON P., Professor, Hôpital St. Joseph, Service de Chirurgie Thoracique, Bd. de Louvain 26, 13008 Marseille, France.

BERTELSEN A.F., Professor, University of Oslo, Department of Mathematics, P.O. Box 1053, Blindern, 0316 Oslo 3, Norway.

CARO C.G., Professor, Imperial College, Physiology Flow Studies Unit, Prince Consort Road, SW7 2AZ London, U.K.

CASSOT F., Dr., C.N.R.S., Laboratoire de Mécanique et Acoustique, Chemin J. Aiguier 31, 13402 Marseille Cedex 09, France.

COLLINS M.W., Dr., The City University, Thermo-Fluids Engineering Research Centre Northampton Square, ECIV OHB London, U.K.

CORIERI P., Ms., Institut von Karman de Dynamique des Fluides, Chaussée de Waterloo 72, 1640 Rhode-St-Genese, Belgique.

DAMAS, A.M., Professor, Universidade do Porto, Instituto Abel Salazar, Sector de Biofisica, Largo da Escola Medica 2, 4000 Porto, Portugal.

DEPLANO V., Ms, Université d'Aix Marseille II, Institut de Mécanique des Fluides, Groupe de Mécanique, Rue Honnorat 1, 13003 Marseille, France.

DROCHON A., Ms, Université de Technologie de Compiègne, Département de Génie Biologique, UA CNRS 858, B.P .649, 60206 Compiègne, France.

GERRARD J.H., Dr., University of Manchester, Department of Engineering, Simon Building Oxford Road, M13 9PL Manchester, U.K.

GRAHAM M., Dr, Imperial College, Department of Aeronautics, SW7 2BY London, U.K.

GREBE R., Dr., RWTH Aachen, Medizinische Facultät, Abteilung Physiologie, Pauwelsstrasse, 5100 Aachen, B.D.R.

GREENWALD S., Dr., London Hospital, Department of Morbid Anatomy, Whitechapel Road, E1 1BB London, U.K.

GRIFFITH M., Dr., University of Wales College of Medicine., Department of Diagnostic Radiology, Heath Park, CF4 4XN Cardiff, U.K.

GUIFFANT G., Dr., Université Paris VI, LBHP, Tour 33-34, 2ème ét., Place Jussieu 2 75251 Paris Cedex 05, France.

HEETHAAR R.M., Professor, Academisch Ziekenhuis Utrecht, Medische Fysica, P.O.Box 85500, 3508 GA Utrecht, Nederland.

HUYGHE J.M., Dr., Rijksuniversiteit Limbourg, Beweging Wetenschappen, P.O.Box 616, 6200 MD Maastricht, Nederland.

HYDON P.E., Mr., Cambridge University, Department of Applied Mathematics and Theoretical Physics, Silver Street, CB3 9EW Cambridge, U.K.

JAFFRIN M.Y., Professor, Université de Technologie de Compiègne, Département de Génie Biologique, B.P. 649, 60206 Compiègne Cedex, France.

JENSEN O.E., Mr., Cambridge University, Department of Applied Mathematics and Theoretical Physics, Silver Street, CB3 9EW Cambridge, U.K.

JONES C.J.H., Dr., Imperial College, Physiological Flow Studies Unit, Prince Consort Road, SW7 2AZ London, U.K.

KISLYAKOV Y.Y., Professor, The USSR Academy of Sciences, Institute for Informatics and Automation, 30, 14th lino, 199178 Leningrad, U.R.S.S.

KITNEY R.I., Dr., Imperial College, Biomedical Systems Group, Prince Consort Road, SW7 2BT London, U.K.

KLOSTERHALFEN B., Dr., RWTH Aachen, Klinikum, Medizinische Facultät, Abteilung Pathologie, Pauwelsstrasse 1, 5100 Aachen, B.D.R.

KRAUSE E., Professor, RWTH Aachen, Aerodynamische Institut, Wülnerstrasse 5-7, 5100 Aachen, B.D.R.

LERCHE D., Professor, Humboldt-Universität, Charité, Institut für Medizinische Physik und Biophysik, Invalidenstrasse 42, 1040 Berlin, D.D.R.

LEVER M.J., Dr., Imperial College, Physiological Flow Studies Unit, Prince Consort Road, SW7 2AZ London, U.K.

LIEPSCH D., Professor, Fachhochschule München, Versorgungstechnik Fachbereich 05, Lothstrasse 34, 8000 München 2, B.D.R.

MARTINS-SILVA J., Professor, Universidade de Lisboa, Faculdade de Medicina, Instituto de Bioquimica, Avenida Egass Moniz, 1699 Lisboa Cedex, Portugal.

MOSORA F., Professor, Univérsité de Liège, Institut de Physique B5, Biophysique, Biomécanique, Sart-Tilman, 4000 Liège, Belgique.

ODDOU C., Professor, Université Paris XII, U.F.R. Sciences et Technologie, Laboratoire de Mécanique Physique, Av. du Général de Gaulle, 94010 Creteil, France.

PAPENFUSS H-D., Professor, Ruhr-Universität Bochum, Institut für Thermo-und Fluiddynamik, Angewandte Strömungsmechanik, Geb Ib 6/49, Postf. 2148, 4630 Bochum 1, B.D.R.

PELISSIER R., Professor, Université. d'Aix Marseille II, Institut de Mécanique des Fluides, Groupe de Mécanique, Rue Honnorat 1, 13003 Marseille, France.

PERONNEAU P., Dr., U 256 I.N.S.E.R.M, Hôpital Broussais, Rue Didot 96, 75674 Paris Cedex14, France.

POCHET Th., Ir., Université de Liège, Institut de Génie Civil C2, L.H.C.N., Quai Banning 6, 4000 Liège, Belgique.

RENEMAN R.S., Professor, Rijksuniversiteit Limburg, Fysiologie Afdeling, Biomedische Centrum, P.O. Box 616, 6200 MD Maastricht, Nederland.

RIEU R., Dr., Univérsité d'Aix Marseille II, Institut de Mécanique des Fluides, Unité Mixte du CNRS 34, Rue Honnorat, 1, 13003, Marseille, France.

SAHLOUL M., Dr. Ir., Université de Liège, Institut de Génie Civil C2, L.H.C.N.,Quai Banning 6, 4000 Liège, Belgique.

SAMPOL J., Professor, Hôpital de la Conception, Service d'Hématologie, Bd. Baine 141, 13385 Marseille, France.

SCHMID-SCHÖNBEIN H., Professor, RWTH Aachen, Klinikum, Medizinische Facultät, Abteilung Physiologie, Pauwelsstrasse 1, 5100 Aachen, B.D.R.

SCHROTER R.C., Dr., Imperial College, Physiological Flow Studies Unit, Prince Consort Road, SW7 2AZ London, U.K.

SEIFFGE D., Dr., Hoechst AG Werk Kalle-Albert, Pharmakologie, P.O.Box 129101, 6200 Wiesbaden 12, B.D.R.

THIRIET M., Dr., Université Paris VII, LBHP, URA 343 CNRS, Place Jussieu 2, 75251 Paris Cedex 05, France.

TSANGARIS S., Dr., National Technical University of Athens, Department of Mechanical Engineering, Fluids Section, P.O.Box 64070, 15710 Zografou, Athens, Greece.

VAN STEENHOVEN A.A., Dr., Technische Universiteit Eindhoven, Mecanische Engineering Afdeling, Postbus 513, 5600 MB Eindhoven, Nederland.

WESTERHOF N, Professor, Vrije Universiteit, Medische Faculteit, Laboratorium voor Fysiologie, Boechorststraat 7, 1081 BT Amsterdam, Nederland.

WICK T.M., Professor, Georgia Institute of Technology, School of Chemical Engineering 30332-0100 Atlanta, Georgia, U.S.A.

YAMAGUCHI T., Dr., National Cardiovascular Center Research. Institute., Department of Vascular Physiology, Vascular Pathophysiology Laboratory, Fujishirodai 5 -7-1, 565 Suita, Osaka, Japan.

AUTHOR INDEX

Cross-sectional area, 208, 349
Cupformers, 231
CW Doppler devices, 293
Cytoplasmic solutes, 239

Damping effect, 115
Darcy's law, 209
Dean parameter, 60
DEC Microvax II computer, 259
DEC-VMS operating system, 259
Dehydrobenzperidol, 16
Diastolic phase, 202
Diazepam, 361
Diffusion, 11,17,
   constant of oxygen, 21
Discharge hematocrit, 206, 214
Dissipative structure, 185-186, 227, 229
      (*see also* Chaos theory)
Duplex scanner, 384
Duromedics valve, 107-108
Dynamic elastic modulus, 347, 349, 352,
   354-355

Echinocyte formers, 231
EDRF, 155-163
Elastic silicon rubber model, 165-167
Electrodes, 71
Electron spin resonance (ESR), 235, 237-
   238
Endocardium, 15, 77
Endothelial cell, 283, 286, 313-319, 323-
   326
   density in arteries, 10, 155
   structure, 168, 283, 285-290
   tearing resistance, 323-327
Endothelium-derived-relaxing factor
   (EDRF), 345
Endotoxaemia, 314
Energy dissipation, 189
Entropy flux, 187
Epicardium, 15
Erythrocyte, 217, 235-241, 274, 283
   aggregation, 366, 370
   aqueous volume, 239-241
   deformability, 185, 189-190, 217-218
      361, 375
   filterability, 361, 366
   intracellular microviscosity, 235, 239-
      240
   labelled, 277
   mean curvature, 223, 229
   membrane, 193
      lamina, 224-226
      model, 223-231
      skeleton, 224
   non-nucleated, 188
   osmotic fragility, 361-363, 367
   rectification, 185-186, 191-194

Erythrocyte (continued)
   shape, 223, 228
   tanktreading, 185, 191, 217, 222
   transport, 376
   velocity, 209, 218, 341, 343
   volume fraction, 243-244
Escherichia coli endotoxin, 314, 316
Extracorporeal circulation, 271, 359, 362-
   370
   continous, 359, 364-367
   pulsatile, 359-360, 364-367
Extravascular shunting, 15-20

Factor VIII-antigen, 345
Fahraeus
   effect, 206, 209
   factor, 188
Fahraeus-Lindqvist effect, 206, 209
Femoral artery model, 265, 329
Femoropopliteal vein graft, 383
Fentanyl, 361
Fibrinogen, 362, 365-366
Fibronectin, 284, 287-289
   fibers, 285-286
   fluorescently labeled, 285
Finite element
   method, 79, 125-126
   model
      axisymmetric two-phase, 23-29 (*see*
      also Left ventricle)
      three dimensional, 52, 129
   Crouzeix-Raviart, 69
Flow
   in artificial heart valve, 87-94, 105-111
   in bifurcation, 62, 63, 115-121, 125-133
      of carotid artery blood, 67-74
      in a dracon graft, 120-121
      in lung vesseels, 143-154
      with stenosis, 67
   in capillaries, 185-194, 205-212, 217-222,
      243
   in collapsible tubes, 33-38
   conductance relationship, 162, 211
   in curved tube, 51-57, 72-74, 268
   diameter relationship, 161-162, 211
   of an incompressible fluid, 78, 79
   instabilities, 37, 38
   mapping, 106, 108, 111-112, 211, 275
      scanner, 267
   in microcirculatory network, 186, 205-214
   non-Newtonian, 165-166, 193, 243-249
   of non-pulsatile fluid, 155-162
   pattern, *see* Flow mapping
   in pipes with varying cross-sections, 41-49
   of Poiseuille, 138, 155, 189
   in porous-media-like tissues, 205, 209
   pressure, 161-162, 308-310
   rate, 36-37, 155, 157-158,